Systems Biology: Modeling and Analysis

Systems Biology: Modeling and Analysis

Edited by Alexis White

SYRAWOOD
PUBLISHING HOUSE

New York

Published by Syrawood Publishing House,
750 Third Avenue, 9th Floor,
New York, NY 10017, USA
www.syrawoodpublishinghouse.com

Systems Biology: Modeling and Analysis
Edited by Alexis White

International Standard Book Number: 978-1-68286-667-2 (Hardback)

Cataloging-in-Publication Data

Systems biology : modeling and analysis / edited by Alexis White.
    p. cm.
Includes bibliographical references and index.
ISBN 978-1-68286-667-2
1. Systems biology. 2. Bioinformatics. 3. Biological systems. 4. Computational biology.
I. White, Alexis.
QH324.2 .S97 2019
572.8--dc23

TABLE OF CONTENTS

PREFACE

Systems biology is defined as the mathematical and computational modeling of biological systems. It is also integral to the field of bioinformatics. The concepts of systems biology are used across several fields of study such as genomics, phonemics, proteomics, etc. The field also involves the study of metabolic and cell signaling networks to understand the properties and functions of cells, tissues, etc. in living organisms. This book elucidates the concepts and innovative models around prospective developments with respect to systems biology. The topics included in this book are of utmost significance and bound to provide incredible insights to the readers. With its detailed analyses and data, this book will prove immensely beneficial to professionals and students involved in this area at various levels.

The information contained in this book is the result of intensive hard work done by researchers in this field. All due efforts have been made to make this book serve as a complete guiding source for students and researchers. The topics in this book have been comprehensively explained to help readers understand the growing trends in the field.

I would like to thank the entire group of writers who made sincere efforts in this book and my family who supported me in my efforts of working on this book. I take this opportunity to thank all those who have been a guiding force throughout my life.

Editor

Selecting high-quality negative samples for effectively predicting protein-RNA interactions

Zhanzhan Cheng[1†], Kai Huang[1†], Yang Wang[5], Hui Liu[2,4], Jihong Guan[3] and Shuigeng Zhou[1,2*]

Abstract

Background: The identification of Protein-RNA Interactions (PRIs) is important to understanding cell activities. Recently, several machine learning-based methods have been developed for identifying PRIs. However, the performance of these methods is unsatisfactory. One major reason is that they usually use unreliable negative samples in the training process.

Methods: For boosting the performance of PRI prediction, we propose a novel method to generate reliable negative samples. Concretely, we firstly collect the known PRIs as positive samples for generating positive sets. For each positive set, we construct two corresponding negative sets, one is by our method and the other by random method. Each positive set is combined with a negative set to form a dataset for model training and performance evaluation. Consequently, we get 18 datasets of different species and different ratios of negative samples to positive samples. Secondly, sequence-based features are extracted to represent each of PRs and protein-RNA pairs in the datasets. A filter-based method is employed to cut down the dimensionality of feature vectors for reducing computational cost. Finally, the performance of support vector machine (SVM), random forest (RF) and naive Bayes (NB) is evaluated on the generated 18 datasets.

Results: Extensive experiments show that comparing to using randomly-generated negative samples, all classifiers achieve substantial performance improvement by using negative samples selected by our method. The improvements on accuracy and geometric mean for the SVM classifier, the RF classifier and the NB classifier are as high as 204.5 and 68.7%, 174.5 and 53.3%, 80.9 and 54.3%, respectively.

Conclusion: Our method is useful to the identification of PRIs.

Keywords: Protein-RNA interactions, Reliable negative samples, Unreliable negative samples

*Correspondence: sgzhou@fudan.edu.cn
†Equal contributors
1 School of Computer Science, Fudan University, Handan Road, 200433 Shanghai, China
2 The Bioinformatics Lab at Changzhou NO. 7 People's Hospital, Changzhou, 213011 Jiangsu, China
Full list of author information is available at the end of the article

Background

Exploring the interactions between proteins and RNAs can help us to understand the mechanisms of life, such as the protein translation process [1–3], gene expression [4, 5], RNA post-transcriptional modification [6–8], cellular regulation [9, 10].

A lot of effort has been put on the identification of PRIs using traditional experimental methods and post-experimental methods. As experimental methods consume more time and money than post-experimental methods, the latter is gaining more and more attention. There are mainly two categories of post-experimental methods: 1)structural & chemical-based methods and 2)computational methods.

The first category of methods attempted to analyze the interacting mechanism of protein and RNA at structural and chemical levels. For example, Jones et al. [11] focused on analyzing protein-RNA complexes, and obtained the physical-chemical properties of RNA-binding residues and the distribution of atom-atom within the complexes. With protein-RNA experimental data, Ellis et al. [12] presented a statistics on properties of binding residues bounding to functional various RNAs. Besides, some function-based works [13, 14] also discussed the protein-RNA interactions.

As for computation-based methods, several machine learning techniques have been employed on identifying PRIs, such as random forest (RF), Naive Bayes (NB) and support vector machine (SVM). Pancaldi et al. [15] used RF and SVM for identifying PRIs by considering more than 100 properties of RNAs and proteins. Instead, Muppirala et al. [16] used only protein and RNA sequence information for predicting interactions. Similarly, Wang et al. [17] improved the Naive Bayes (ENB) classifiers for predicting PRIs with only sequence data. Recently, we also proposed learning method [18] with only positive and unlabeled samples on PRIs prediction.

Compared with structural & chemical-based methods, computational methods are more efficient and effective. However, the performance of computational methods heavily depends on the quality of training datasets, which usually consist of positive samples and negative samples. Here, positive samples are not the problem. The difficulty lies in that we do not have experimentally-validated negative samples. Current works [16, 17] addressed this problem by randomly pairing RNAs and proteins and then removing these pairs included in the positive set. In this paper, we call this method *random method* or *traditional method*. Obviously, random negative samples must not be real negative samples. So the quality of random negative sets cannot be guaranteed. This will unavoidably impact prediction performance of classifiers trained on datasets with random negative samples.

This paper addresses how to select highly reliable negative samples to improve PRI prediction. To this end, we present an effective method *FIRE* — the abbreviation of *FI*nding *R*eliable n*E*gative samples). The basic idea of our method is like this: given a known PRI of protein i and RNA j, for a protein k, the more difference between protein i and protein k, the less possibility that protein k interacts RNA j.

We first construct positive sets using known PRIs. Given a positive set, we establish two negative sets: one is by random method and the other by our method. And the positive set is combined with each of the two negative sets to form a dataset for model training and performance evaluation. In such a way, we construct 18 datasets of different species and different ratios of negative samples to positive samples. Then, we extract the features of each pair of protein and RNA. Here, each feature is composed of a conjoint triad of vicinal amino acids and a k nucleotide acids. To cutoff computational cost, a filter-based feature selection method is employed to reduce the dimensionality of feature vectors. Finally, we conduct extensive experiments to evaluate the proposed method by training and testing SVM, RF and NB classifier on the 18 datasets. The experimental results show that these classifiers perform much better using the negative samples generated by our method than using random negative samples.

Methods

We collected non-redundant known PRIs as positive samples, and generated 18 datasets based on our method and the random method, which were used to evaluate the performance of PRI prediction by SVM, RF and NB classifiers. Figure 1 is the procedure of our method, which contains five steps: 1) Generating negative datasets by using our method *FIRE* and the random method; 2) Constructing feature vectors for each pair of protein-RNA; 3) Reducing the dimension of feature vectors; 4) Training classifiers; 5) Performance evaluation.

Datasets

We constructed 9 non-redundant positive PRI sets from PRIDB [19], NPInter [20], 9 reliable negative sets based on the positive sets and the STRING [21] database by our method, and 9 random negative sets with the random method. The procedure for negative sample construction will be detailed later. Each positive set is merged with a negative set to construct a PRI dataset, consequently 18 PRI datasets in total are constructed. PRIDB is a database of protein-RNA interfaces calculated from protein-RNA complexes in PDB [22]. NPInter is a complete database covering eight-category functional interactions between proteins and noncoding RNAs of six model organisms, including *Caenorhabditis elegans*, *Drosophila*

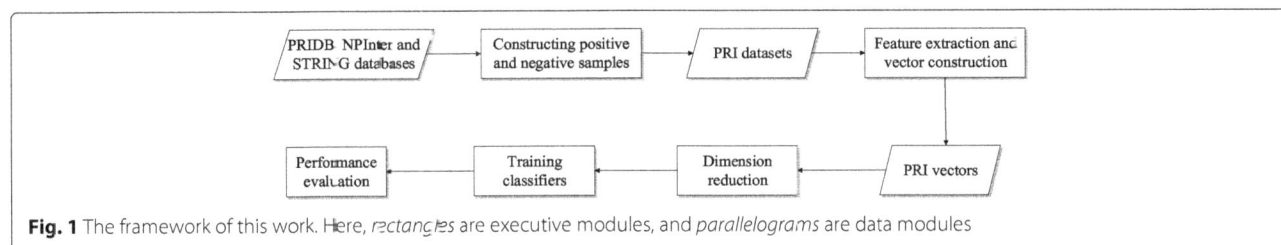

Fig. 1 The framework of this work. Here, *rectangles* are executive modules, and *parallelograms* are data modules

melanogaster, Escherichia coli, Homo sapiens, Mus musculus and *Saccharomyces cerevisiae*. STRING is an updated online database resource Search Tool for the Retrieval of Interacting Genes, it provides uniquely comprehensive coverage and ease of access to both experimental and predicted protein-protein interaction (PPI) information.

The 18 datasets are divided to 3 groups. The first group of datasets (denoted group 1) contain 336 experimental-validated PRIs that are used as positive samples, which are related to the six organisms above and constructed from the NPInter and STRING databases. This group consists of six sub-datasets (named by SO) as follows:

1. The first sub-dataset ($SO\_reliable_{1:1}$) contains 168 positive samples and 168 reliable negative samples generated by our method, the ratio of positives to negatives is 1 : 1;
2. The second sub-dataset ($SO\_reliable_{2:1}$) contains 336 positive samples and 168 reliable negative samples, the ratio is 2 : 1;
3. The third sub-dataset ($SO\_reliable_{1:2}$) contains 168 positive samples and 336 reliable negative samples, the ratio is 1 : 2;
4. The fourth sub-dataset ($SO\_random_{1:1}$) contains 168 positive samples and 168 random negative samples generated by the random method, and the ratio of positives to negatives is 1 : 1;
5. The fifth sub-dataset ($SO\_random_{2:1}$) contains 336 positive samples and 168 random negative samples, the ratio is 2 : 1;
6. The last sub-dataset ($SO\_random_{1:2}$) contains 168 positive samples and 336 random negative samples, the ratio is 1 : 2.

The second group of datasets (denoted as group 2) includes 1320 experimental-validated homo species PRIs used as positive samples, which are extracted from the PRIDB and STRING databases, it also consists of six sub-datasets. Following the nomenclature of the first group of datasets, these PRI datasets are named as $HOMO\_reliable_{1:1}$, $HOMO\_reliable_{2:1}$, $HOMO\_reliable_{1:2}$, $HOMO\_random_{1:1}$, $HOMO\_random_{2:1}$, $HOMO\_random_{1:2}$.

The third group of datasets (denoted as group 3) has 114 experimental-validated mouse PRIs as positive samples,

which also consists of six sub-datasets: $MUS\_reliable_{1:1}$, $MUS\_reliable_{2:1}$, $MUS\_reliable_{1:2}$, $MUS\_random_{1:1}$, $MUS\_random_{2:1}$, $MUS\_random_{1:2}$.

Table 1 gives the statistics of the total 18 PRI datasets.

Construction of random negative samples

Previous works [16, 17] randomly select negative samples, the underlying hypothesis is: if there is no validated interaction between a protein and a RNA, then the protein and the RNA constitute a negative sample. Obviously, the hypothesis is not completely reasonable. The flowchart for generating random negative samples is shown in Fig. 2.

In Fig. 2, the major steps of the random method are as follows:

1. Each PRI extracted from PRIDB and NPInter is included in the positive set. From the positive set, we can get a set P of proteins and a set R of RNAs, each protein/RNA in P/R is involved in at least a positive PRI.

Table 1 The 18 PRI datatsets used in this paper

Datesets	# Positive samples	# Negative samples
$SO\_reliable_{1:1}$	168	168
$SO\_reliable_{2:1}$	336	168
$SO\_reliable_{1:2}$	168	336
$SO\_random_{1:1}$	168	168
$SO\_random_{2:1}$	336	168
$SO\_random_{1:2}$	168	336
$HOMO\_reliable_{1:1}$	660	660
$HOMO\_reliable_{2:1}$	1320	660
$HOMO\_reliable_{1:2}$	660	1320
$HOMO\_random_{1:1}$	660	660
$HOMO\_random_{2:1}$	1320	660
$HOMO\_random_{1:2}$	660	1320
$MUS\_reliable_{1:1}$	57	57
$MUS\_reliable_{2:1}$	114	57
$MUS\_reliable_{1:2}$	57	114
$MUS\_random_{1:1}$	57	57
$MUS\_random_{2:1}$	114	57
$MUS\_random_{1:2}$	57	114

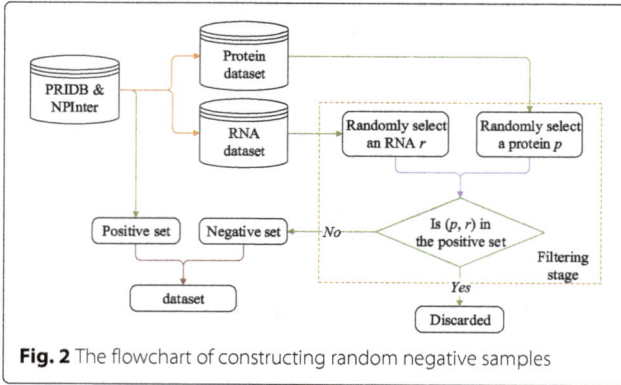

Fig. 2 The flowchart of constructing random negative samples

2. For each protein p in P and each RNA r in R, there is a corresponding protein-RNA pair (p, r).
3. If (p, r) is not included in the positive set, it is a negative sample.
4. The positives and negatives are merged to a PRI dataset.

Construction of reliable negative samples

The basic idea of our method is like this: for an experimentally-validated PRI of protein p and RNA r, r is highly possible to interact with any protein p' similar to p. On the contrary, if protein p' is dissimilar to p, there is low possibility that p' interacts r. Based on this idea, we propose the method *FIRE* to construct reliable negative PRIs. The flowchart of FIRE is shown in Fig. 3. Concretely, for each positive PRI (p, r), we try to find any protein p' that is as much dissimilar as possible to p. If (p', r) is not an experimentally-validated PRI, then it is selected as a negative PRI.

We first compute the similarity between each pair of proteins based on three different data sources, then we combine these similarity scores as a final score to measure the similarity between the two proteins. Detail is delayed to "Protein-protein similarity computation" section.

The procedure of our method FIRE is as follows:

1. Construct the positive set PS of PRIs based on the PRIDB and NPInter databases, and compute the similarity matrix SP of proteins involved in PS as in "Protein-protein similarity computation" section.

2. For protein p_i and RNA r_j that do not form a positive PRI in PS, i.e., $(p_i, r_j) \notin PS$, compute a score between p_i and r_j as follows:

   (a) If protein p_k $(k \neq i)$ and r_j forms a PRI in the positive PRI set PS, then the score SPR_{ijk} indicating the confidence of (p_i, r_j) being a positive PRI via protein p_k can be evaluated via SP_{ik}, which is the similarity between p_i and p_k.

   (b) As there may be multiple (say n) positive PRIs involving r_j in PS, we aggregate the scores SPR_{ijk} over all positive PRIs (p_k, r_j) $(k \neq i$ and $k = 1..n)$ as follows:

   $$SPR_{ij} = \sum_{k=1}^{n} SPR_{ijk} = \sum_{k=1}^{n} \delta(k, j) \times SP_{ik}, \quad (1)$$

   SPR_{ij} indicates the confidence of (p_i, r_j) being a positive PRI, $\delta(i, j) = 1$ if (p_k, r_j) is a positive PRI, otherwise 0.

3. As $(p_i, r_j) \notin PS$, it is a potential negative PRI. Sorting all generated potential PRIs (p_i, r_j) via their scores SPR_{ij} in increasing order, the top-m protein-RNA pairs in the sorted list are taken as negative PRIs if m negative PRIs are to be generated.

Protein-protein similarity computation

We compute the similarity between any two proteins involved in the positive set based on three types of data sources: sequence information, functional annotations and protein domains, these computed similarities are called *sequence similarity*, *functional annotation semantic similarity* and *protein domain similarity*, which are merged to get the final similarity of the two proteins.

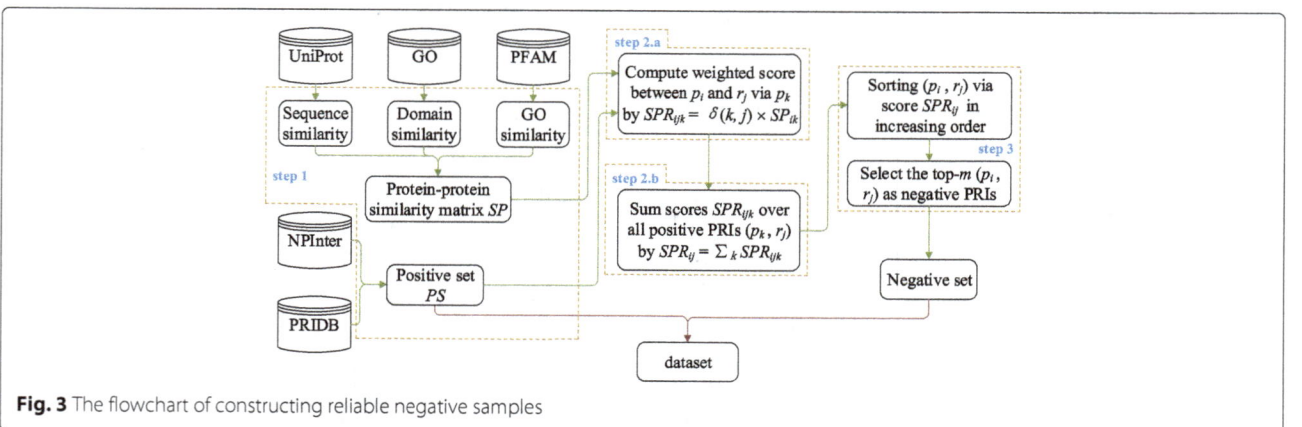

Fig. 3 The flowchart of constructing reliable negative samples

Sequence similarity (SS). Protein sequences are obtained from the UniProt database [23]. We compute sequence similarity between two proteins using a normalized version of Smith-Waterman score [24]. The normalized Smith-Waterman score between two proteins p_i and p_j is $nsw(p_i,p_j)=sw(p_i,p_j)/\sqrt{sw(p_i,p_j)}\sqrt{sw(p_j,p_j)}$ where $sw(.,.)$ means the original Smith-Waterman score. By applying this operation to protein pair p_i and p_j, we can obtain their sequence similarity $SS(p_i,p_j)=(nsw(p_i,p_j)+nsw(p_j,p_i))/2$.

Functional annotation semantic similarity (FS). GO annotations are downloaded from the GO database [25]. Semantic similarity between each pair of proteins is calculated based on the overlap of the GO terms associated with the two proteins [26]. All three types of GO are used in the computation as similar RNAs are expected to interact with proteins that act in similar biological processes, or have similar molecular functions or reside in similar cell compartments. We compute the Jaccard value [27] with respect to the GO terms of each pair of proteins as their similarity. The Jaccard score between term sets t_i and t_j of proteins p_i and p_j is defined as $|t_i \cap t_j|/|t_i \cup t_j|$, which is the ratio of the number of common terms between proteins p_i and p_j to the total number of terms of p_i and p_j, which is used as the functional annotation semantic similarity $FS(p_i,p_j)$ of proteins p_i and p_j.

Protein domain similarity (DS). Protein domains are extracted from Pfam database [28]. Each protein is represented by a domain fingerprint (binary vector) whose elements encode the presence or absence of each retained Pfam domain by 1 or 0, respectively. We compute the Jaccard value of any two proteins p_i and p_j with their domain fingerprints as their similarity $DS(p_i,p_j)$.

For proteins p_i and p_j, we compute the aggregated similarity (AS) by merging the three different similarity measures above as follows:

$$AS(p_i,p_j) = (SS(p_i,p_j)+FS(p_i,p_j)+DS(p_i,p_j))/3. \quad (2)$$

PRI feature vectors

Existing works [29–31] found that properties of amino acids are effective in protein classification. To reduce the dimensionality of protein representation, Shen et al. [32] classified the 20 amino acid residues as seven classes according to their physicochemical properties, meanwhile the concept of conjoint triads were also proposed to represent the protein properties. Wang et al. [17] further reduced the dimension of feature vector by encoding the 20 amino acids residues into four classes: {DE}, {HRK}, {CGNQSTY}, and {AFILMPVW}. In this work, we use the same strategy for encoding protein sequences.

Feature construction

To compute protein feature vectors, we used conjoint triads as protein properties as in [16, 17, 32]. 3 continuous

amino acids constitute a conjoint triad, we can get 64 ($4 \times 4 \times 4$) classes of conjoin triads. Note that two triads are treated as the same class if their residues in the corresponding positions belong to the same class. For RNA sequences, we used k-nucleotide acids (k-NAs) as properties. A k-NAs refers to a unit of k continuous nucleotide acids. k-NAs of size 1 (i.e. $k = 1$) are called "uniNAs", size 2 (i.e. $k = 2$) are called "biNAs", size 3 (i.e. $k = 3$) are called "triNAs", size 4 or more (i.e. $k \geqslant 4$) are simply called "k-NAs". Because RNA sequences contain only the four bases A, U, C, G, we have 4 unique uniNAs, 4^2 unique biNAs and 4^3 unique triNAs. Finally, by pairing the k-NAs ($k = 1, 2, 3$) and triads, we can get at most 256 (64×4) 4-mers, 1024 (64×4^2) 5-mers and 4096 (64×4^3) 6-mers, each of which is composed of a conjoint triad and a uniNA, biNA and triNA respectively. In the sequel, we also call 4-mers, 5-mers, 6-mers as type 1, 2, 3 ($k+3$)-mers.

Table 2 gives the combination of triads and k-NAs examples. For a pair of hypothetical amino acid sequence *DPPVPPPPV* and nucleotide acids sequence *CCUCU*, two classes of triads {DPP}, {PPV, PVP, VPP, PPP} (note that 'P' and 'V' belong to the same class), three classes of 3-NAs {CCU}, {CUC} and {UCU}, three classes of 2-NAs {CC}, {CU} and {UC} and two classes of 1-NAs {C} and {U} are generated. Hence, we can get the following 15 6-mers by matching the 3-NAs and triads: CCU-DPP, CCU-PPV, CCU-PVP, CCU-VPP, CCU-PPP, CUC-DPP, CUC-PPV, CUC-PVP, CUC-VPP, CUC-PPP, UCU-DPP, UCU-PPV, UCU-PVP, UCU-VPP and UCU-PPP, and 15 5-mers by matching the 2-NAs and triads: CC-DPP, CC-PPV, CC-PVP, CC-VPP, CC-PPP, CU-DPP, CU-PPV, CU-PVP, CU-VPP, CU-PPP, UC-DPP, UC-PPV, UC-PVP, UC-VPP and UC-PPP, and 10 4-mers by matching the 1-NAs and triads: C-DPP, C-PPV, C-PVP, C-VPP, C-PPP, U-DPP, U-PPV, U-PVP, U-VPP, U-PPP.

Feature value computation

In order to discriminate the significance of different types of features in a feature vector, we introduce the concept of concentration of different features. Denote the number of unique ($k + 3$)-mers of type i as N_i. The concentration of type i is the ratio of N_i to the total number of unique ($k + 3$)-mers, that is,

Table 2 An example of feature extraction for a pair of protein and RNA sequences

Protein sequence	*D P P V P P P P V*
RNA sequence	*C C U C U*
Triads	{*DPP*}, {*PPV, PVP, VPP, PPP*}
3-NAs	{*CCU*}, {*CUC*}, {*UCU*}
2-NAs	{*CC*}, {*CU*}, {*UC*}
1-NAs	{*C*}, {*U*}

$$C_i = \frac{N_i}{\sum_{j=1}^{3} N_j}, \qquad i = 1, 2, 3. \qquad (3)$$

For example, the number of unique 6-mers is 64×4^3. The total number of unique $(k+3)$-mers used in this study is 5376, therefore the concentration of 6-mers is $C_3 = 4096/5376 = 0.762$. Then, the elements of a feature vector are calculated by

$$f_j = t_j \times C_i, \qquad 1 \le j \le 5376 \qquad (4)$$

Above, t_j is the occurrence frequency of a certain unique $(k+3)$-mer of type i. A feature vector contains 5376 dimensions, each of which corresponds to a unique $(k+3)$-mer of a certain type i ($i = 1, 2$ and 3). Within a vector, the dimensions are arranged in the order of 6-mers, 5-mers and 4-mers. Then f_i is further normalized to ff_i as follows:

$$ff_j = \frac{f_j - f_{min}}{f_{max} - f_{min}} \qquad (5)$$

where f_{max} and f_{min} denote the maximum and the minimum of all f_j ($j = 1, 2, \ldots, 5376$), respectively.

Feature reduction

In order to reduce the computational cost, we employed a filter-based method for cutting down the dimension of feature vectors.

For the i-th feature $ff_j(i)$ of the j-th vector, let $F(i)_p$ and $F(i)_n$ denote its occurrence frequency in the positive and negative sample set respectively, which are calculated by

$$F(i)_p = \sum_{j=1}^{N} ff_j(i), \; vector \; j \; \in \; the \; positive \; set, \qquad (6)$$

$$F(i)_n = \sum_{j=1}^{M} ff_j(i), \; vector \; j \; \in \; the \; negative \; set, \qquad (7)$$

where N and M are the numbers of positives and negatives in the dataset.

$F(i)_p$ and $F(i)_n$ are further normalized to $FF(i)_p$ and $FF(i)_n$ as in Eq. (5), and then the final score of each feature is defined as follows:

$$FScore(i) = \frac{FF(i)_p}{FF(i)_n}, \qquad i = 1, 2, \ldots, 5376. \qquad (8)$$

Our objective is to choose those discriminative features that either frequently occur in the positive set but seldom occur in the negative set, or frequently occur in the negative set but rarely occur in the positive set. In such a way, we choose the features that help us to distinguish positive samples from negative samples.

As $FScore(i)$ measures the relative enrichment of the i-th feature in the positives over the negatives, it can be regarded as an indicator of the usefulness of the i-th feature. Based on the calculated $FScore$ values, the most "useful" features that have the largest or smallest $FScore$

values are selected to represent the PRI pairs. Suppose that we reduce the PRI vectors to k dimensions, we select the $\frac{k}{2}$ features with the largest $FScore$ values and the $\frac{k}{2}$ features with the smallest $FScore$ values to represent the k-dimension PRI vectors. In our work, k is set to 1000.

The classifiers and performance metrics

As several studies have successfully used random forest (RF), naive Bayes (NB) and support vector machine (SVM) to predict PRIs [15–17], we also use them to evaluate our method by 10-fold cross validation.

Four widely-used performance metrics, *sensitivity* (SE), *specificity* (SP), *accuracy* (ACC) and *geometric mean* (GM) are used in this paper. GM is commonly used for class-imbalance learning [33] because it can give a more accurate evaluation on imbalanced data. Therefore, for the imbalance datasets, we pay more attention to GM rather than ACC. These metrics are evaluated as follows:

$$SE = \frac{TP}{TP + FN}, \qquad (9)$$

$$SP = \frac{TN}{TN + FP}, \qquad (10)$$

$$GM = \sqrt{SE \times SP}, \qquad (11)$$

$$ACC = \frac{TP + TN}{TP + FN + TN + FP}, \qquad (12)$$

where TP is the number of true positives, TN is the number of true negatives, FP is the number of false positives, and FN is the number of false negatives.

In addition, we also use AUC (Area Under the receiver operating characteristic (ROC) Curve) to evaluate prediction performance in some experiments. AUC falls between 0 and 1. The maximum value 1 means a perfect prediction. For a random guess, the value of AUC is close to 0.5.

Results and Discussion

In our experiments, eighteen PRI datasets are used, these datasets either contain PRI data of different species or have different ratios of positive PRIs to negative PRIs. For each dataset, 10-cross validation is performed on SVM, RF and NB classifiers respectively, and the performance metrics of SE, SP, GM and ACC as well as AUC are used.

In the sequel, for the simplicity of notation, we denote the ratio of positive samples to negative samples as PNR, and remove the words "reliable" and "random" from the dataset names in Table 1. For example, both $SO\_reliable_{1:1}$ and $SO\_random_{1:1}$ are simplified to $SO_{1:1}$. In other words, $SO_{1:1}$ represents both $SO\_reliable_{1:1}$ and $SO\_random_{1:1}$.

Performance comparison

Figures 4, 5 and 6 respectively show the performance comparison between using our reliable negative samples

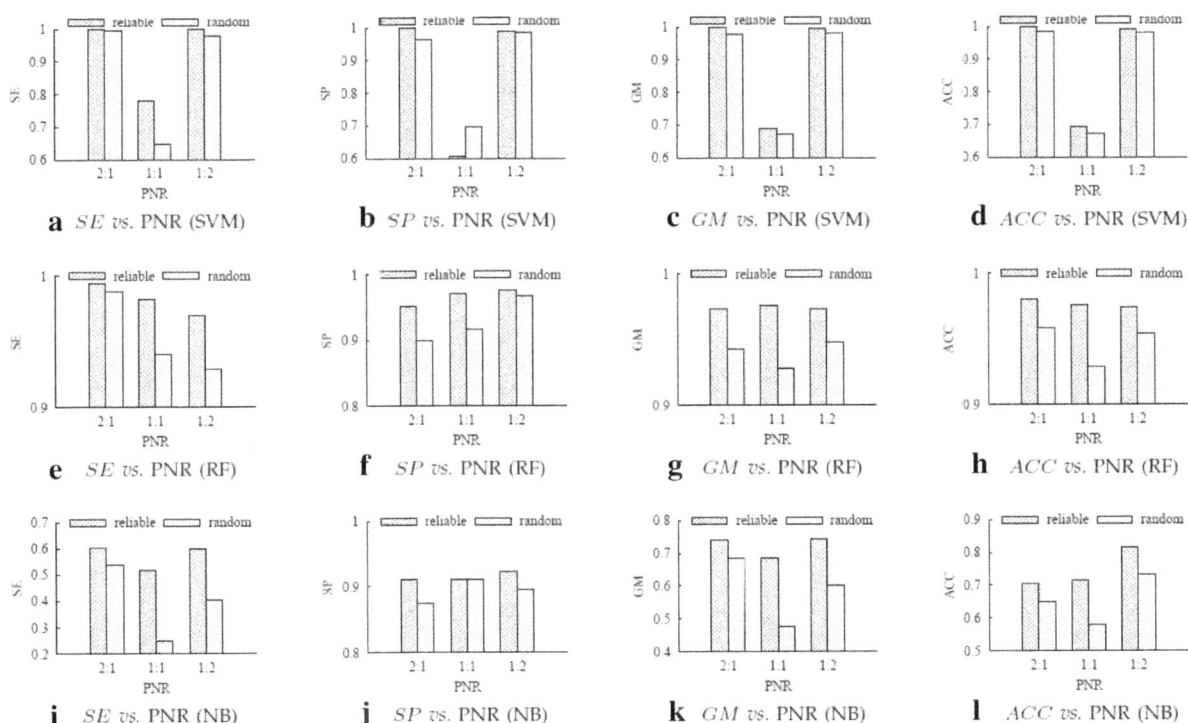

Fig. 4 Experimental results on SO datasets. **a–d** are the *SE*, *SP*, *GM* and *ACC* values of SVM classifiers; (**e**)–(**h**) are the *SE*, *SP*, *GM* and *ACC* values of RF classifiers; and (**i**)–(**l**) are the *SE*, *SP*, *GM* and *ACC* values of NB classifiers

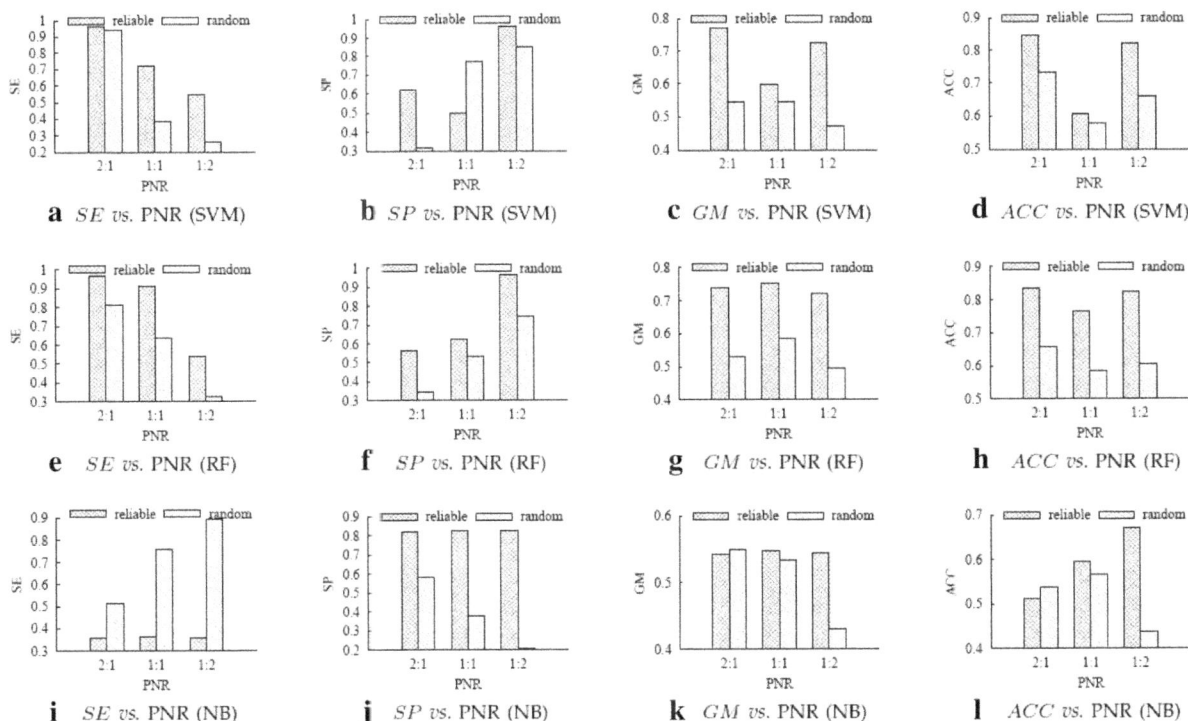

Fig. 5 Experimental results on HOMO datasets **a–d** are the *SE*, *SP*, *GM* and *ACC* values of SVM classifiers; (**e**)–(**h**) are the *SE*, *SP*, *GM* and *ACC* values of RF classifiers; and (**i**)–(**l**) are the *SE*, *SP*, *GM* and *ACC* values of NB classifiers

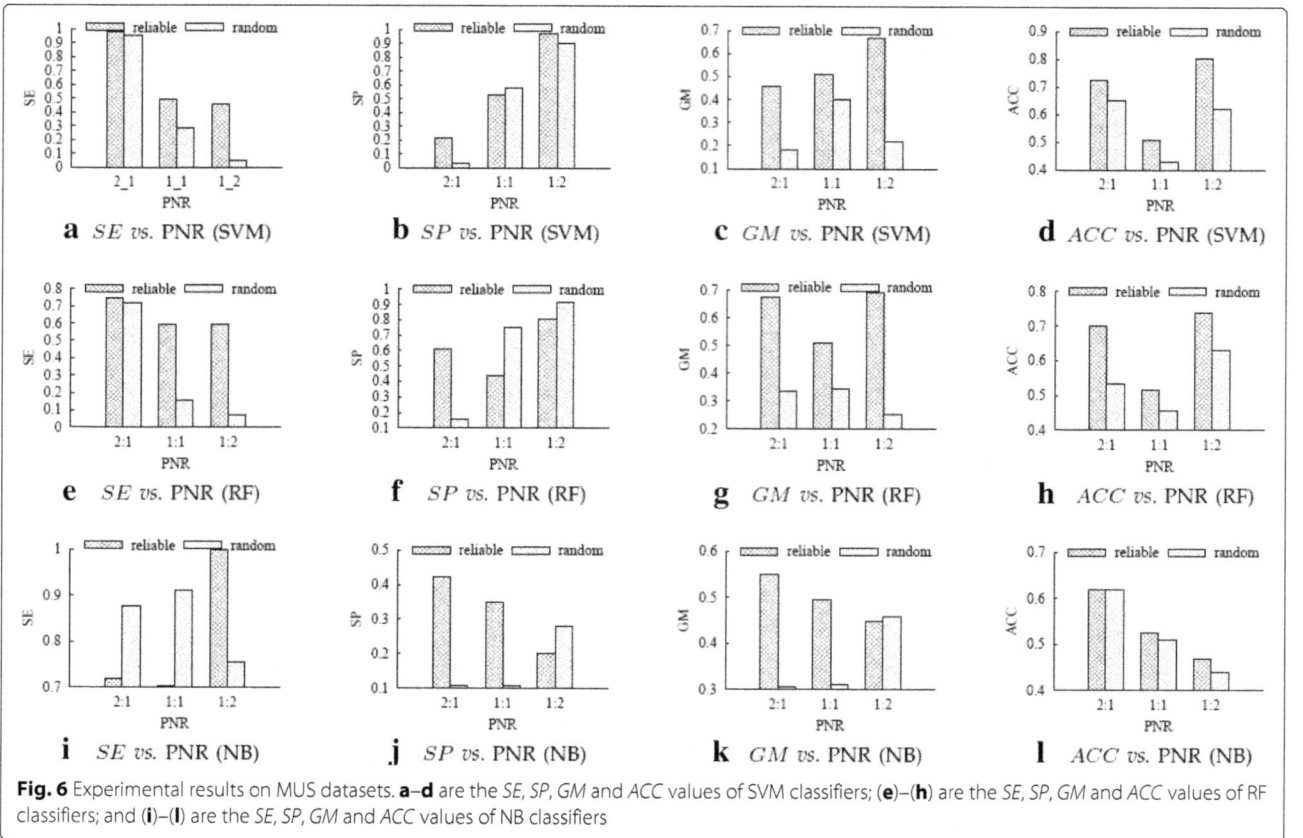

Fig. 6 Experimental results on MUS datasets. **a**–**d** are the *SE*, *SP*, *GM* and *ACC* values of SVM classifiers; (**e**)–(**h**) are the *SE*, *SP*, *GM* and *ACC* values of RF classifiers; and (**i**)–(**l**) are the *SE*, *SP*, *GM* and *ACC* values of NB classifiers

and using random negative samples on the *SO* datasets, *HOMO* datasets and *MUS* datasets.

To more clearly evaluate the advantage of reliable negative samples over random negative samples, we define the performance *improvement ratio* (*IR*) of using our reliable negatives over using random negatives as follows:

$$IR = \frac{result_{reliable} - result_{randm}}{result_{random}} \times 100\%, \qquad (13)$$

where $result_{reliable}$ and $result_{random}$ denote the performance measure (any of SE, SP, GM and ACC) of using our reliable negatives and using random negatives, respectively. A positive IR means using our reliable negatives achieves better performance than using random negatives. Table 3 shows the *IR* values calculated based on the results in Figs. 4, 5 and 6.

From Table 3, we can see that out of the 108 IR values, only 14 IRs are negative, one is 0, the other 93 (93/108≈86%) values are positive. As *GM* and *ACC* are more comprehensive than *SE* and *SP* in measuring classification performance, we check their IR values more carefully. Of the 54 IR values for *GE* and *ACC*, 51 (51/54≈94%) values are positive. Therefore, in most cases performance measure of our method is better than the random method. The largest IR is 760.4%, which is achieved for *SE* by SVM

on dataset $MUS_{1:2}$. We can also see that SVM and RF perform better than NB on these datasets.

The results above show that using the reliable negative samples selected by our method indeed boosts the performance of PRI prediction, and our method can serve as a practical and effective method for computationally predicting PRIs.

The effect of score threshold

To select negative samples, we have to set a score threshold, and require that all candidate negative samples (protein-RNA pairs) have scores (defined in Eq. (1)) no larger than the threshold. So the value of threshold will impact the quality of selected negative samples, and will subsequently impact the prediction performance. The smaller the threshold, the higher the quality of selected negatives, and the smaller the number of negatives that can be selected. So there is a tradeoff between the quality and the number of selected negatives. In this part, we check the impact of score threshold on prediction performance and thus suggest proper values for the threshold. Here, we use *AUC* to evaluate prediction performance.

We randomly select 908 nonredundant positive PRIs of *Homo sapiens* from PRIDB and NPInter, then construct an equal number of negative samples by our method with different score threshold values. Concretely, we generate

Table 3 The improvement ratio (IF) values of different classifiers on different datasets

Dataset	SVM				RF				NB			
	SE	SP	GM	ACC	SE	SP	GM	ACC	SE	SP	GM	ACC
$SO_{1:1}$	20.2	−12.8	2.4	3.1	4.5	5.8	5.1	5.1	107.2	0	43.9	23.1
$SO_{1:2}$	2.5	0.23	1.4	1.0	4.4	1.0	2.7	2.1	48.4	3.0	23.6	11.4
$SO_{2:1}$	0.6	3.7	2.1	1.6	0.6	6.0	3.2	2.3	12.1	4.2	8.1	8.5
$HOMO_{1:1}$	86.5	−35.4	9.8	5.3	42.1	17.2	29.0	30.8	−52.2	120.4	2.6	5.0
$HOMO_{1:2}$	109.2	12.3	53.3	25.2	65.4	29.0	46.1	35.5	−59.9	302.2	26.9	54.3
$HOMO_{2:1}$	2.3	94.5	41.1	15.7	19.1	63.1	39.4	26.8	−30.3	40.3	−1.1	-4.6
$MUS_{1:1}$	249.5	−63.4	13	68.7	372.2	−18.7	95.9	53.9	−21.1	299.5	77.5	21.8
$MUS_{1:2}$	760.4	7.8	204.5	29.3	751.4	−11.5	174.5	16.7	32.6	−28.1	−2.4	6.7
$MUS_{2:1}$	2.7	497.8	147.8	11.7	3.8	286.9	100.4	31.9	−18.0	299.1	80.9	0

negative samples like this: give a threshold value st (st is set to 0, 0.2, 0.4, 0.7 and 1.0 respectively), we select 908 protein-RNA pairs whose scores are closest to st. Thus, we construct five PRI datasets. Finally, we evaluate the AUC values of three classifiers RF, SVM and NB on the five constructed datasets by 10-fold cross validation. Figure 7 shows the results. As we can see, for all the three classifiers, with the increase of *threshold* value, the AUC value shows a decreasing trend, which conforms to our expectation. And when the score threshold is less than 0.7, the prediction performance is stable.

Capability of finding new positive PRIs

In this paper, we define a score (Eq. (1)) to measure the relationship between each protein and each RNA. The smaller the score, the more possible this protein-RNA pair is a negative PRI. Otherwise, the more possible it is a PRI. So the merits of our method are two-fold. On the one hand, we can use it to select highly credible negative PRIs; On the other hand, it can be used to directly predict positive PRIs.

We randomly select 908 nonredundant positive PRIs of *Homo sapiens* from PRIDB and NPInter, and compute the score of any protein-RNA pair not included in the positive set by our method. Among the screened protein-RNA pairs, for each RNA we extract the top 4 protein-RNA pairs in terms of the aggregated score AS defined in Eq. (1) and requiring $AS > 1$, then we get 397 protein-RNA pairs involving 107 unique RNAs and 96 unique proteins. We search each protein-RNA pair against the NPInter and PRIDB datasets, and find that 22 pairs have been validated by biological experiments.

Furthermore, from the 397 protein-RNA pairs gotten above, we filter out those pairs whose proteins appear in PRIs of the NPInter and PRIDB datasets, and get 256 protein-RNA pairs involving 56 unique RNAs and 74 unique proteins. Then we annotate manually the 74 proteins in the 256 protein-RNA pairs by the Gene Ontology database, and we find that 64 (64/74≈86.5%) proteins have RNA binding, chromatin binding or nucleotide binding functions, which play important roles in positive or negative regulation of transcription, gene expression and RNA processing.

Figure 8 is a protein-RNA interaction network constructed by the true positive PRIs and the predicted ones. The network includes 908 true PRIs represented by solid line and 256 highly credible predicted PRIs represented by dotted line. Based on our experimental results, we can believe that these predicted PRIs are very possibly true PRIs.

Conclusion

In this paper, we present a novel method *FIRE* for boosting the performance of protein-RNA interaction prediction by selecting high-quality negative protein-RNA pairs to construct high-performance classifiers. Experiments over 18 PRI datasets show that the three compared classifiers, including SVM, RF and NB all achieve better performance on the negative sets selected by our method than on the random negative sets. This means that our

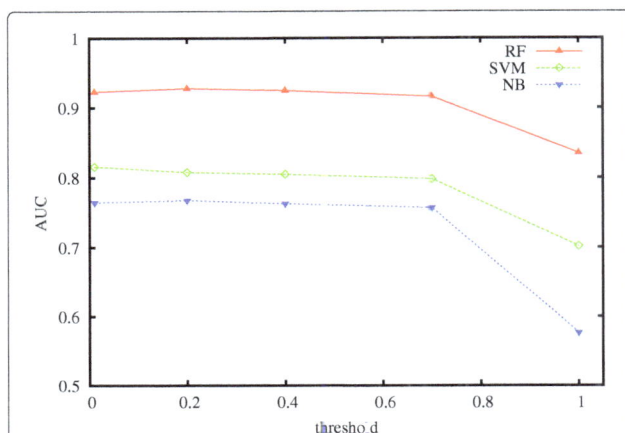

Fig. 7 AUC vs. score threshold (RF, SVM and NB)

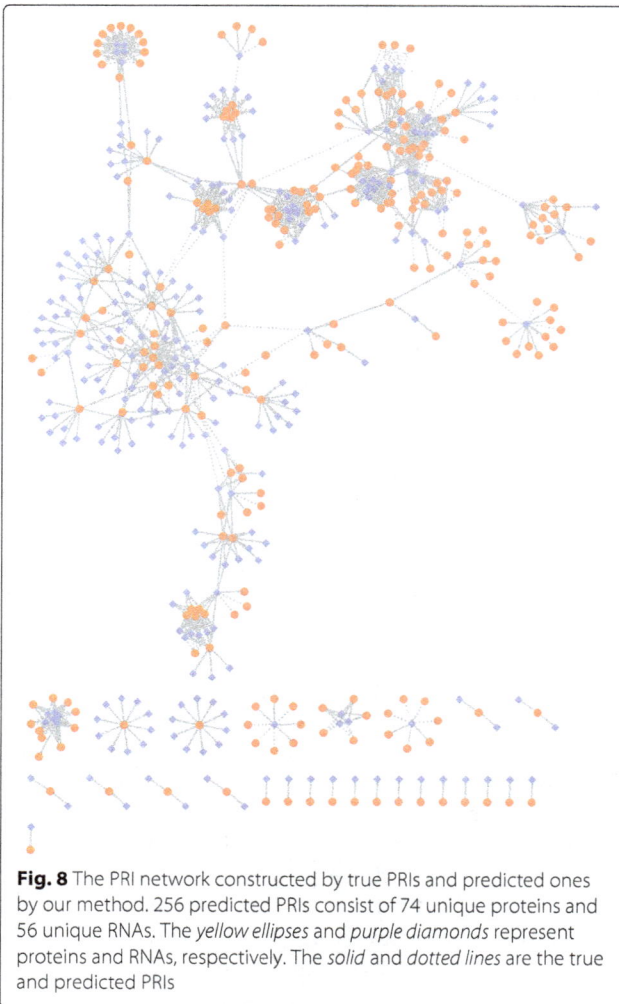

Fig. 8 The PRI network constructed by true PRIs and predicted ones by our method. 256 predicted PRIs consist of 74 unique proteins and 56 unique RNAs. The *yellow ellipses* and *purple diamonds* represent proteins and RNAs, respectively. The *solid* and *dotted lines* are the true and predicted PRIs

method can screen highly-credible negative PRIs, and thus can improve PRI prediction performance. As for future work, we will further explore the interacting mechanism between protein and RNA, and propose new and more effective methods to select reliable negative samples.

Acknowledgments
Not applicable.

Funding
National Natural Science Foundation of China (NSFC) (grant No. 61272380) for the design of the study, data generation and analysis, manuscript writing, and publication cost; The National Key Research and Development Program of China (grant No. 2016YFC0901704) for data collection and analysis; NSFC (grant No. 61672113) and the Program of Shanghai Subject Chief Scientist (15XD1503600) for data interpretation and manuscript writing.

Authors' contributions
SZ conceived and supervised the research, and revised the manuscript. ZC implemented the proposed method, carried out the experiments, did most data analysis, and drafted the manuscript. KH did some data analysis. YW prepared datasets and implemented some compared methods. HL and JG participated discussions and manuscript revision. All authors read and approved the final manuscript.

Competing interests
The authors declare that they have no competing interests.

Author details
[1]School of Computer Science, Fudan University, Handan Road, 200433 Shanghai, China. [2]The Bioinformatics Lab at Changzhou NO. 7 People's Hospital, Changzhou, 213011 Jiangsu, China. [3]Department of Computer Science and Technology, Tongji University, 201804 Shanghai, China. [4]Lab of Information Management, Changzhou University, 213164 Changzhou, China. [5]School of Computer Science, Jiangxi Normal University, 330022 Nanchang, China.

References
1. Moore PB. The three-dimensional structure of the ribosome and its components. Annu Rev Biophys Biomol Struct. 1998;27(1):35–58.
2. Moras D. Structural and functional relationships between aminoacyl-tRNA synthetases. Trends Biochem Sci. 1992;17(4):159–64.
3. Ramakrishnan V, White SW. Ribosomal protein structures: Insights into the architecture, machinery and evolution of the ribosome. Trends Biochem Sci. 1998;23(6):208–12.
4. Mata J, Marguerat S, Bähler J. Post-transcriptional control of gene expression: A genome-wide perspective. Trends Biochem Sci. 2005;30(9): 506–14.
5. Siomi H, Dreyfuss G. RNA-binding proteins as regulators of gene expression. Curr Opin Genet Dev. 1997;7(3):345–53.
6. Frank DN, Pace NR. Ribonuclease P: Unity and diversity in a tRNA processing ribozyme. Ann Rev Biochem. 1998;67(1):153–80.
7. Ramos A, Varani G. A new method to detect long-range protein-RNA contacts: NMR detection of electron-proton relaxation induced by nitroxide spin-labeled RNA. J Am Chem Soc. 1998;120(42):10992–10993.
8. Singh D, Febbo PG, Ross K, Jackson DG, Manola J, Ladd C, Tamayo P, Renshaw AA, D'Amico AV, Richie JP, et al. Gene expression correlates of clinical prostate cancer behavior. Cancer Cell. 2002;1(2):203–9.
9. Hall KB. RNA–protein interactions. Curr Opin Struct Biol. 2002;12(3):283–8.
10. Tian B, Bevilacqua PC, Diegelman-Parente A, Mathews MB. The double-stranded-RNA-binding motif: Interference and much more. Nat Rev Mol Cell Biol. 2004;5(12):1013–23.
11. Jones S, Daley DT, Luscombe NM, Berman HM, Thornton JM. Protein–RNA interactions: A structural analysis. Nucleic Acids Res. 2001;29(4):943–54.
12. Ellis JJ, Broom M, Jones S. Protein–RNA interactions: Structural analysis and functional classes. Proteins Struct Funct Bioinforma. 2007;66(4): 903–11.
13. Auweter SD, Oberstrass FC, Allain FH-T. Sequence-specific binding of single-stranded RNA: Is there a code for recognition? Nucleic Acids Res. 2006;34(17):4943–959.
14. Hermann T, Westhof E. Simulations of the dynamics at an RNA–protein interface. Nat Struct Mol Biol. 1999;6(6):540–4.
15. Pancaldi V, Bähler J. In silico characterization and prediction of global protein–mRNA interactions in yeast. Nucleic Acids Res. 2011;39(14): 5826–836.
16. Muppirala UK, Honavar VG, Dobbs D. Predicting RNA-protein interactions using only sequence information. BMC Bioinforma. 2011;12(1):489.
17. Wang Y, Chen X, Liu ZP, Huang Q, Wang Y, Xu D, Zhang XS, Chen R, Chen L. De novo prediction of RNA–protein interactions from sequence information. Mol BioSyst. 2013;9(1):133–42.
18. Cheng Z, Zhou S, Guan J. Computationally predicting protein-RNA interactions using only positive and unlabeled examples. J Bioinforma Comput Biol. 20151541005. doi:10.1142/S021972001541005X.

19. Lewis BA, Walia RR, Terribilini M, Ferguson J, Zheng C, Honavar V, Dobbs D. PRIDB: a protein–RNA interface database. Nucleic Acids Res. 2011;39(suppl 1):277–82.

20. Yuan J, Wu W, Xie C, Zhao G, Zhao Y, Chen R. NPInter v2. 0: an updated database of ncRNA interactions. Nucleic Acids Res. 2014;42(D1):104–8.

21. Szklarczyk D, Franceschini A, Kuhn M, Simonovic M, Roth A, Minguez P, Doerks T, Stark M, Muller J, Bork P, et al. The STRING database in 2011: functional interaction networks of proteins, globally integrated and scored. Nucleic Acids Res. 2011;39(suppl 1):561–8.

22. Berman HM, Westbrook J, Feng Z, Gilliland G, Bhat T, Weissig H, Shindyalov IN, Bourne PE. The protein data bank. Nucleic Acids Res. 2000;28(1):235–42.

23. Consortium TU. Update on activities at the Universal Protein Resource (UniProt) in 2013. Nucleic Acids Res. 2013;41(D1):43–7.

24. Smith TF, Waterman MS. Identification of common molecular subsequences. J Mol Biol. 1981;147(1):195–7.

25. Consortium TGO. Gene Ontology Annotations and Resources. Nucleic Acids Res. 2013;41(D1):530–5. doi:10 1093/nar/gks1050.

26. Couto FM, Silva MJ, Coutinho PM. Measuring semantic similarity between Gene Ontology terms. Data Knowl Eng 2007;61(1):137–52.

27. Jaccard P. Nouvelles recherches sur la distribution florale. Bul Soc Vaudoise Sci Nat. 1908;44:223–70.

28. Finn RD, Bateman A, Clements J, Coggill P, Eberhardt RY, Eddy SR, Heger A, Hetherington K, Holm L, Mistry J, Sonnhammer ELL, Tate J, Punta M. Pfam: the protein families database. Nucleic Acids Res. 2013. doi:10.1093/nar/gkt1223.

29. Han L, Cai C, Ji Z, Cao Z, Cui J, Chen Y. Predicting functional family of novel enzymes irrespective of sequence similarity: A statistical learning approach. Nucleic Acids Res. 2004;32(21):6437–444.

30. Liu ZP, Wu LY, Wang Y, Zhang XS, Chen L. Prediction of protein–RNA binding sites by a random forest method with combined features. Bioinformatics. 2010;26(13):1616–22.

31. Terribilini M, Lee JH, Yan C, Jernigan RL, Honavar V, Dobbs D. Prediction of RNA binding sites in proteins from amino acid sequence. RNA. 2006;12(8):1450–62.

32. Shen J, Zhang J, Luo X, Zhu W, Yu K, Chen K, Li Y, Jiang H. Predicting protein–protein interactions based only on sequences information. Proc Natl Acad Sci. 2007;104(11):4337–341.

33. Akbani R, Kwek S, Japkowicz N. In: Boulicaut J-F, Esposito F, Giannotti F, Pedreschi D, editors. Applying Support Vector Machines to Imbalanced Datasets. Berlin: Springer; 2004, pp. 39–50.

2

Modeling *de novo* granulation of anaerobic sludge

Anna Doloman[1†], Honey Varghese[2†], Charles D. Miller[1] and Nicholas S. Flann[2*]

Abstract

Background: A unique combination of mechanical, physiochemical and biological forces influences granulation during processes of anaerobic digestion. Understanding this process requires a systems biology approach due to the need to consider not just single-cell metabolic processes, but also the multicellular organization and development of the granule.

Results: In this computational experiment, we address the role that physiochemical and biological processes play in granulation and provide a literature-validated working model of anaerobic granule *de novo* formation. The agent-based model developed in a *cDynoMiCs* simulation environment successfully demonstrated a *de novo* granulation in a glucose fed system, with the average specific methanogenic activity of 1.11 ml CH_4/g biomass and formation of a 0.5 mm mature granule in 33 days. The simulated granules exhibit experimental observations of radial stratification: a central dead core surrounded by methanogens then encased in acidogens. Practical application of the granulation model was assessed on the anaerobic digestion of low-strength wastewater by measuring the changes in methane yield as experimental configuration parameters were systematically searched.

Conclusions: In the model, the emergence of multicellular organization of anaerobic granules from randomly mixed population of methanogens and acidogens was observed and validated. The model of anaerobic *de novo* granulation can be used to predict the morphology of the anaerobic granules in a alternative substrates of interest and to estimate methane potential of the resulting microbial consortia. The study demonstrates a successful integration of a systems biology approach to model multicellular systems with the engineering of an efficient anaerobic digestion system.

Keywords: Agent-based modeling, Multicellular modeling, Anaerobic granulation, Wastewater treatment

Background

An efficient anaerobic digestion (AD) of organic matter is a result of a complex microbial interaction inside a bioreactor. For the high-rate anaerobic digestion of a feedstock, an up-flow anaerobic sludge blanket reactor (UASB) is a common choice. The superior performance of this reactor is due to the particular organization of microorganisms into spherical granular structures. The process of granulation was first noticed and documented in the early 1980s [1, 2] and since then a number of anaerobic granulation theories have been presented. The main reasoning for the granulation *per se* is the up-flow velocity inside sludge bed of a UASB reactor. Microbial cells moving up with the flow of the feed tend to stick to the other microbial cells. Such sticking behavior prevents a washout of the microbial inoculum from a reactor since the outlet for the digested feed is located in the top of the reactor [3, 4] (see Fig. 1). The most widely accepted theory states that granulation starts with a formation of a future granule's core, comprised of filamentous methanogenic bacteria *Methanothrix*, together with *Methanosarcina*, which secrete extracellular polymers (ECP) [5–7]. The surface charge of this core changes and become attractive for the oppositely charged anaerobic bacteria that are present in the dispersed inoculum of a UASB rector [8–10]. Chemo-attractance of other bacteria towards ECPs and substrate around the granule core may also play a major role in the further aggregation and formation of mature granules [11, 12]. Despite these possible explanations of the granulation

*Correspondence: nick.flann@usu.edu
†Equal contributors
[2]Department of Computer Science, Utah State University, Old Main Hill 420, 84322-4205 Logan, UT, USA
Full list of author information is available at the end of the article

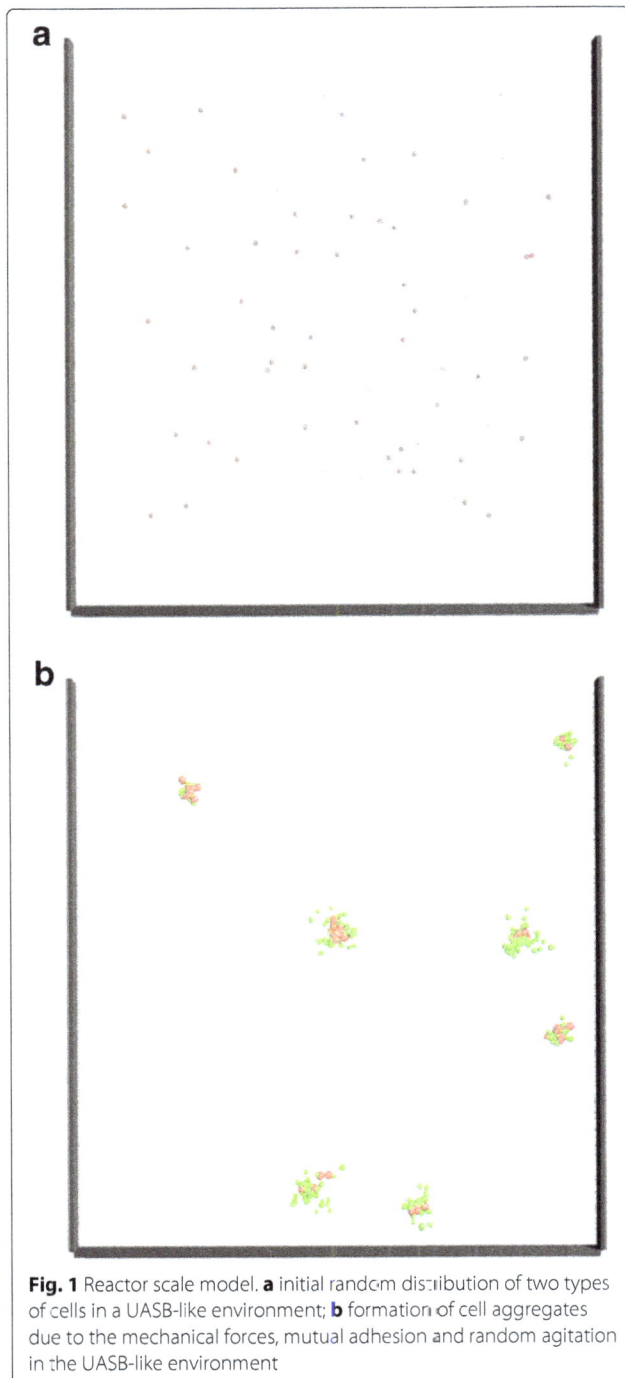

Fig. 1 Reactor scale model. **a** initial random distribution of two types of cells in a UASB-like environment; **b** formation of cell aggregates due to the mechanical forces, mutual adhesion and random agitation in the UASB-like environment

An effective means to get a better understanding the granulation process is through the construction of a computational granulation model. This model must incorporate testing of different key granulation factors. There are already some granulation models available in the literature, but they do not describe a process of *de novo* granulation and only describe the kinetics of anaerobic digestion with an already mature granular consortia. For example, one of the earliest models [13] assumes a layered granule structure with a homogeneous distribution of microbial groups from the very beginning of the simulation. Authors describe the kinetics of substrate transformation in a mature granule that reached a steady state. Using the same assumption [14] they successfully predicted the substrate distribution inside a granule, based on diffusivity gradient inside a biomass. Authors of another study [15] took the substrate kinetics in the granule one step further, incorporating behavior of granular agglomerates into the operation predictions of the whole UASB reactor. The mass of granules in a reactor, rates of granule decline and general bacterial growth kinetics were used as a basis for the model. In another study [16], researchers have applied a cellular automata theory, developed by Wimpenny et al., [17], to model granulation during anaerobic digestion. However, authors assumed a homogeneous layered structure of a granule and obtained calculated values of substrate utilization rates that do not agree with the experimental data they used as a reference.

A commonly applied assumption of a homogenous-layered structure of anaerobic granule does not conform with experimental data. In particular, data suggests a spatially organized granule containing a mixed composition of bacterial groups inside the granule. In models lacking this property, there is no strict compartmentalization of trophic groups, like methanogens and acidogens, in the core and outer layer, respectively. Strict anaerobes, like methanogens, can also be found in the outer layer of the granule, as visualized with fluorescent probing experiments and scanning electron microscopy [18–21]. A non-homogeneous bacterial distribution is investigated in a model described in [22]. However, the study does not address the process of granulation itself, and an entirely formed granule is employed as an initial condition and seed of a model. The model, therefore, predicts a mature granule's further development, growth, and formation of an inert core insie it.

An enormous amount of knowledge has been developed on predicting the rates of anaerobic digestion in UASB reactors with mature granules. However, these models are not complete and do not represent the actual input for large scale applications, specifically those of the widely accepted biochemical model of the anaerobic digestion process (ADM1) [23]. The most recent review of a current status of ADM1 clearly states the need to

process, there is still no agreement on which of the possible theories correctly explain this most important and crucial role of granulation. The key factors of granulation are still to be determined, whether they are physical, biochemical or a combination of physicochemical properties of the cells and the way the organic matter transforms over space and time.

thoroughly address the application of ADM1 to various types of anaerobic reactors, UASB in particular. Thus, a complete and trustful model of anaerobic digestion in UASB must take into account both granulation in general and initial *de novo* granulation [24]. Knowledge of the critical parameters facilitating *de novo* granule formation will aid in robust UASB reactor operation and production of increased methane yields with high organic matter transformation rates.

To model *de novo* anaerobic granulation, a number of computational platforms has been reviewed to find the best fit. The cellular Potts model was a pioneer [25] in biofilm modeling and has been extensively implemented in modeling of biofilms of the eukaryotic origin [26, 27]. To effectively apply this approach to the microbial liquid-based environment (thus without influence of attachment/detachment to the substratum), this model needs a lot of improvements, to prevent formation of artifacts [28, 29]. To model *de novo* anaerobic granulation, a number of computational platforms has been reviewed to find the best fit. The cellular Potts model was a pioneer [25] in biofilm modeling and has been extensively implemented in modeling of biofilms of the eukaryotic origin [26, 27]. To effectively apply this approach to the microbial liquid-based environment (thus without influence of attachment/detachment to the substratum), this model needs a lot of improvements, to prevent formation of artifacts [28, 29]. A simulator framework *cDynoMics* [30, 31], on the other hand, is more quantitative and is very flexible to adjust for modeling of bacterial aggregates. This framework has built-in functions to specify all the necessary substrate limiting kinetics for cell growth and biomass decay due to the starvation, which are absent in other previously described platforms. Absence of a solid substratum in the anaerobic digestion system excludes need for the use of attractive van der Waals force in the model, unlike in other reported biofilm developing tools [32].

A model of *de novo* granulation proposed in this paper addresses some of the key aspects that influence aggregation of microbial biomass into defined granular structures. Those key elements include: initial concentrations of the substrate used as a feedstock for anaerobic digestion; ratio of methanogenic and acidogenic cells at the start of the reactor; the role of chemotactic attractions and cell-to-cell adhesion properties. This study addresses all these factors. Additionally, an extensive computational search of the initial parameter values is made to determine an optimal initial combination that yields the highest start-up methane production rates.

Results and discussion

Simulation experiments were conducted on the computational granulation model to give insights into different stages in the development of granules in aerobic sludge reactors. Where available, literature supported model parameters were employed. Other parameters, such as those that influence particle aggregation and mechanical sorting, were fine tuned based on correspondence between observations made from simulations and comparisons with reported granule images. The resulting granule spatial organization and product production of model simulations are analyzed and compared with values from real biological systems. Another objective of the study was to employ a search engine to find the amount of initial glucose concentration and populations of methanogens and acidogens that lead to optimal methane production.

Study I: reactor scale model

In the reactor scale phase of modeling, randomly distributed acidogens and methanogens (illustrated in Fig. 1a) interact with each other in a simulated UASB reactor environment, where upflow velocity and agitation play key roles to promote granulation of sludge. In the simulated environment microbial cells move around the system due to agitation and cells are bound together due to biomechanical adhesive forces, allowing formation of cell agglomerates (illustrated in Fig. 1b).

Study IIa: stages of granule formation

To investigate the development of a mature granule and dynamic changes in the cell growth, consumption of glucose, a series of simulator output snapshots were performed (Fig. 2). At the initial stage (t=0 h), single cell aggregate appear as a small cluster of acidogens and methanogens (zoomed from Reactor scale model, Fig. 1). As time proceeds (t=300, t=480 and t=700 h) cells grow and corresponding solute gradients demonstrate accumulation of acetate and methane in the system. Methane, being a volatile compound, is slowly diffused out of the system and depicted values on the scale of gradient images are not the cumulative values, as in the case of the glucose and acetate. At 480 h of granule development, a black "dead" core of cells start to emerge in the middle of the granule sphere. Appearance of a "dead" core is due to the diffusion boundaries of glucose or acetate inside granular cluster. Thus, cells of both types (acidogens and methanogens) are not getting enough energy supply and are forced to transition into the inert biomass. This transition is set to be irreversible in the model, thus leading to a formation of a "dead core". A similar core can be seen on the Fig. 4a of the laboratory-observed granule, which is used as evaluation criterion in current study and is descried later in detail. The final stage of granule development simulation (t=650 h) demonstrates a mature granule with 0.5 mm in diameter.

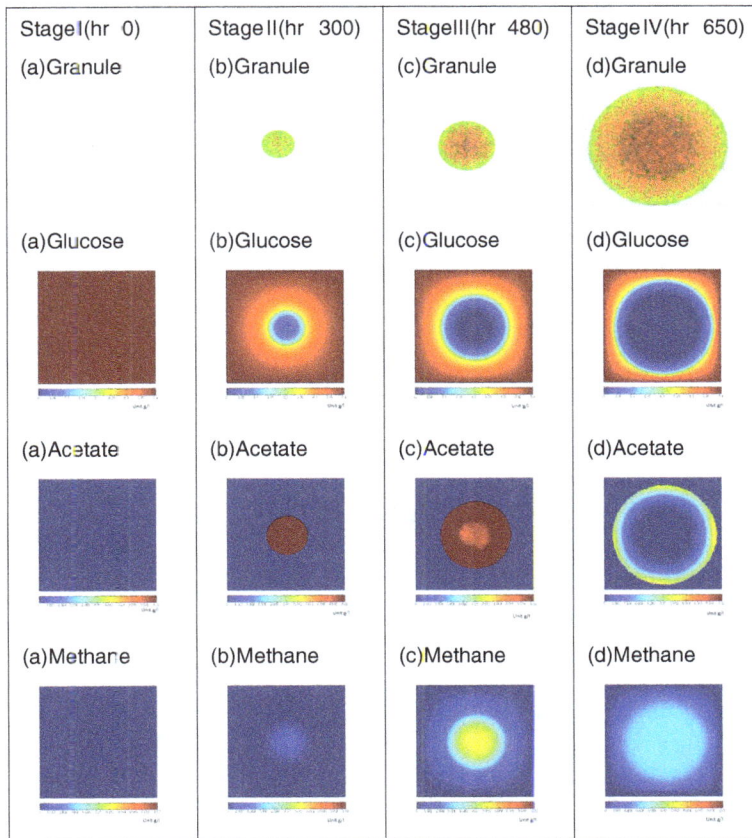

Fig. 2 Simultion of 0.5 mm granule formation. Stages of simulated *de novo* granulation and associated dynamic changes in the solutes concentrations (glucose, acetate and methane). Only the critical time points of simulation are depicted through stages I-IV (t=0 h through t=650 h)

Study IIb: analysis of granule growth dynamics

In addition to visual (qualitative) investigation of *de novo* granulation, a close up quantitative study was performed on dynamic changes in solute amounts and cell biomass accumulation (both in values of cell numbers and cell biomass numbers). Graphs for dynamic changes are provided in Fig. 3. Figure 3a demonstrates changes in the total number of two types of cells (acidogens and methanogens) with regard to the simulation time. Simulation was initiated with 100 cells of each type. Due to the fast growth of the acidogens (see the Table 1 with growth kinetics parameters), we can see an exponential growth of acidogens from t=80 h to t=360. A similar dynamic is depicted in Fig. 3b. Due to the product inhibition by the produced acetate and lack of diffused glucose, acidogens decrease their relative growth rate and reach the stationary phase of growth at around t=600 h. Dynamics of methanogens growth is slightly different, mainly due to the lack of available acetate from the start-up of the system and a lower growth rate, contrary to acidogens (Table 1 with model parameters). Methanogen growth goes through a long lag phase (t=0 h until t=220 h), where biomass is

accumulated at a very slow rate (Fig. 3b). At this lag phase methanogen cells are waiting for the supply of acetate from acidogens. As soon as enough acetate is accumulated in the system (around t=220 h), methanogens start exponential growth and decrease their relative growth rate at about t=520 h. This decrease is in direct correspondence with the amount of available acetate in the system at the same time period (t=480–500 h), (Fig. 3c) when acidogens are inhibited by the produced acetate and are not provided with a high flow of glucose (due to the slow diffusion into the center of the granular biomass). Kinetics of acetate accumulation/conversion and methane production are in a good correlation with experimental data reported by Kalyzhnyy et al. and others [33–36].

Study III: formation of a mature granule

Figure 4 shows images of a 1 mm in diameter granule, obtained from both a laboratory experiment reported by Sekiguchi et al. [19] (Fig. 4a) and an image from our simulated model (Fig. 4b). Simulation of 1 mm in diameter granule formation took 800 h (around 33 days), which corresponds to the published studies observing

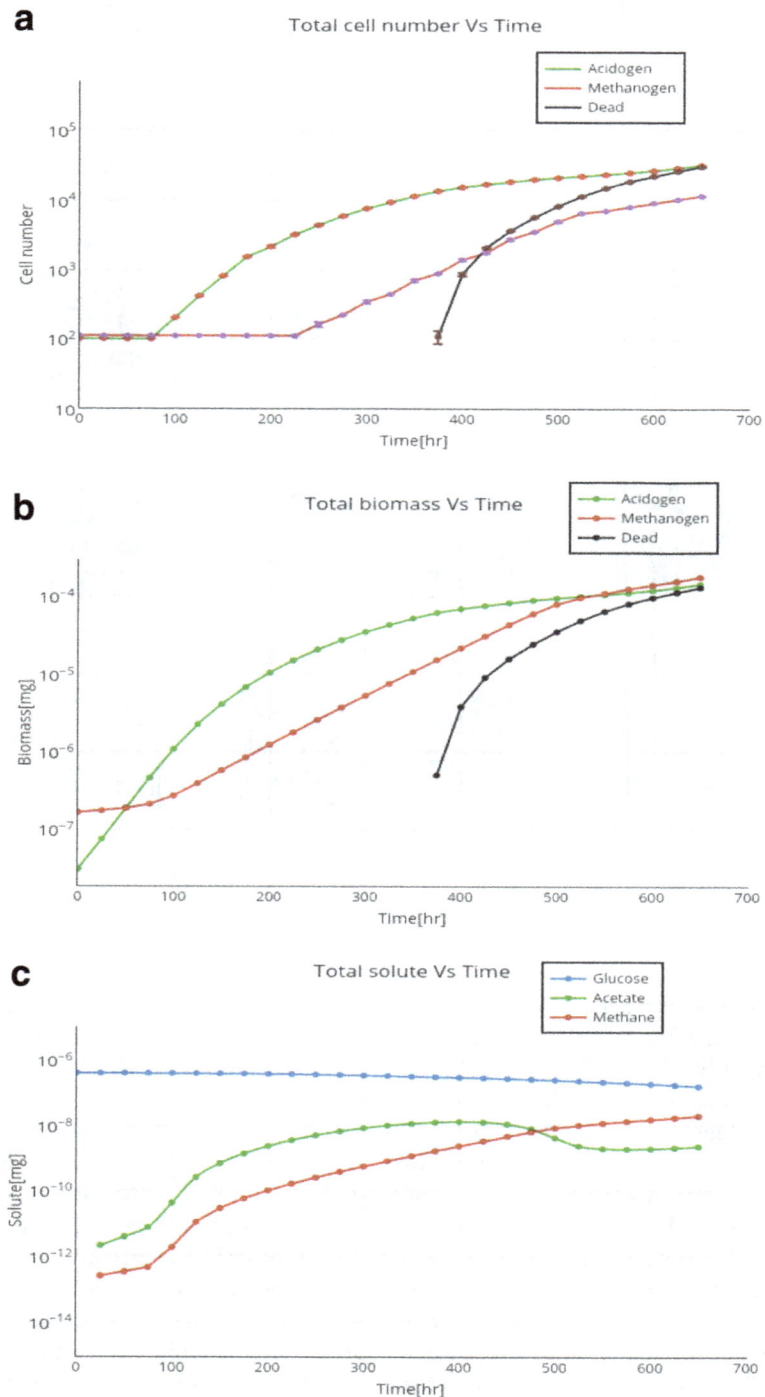

Fig. 3 Simulation related changes in solute concentrations and cell biomass. A close-up of the dynamic changes in the **a** cell number over simulation time, **b** cell biomass over simulation time and **c** solutes concentrations over simulation time. All the changes are graphed for each type of the cell (acidogens, methanogens, inert dead type) and each type of the solute (glucose, acetate, methane). Ten simulations with different random seeds were graphed to demonstrate standard deviation in the monitored values

granulation in UASB reactors [20, 37]. Figure 4c, d and e depict distribution of solutes (glucose, acetate, and methane) at the final stage of simulated granule growth (t=800 h). One can note a sharp decrease in the glucose diffusion inside the granule, with regard to the biofilm diffusivity capacity. Since acetate is consumed

Table 1 Parameters used in model and their correspondent values

Parameter summary

Model parameter	Symbol	Value	Unit	References
Solutes				
Diffusion of glucose in liquid	D_g	5.8×10^{-6}	m^2/day	[59]
Diffusion of acetate in liquid	D_a	1.05×10^{-4}	m^2/day	[59]
Diffusion of methane in liquid	D_m	1.29×10^{-4}	m^2/day	[60]
Biofilm Diffusivity	γ	30	%	[42]
Acidogens				
Cell mass	B_a	300	fg	[61]
Division radius		3	μm	[62]
Maximum growth rate	$\hat{\mu}_a$	0.208	h^{-1}	[61], [56, 63]
Substrate saturation constant	Ks	0.26	g/L	[35, 56]
Product inhibition constant	Ki	0.1	g/L	[56, 63]
Biomass conversion rate	α_{bg}	0.3	$\frac{g_{biomass}}{g_{glucose}}$	[56, 57]
Substrate conversion rate	α_{ag}	0.82	$\frac{g_{acetate}}{g_{glucose}}$	[56, 63]
Death delay		48	h	Estimated
Death threshold		0.02	g/L	Estimated
Methanogens				
Cell mass	B_m	1500	fg	[62]
Mass of EPS capsule		10	fg	[54]
Division radius		3	μm	[62]
Maximum growth rate	$\hat{\mu}_m$	0.1	h^{-1}	[33, 54]
Substrate saturation constant	Ks	0.005	g/L	[54]
Biomass conversion rate	α_{ba}	0.15	$\frac{g_{biomass}}{g_{acetate}}$	[33, 35]
Substrate conversion rate	α_{ma}	0.26	$\frac{g_{methane}}{g_{acetate}}$	[33]
Death delay		48	h	Estimated
Death threshold		0.00001	g/L	Estimated

by methanogens during their growth and converted to methane, there is a low concentration gradient of both chemicals on the final images (Fig. 4c, d e). Overall, solute distributions for 1mm granule follow a similar pattern as for the 0.5 mm granule, described earlier. Key point in conducting simulation of a 1mm granule development is to demonstrate radial growth, without substantial changes in the overall morphology. Thus, initial stages of granule formation are the key factors for granulation *per se*.

Validation of the model

Validation of the model performance was conducted both qualitatively (Fig. 4a, b) and quantitatively (Fig. 5). Visual comparison of a published fluorescent-labeled image of granule with simulated granule image demonstrates a striking similarity in spatial distribution of main trophic groups of microorganisms - acidogens, methanogens and "dead" biomass. Irregularities and hollow parts (black color) in the published granule image (Fig. 4a) are possibly

caused by the upflow velocity of the liquid and particulate matter in a UASB reactor, where the granule was developed [19], which might have damaged spherical shape of the immature granule, causing mature granule to change its shape and grow further with hollow compartments. Another possible explanation might be granule division. It is well documented [8–10] that due to the shear stress in a UASB reactor, granules cannot grow uncontrollably and will eventually split into "daughter" granules. Those "daughter" granules are susceptible to attachments of additional microbial cells, floating in UASB sludge bed. Those newly attached cells might cause irregularities in future mature granules in forms of randomly distributed cell clusters in a presumably inert ("dead") core (red-labeled cell clusters on Fig. 4a). To validate our simulated model quantitatively, we conducted image processing of the published data and used an algorithm to count the number of distinctly colored pixels/cells at the different distances from the center of the granule image (Fig. 5).

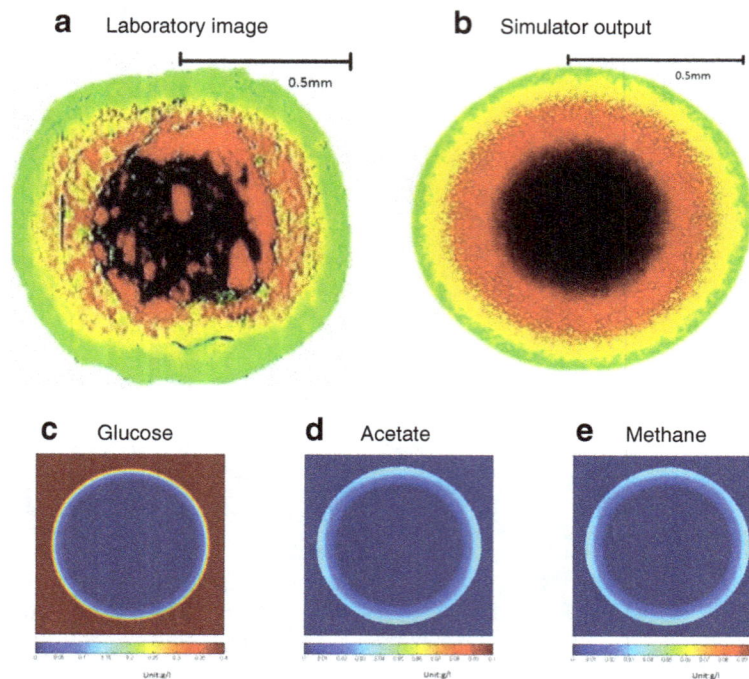

Fig. 4 Validation of the *de novo* granulation model via qualitative analysis. **a** Laboratory image courtesy of Sekiguchi et al. (1999), where *green fluorescence* label was used for Bacteria (represented by a single group of acidogens in current study), *red fluorescence* was emitted by Archaea (represented by a single group of methanogens in current study), *yellow color* correlates with overlapped *red* and *green fluorescence* and *black color* represents absence of fluorescence hybridization, and thus, absence of cell biomass (denoted as dead core here). **b** An image of granule simulated with current model. Same color labeling of the cell types is applied. **c**, **d** and **e** Distribution of the three solutes defining simulation of granulation (glucose, acetate, methane) at the final time point (t=800 h) of the simulation

We used 4 quarters of a spherical granule in the analysis to provide standard deviations of spatial distribution of three distinct cell groups – acidogens, methanogens and inert ("dead") biomass. Results of quantitative distribution of three main cell types in both simulated and real images are in a good correlation, accept for the radial section "3". Such slight discrepancy is due to the possible "division to daughter granules" history of the laboratory granule.

Parameter scan for optimized methane production

Main objective of the parameter scan is to estimate a combination of cell ratio (acidogens:methanogens) and glucose supply needed to start anaerobic system to achieve a desired (maximum) methane yield. The corresponding protocol parameter for glucose value is "SBulk" in world section. The "init area number" for acidogens and methanogens in the species section is used to determine the initial cell ratio for the simulations. The minimum and maximum value of the interval in which the search should be performed is given as an input to the search engine. The methane productivity (calculated from the solute concentration file output from simulator) is given as fitness function for the engine. The search engine simulated granule formation for several combinations of parameter values

within the input interval and calculated total methane produced. The result is produced as a heatmap in Fig. 6.

Figure 6 depicts amount of methane produced (in milliliters) per gram of biomass with varying amount of glucose supplied initially into the system (0.1 to 0.4 g/l). Figure 6a has a constant initial acidogen count of 100 cells, and heatmap demonstrates varying amounts of methane produced with different glucose concentrations and different numbers of initial methanogen cells (from 1 to 900 cells). Same scheme is followed on Fig. 6b, but with varying initial numbers of acidogens (from 1 to 400) and constant initial methanogen count of 100 cells.

One can note from both Fig. 6a and b that increased amount of glucose correlates with increased amount of methane produced in the system. Also, in general increased number of starting cells of acidogens (Fig. 6b) let to the higher amounts of methane produced. This correlates with the earlier explored kinetics of methanogen/acidogen growth, when methanogens are waiting for acetate supply until they start to grow and produce methane. Parameter scan also helped to identify an important observation that a ratio of methanogen cells to acidogens should not be in a high favor of methanogens (100 acidogens and 900 methanogens on Fig. 6a), since

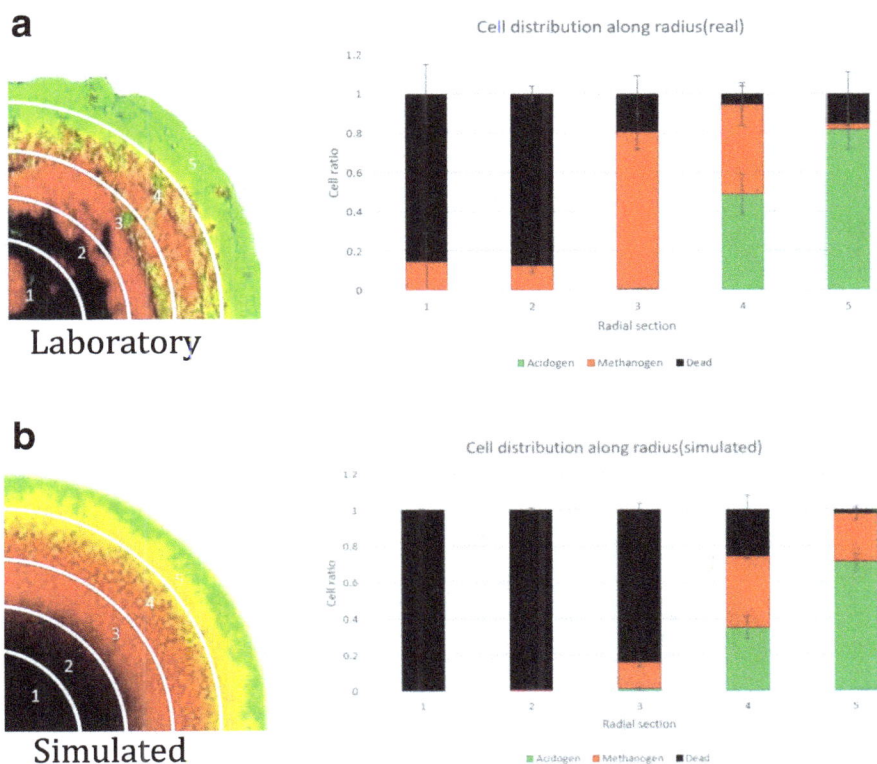

Fig. 5 Validation of the *de novo* granulation model via quantitative analysis. Validation was done via analysis of the three cell type radial distribution in the both laboratory (**a**) and simulated granules (**b**). Both granules were divided into four quarters and each quarter was analyzed for cell distribution. Differences in the cell numbers at the same radial distance in four quarters are depicted in a form of standard deviation. *Red, green* and *black colors* of the bars on bar chart represent acidogen, methanogen and dead cells respectively

this leads to a decreased amount of methane production. The reason for such correlation is lack of acetate in the system to support growth of such a big number of methanogenic cells, which are forced to starve and die off.

Conclusions

A model of anaerobic granulation from digestion of glucose to methane has been successfully implemented in an agent-based simulator framework, *cDynoMiCs*. Simulation studies incorporated modeling of both reactor and single agglomerate scale granule development. Utilized growth mechanisms for generalized glucose-consuming/acetate-producing bacteria and acetate-consuming/methane-producing bacteria resulted in a well-correlated kinetic patterns of substrate conversions and biomass growth (Fig. 3). We were able to successfully qualitatively and quantitatively validate the architecture of the developed simulated anaerobic granule with the granule images and cell distribution from experimental literature studies (Figs. 4 and 5). The described granulation model has direct applications for designs of experiments, to predict yields of methane gas from substrates of interest. One application of the

model was successfully demonstrated in this paper via parameter scan algorithm, searching through different acidogens:methanogens cell ratios and glucose feed that is needed to start anaerobic system to achieve a desired (maximum) methane yield. By changing the parameters of microbial growth to fit bacteria of a specific interest (the bacteria one is targeting to explore in an AD experiment), researchers can apply this model to predict efficiencies of anaerobic digestion in a system. The tested parameter scan is directly applicable to the studies with low-strength feed streams to UASB reactors, such as AD of brewery wastewater (COD=100-800 mg/L) [38], some municipal and industrial wastewaters (COD=100-400 mg/L) [39, 40] and effluents from petroleum refineries (COD from 68 mg/L) [41]. Further development of the model will include a parameter search to investigate methane production from medium and high strength wastewaters. The current model of anaerobic granulation and methane production from simple feed sources (glucose) can be expanded to accommodate microbial conversion of more substrates, such as a mixture and proteins and carbohydrates. This expansion will make it possible to study granulation and methane potential from a more realistic

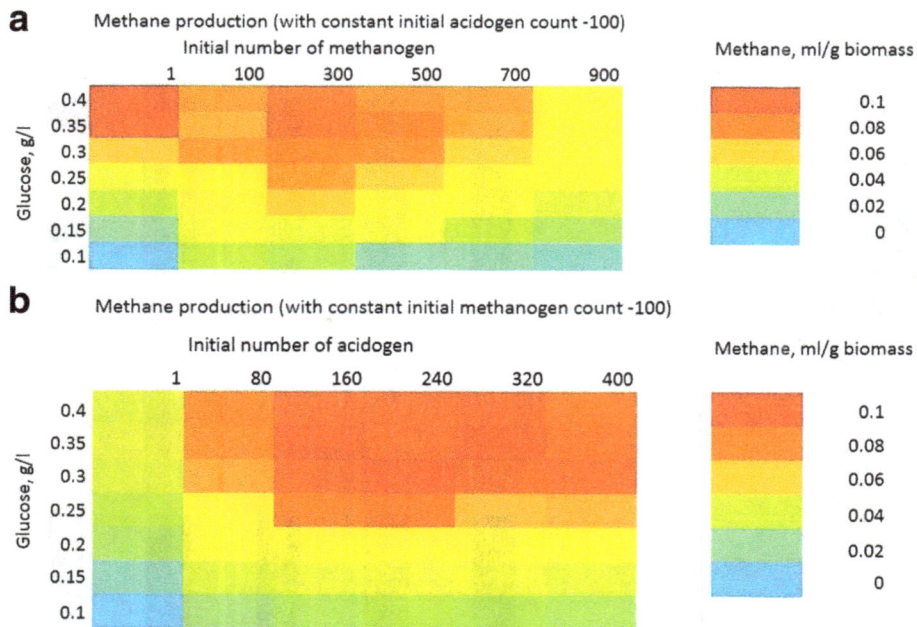

Fig. 6 Parameter scan for the methane production in simulated granule. Parameter scan for the methane production in simulated granule with **a** varying initial number of methanogen cells (constant initial acidogen cell count) and **b** varying initial number of acidogen cells (constant initial methanogen cell count). *Red color* of the heatmap section has the highest value of methane produced (in milliliters of methane per gram of biomass), while *blue* heatmap section has the lowest value of produced methane. Parameter scan was conducted for 0.5 mm granule size and for the period of 650 simulation hours

scenario of wastewater feed, such as dairy and municipal wastewaters. A granulation model from a complex feed should result in a less stratified granule, due to the differential diffusions of the main feed components and a more complex patterns of microbial growth kinetics [18].

In addition, a model framework (*iDynoMiCs*) can be further modified to simulate detachment of excessive biomass from granular surface (simulating sheer stress described in the UASB reactor environment [4, 42–44]) and breakage of a granule into daughter clusters, that subsequently give rise to mature granules with a more complex morphology [18, 21, 45]. Since current model assumes spherical types of cells, exploration of filamentous type of methanogenic bacteria influencing *de novo* granulation based on the "spaghetti theory" is something of future interest [32, 46]. Another possible realm to expand development and application of current granulation model is to explore the mechanisms of enhancing anaerobic granulation, such as addition of positively charged ions and particles of polymers into the UASB system [47, 48]. To converge granulation model with reactor-like environment, a *Biocellion* modelling environment can be used [49, 50]. Possibility to parallelize computation load in Biocellion would eliminate the main bottleneck of the *cDynoMics* and allow development of a whole reactor model with simultaneous substrate conversion and anaerobic granule development. The current model of the

de novo anaerobic granulation and its immediate applications will aid future discoveries in the field of anaerobic digestion, which is regaining its value and popularity in sustainable energy.

Methods

The process of granulation is modeled at two spatial scales in the simulation. At the macroscale, the reactor process is simulated where the cells are introduced into an agitated system (due to the upflow velocity in UASB reactor), cells interact and form multiple agglomerates (centers of granulation). At the mesoscale, simulations are performed that focus on the growth and development of one such agglomerate into a mature granule.

In the macroscale, randomly distributed acidogenic (further referred to as "acidogens") and methanogenic cells (further referred to as "methanogens") are introduced into random positions within the reactor. The particles experience mechanical forces due to agitation in the system as well as biomechanical forces due to homogeneous and heterogeneous adhesion and formation of EPS-driven interactions. As a cumulative effect of these forces, cells come close to each other and form several agglomerates.

To closely monitor the growth patterns in the formation of a granule, the mesoscale simulation is designed to focus on the development of a single granule (from the initial agglomerate of acidogens and methanogens formed during the macro studies). In UASB bioreactors, granules move freely in an agitated system, where the supplied solutes are relatively mixed. To simulate such a mixed environment for the granule growth, we provide a continuous supply of one solute (glucose) from all the sides of the simulation domain with diffusivity as defined in Table 1. The model executes growth reactions that represent the consumption of the supplied glucose by the acidogens, the secretion of the acetate as a metabolite of acidogens and the consumption of acetate by methanogens, which is converted into the methane gas.

An agent-based simulator framework, *cDynoMiCs* [31] is used in this experiment. *cDynoMiCs* is an extension of *iDynoMiCS* framework developed by the Kreft group at University of Birmingham [51] specifically for modeling biofilms. *cDynoMiCs* includes eucaryotic cell modeling processes with the addition of extracellular matrix and cellular mechanisms such as tight junctions and chemotaxis. Each cell is represented as a spherical particle, which has a particular biomass, and implements type and species-specific mechanisms to reproduce cellular physiology. Biochemically, particles can secrete or uptake chemicals that are diffused through the domain by executing reactions. Biomechanically, particles exhibit homogeneous and heterogeneous adhesion, and the formation of tight junctions. Particles model growth by increasing their biomass according to metabolic reactions and split into two particles once a maximum radius threshold is reached. They can also switch from one type of particle to another based on specific microenvironmental conditions and internal states. The simulation process interleaves biomechanical stress relaxation where the particles are moved in response to individual forces, along with the resolution of biochemical processes such as secretion, uptake, and diffusion by a differential equation solver. We assume that the solute fields are in a pseudo steady-state with respect to biomass growth [51].

Particle growth and division can cause particles to overlap, creating biomechanical stress. To resolve this problem a process called shoving is implemented. When the distance between two particles is less than a fixed threshold set by the particle size, a repulsive force is generated to push them apart, proportional to the overlap distance between the two particles. Then the relaxation process commences that iteratively moves each particle in response to its net force, then recalculates the forces due to the movement. The process terminates when only negligible forces remain, and the system has reached a pseudo steady state.

cDynoMiCs adds new functionality to the Java code of *iDynoMiCS* and extends the XML protocol, used to specify many different types of simulations. *iDynoMiCS* writes plain-text XML files as output, and these may be processed using any number of software tools, such as Matlab

and R. In addition to XML files, *iDynoMiCS* also writes files for POV-Ray that is used to render 3-D ray-traced images of the simulation. For the experiment to form the 1mm granule a 1.16 mm × 1.16 mm domain size was used. For all other experiments, a 508 μm × 508 μm domain size (2D) is used. A summary of the protocol parameter values can be found in Table 1.

Three solutes glucose (S_g), acetate (S_a) and methane (S_m) exist within the reactor model. The distribution of these solutes is controlled by Eqs. 1, 2, and 3 respectively. The diffusion coefficients and reaction rates take different forms for each region depending upon the spatial distribution of acidogen biomass (B_a), methanogen biomass (B_m) and dead biomass (B_d) described in Eq. 4. The effective diffusion coefficient is decreased within the granule compared with the liquid value in order to account for the increased mass transfer resistance. The diffusivity values used for the model (specified in Table 1) are taken from literature related to biofilm diffusivity studies [42, 52]. The growth rate of acidogens is $\mu_a(S_g, S_a)$, defined in Eq. 8, and the growth rate of methanogens is $\mu_m(S_a)$ defined in Eq. 9.

$$\frac{\partial S_g}{\partial t} = B(x,y).D_g.\frac{\nabla^2 S_g}{\partial x \partial y} - \mu_a(S_g, S_a).\frac{B_a}{\alpha_{bg}} \quad (1)$$

$$\frac{\partial S_a}{\partial t} = B(x,y).D_a.\frac{\nabla^2 S_a}{\partial x \partial y} + \mu_a(S_g, S_a).\frac{\alpha_{ag}.B_a}{\alpha_{bg}} \quad (2)$$

$$\frac{\partial S_m}{\partial t} = B(x,y).D_m.\frac{\nabla^2 S_m}{\partial x \partial y} + \mu_m(S_a).\frac{B_m}{\alpha_{ba}} \quad (3)$$

where,

$$B(x,y) = \begin{cases} 1.0 & \text{if location } x, y \text{ contains no biomass} \\ \gamma & \text{if location } x, y \text{ contains biomass} \end{cases}$$

$$(4)$$

Equations 5 and 6 describe acidogen and methanogen biomass changes as a function of local acetate and glucose concentration. Cell death due to lack of food is modeled using a discrete switching mechanism defined as the function $die(B_i)$ in the equations. Acidogen cells are converted to dead cells when the amount of glucose is below a threshold value (death threshold in Table) for a period of 48 h. Similarly, the methanogen cells are converted to dead cells when the amount of glucose is below a threshold value (death threshold in Table 1) for a period of 48 h. The rate of increase in dead cell mass is define in Eq. 7. The parameter values for controlling cell death are estimated due to the lack of studies quantifying the response of acidogen and methanogen cells to nutritional stress.

$$\frac{\partial B_a}{\partial t} = \mu_a(S_g, S_a)B_a - die(B_a) \quad (5)$$

$$\frac{\partial B_m}{\partial t} = \mu_a.S_a.B_m - die(B_m) \quad (6)$$

$$\frac{\partial B_d}{\partial t} = die(B_a) + die(B_m) \quad (7)$$

Acidogens grow by consuming glucose and producing acetate described by the Monod-kinetic Eq. 8, where $\hat{\mu}_a$ is the maximum growth rate for acidogens. Similarly, methanogen growth by consuming acetate and producing methane described by Monod-kinetic Eq. 9, where $\hat{\mu}_m$ is the maximum growth rate for mathanogens. Values for growth constants, such as biomass yield and substrate conversion rate, for both acidogens and methanogens were taken from literature and averaged. Thus, maximum growth rate for acidogens was twice as high as that that of methanogens, see [3, 35, 53–58]. Biomass decay rate is not taken into account for both cell types, since decay for anaerobic type of growth is usually less or equal to 1% of specific growth rate and thus can be ignored [58]. Non-competitive product inhibition is considered for growth of acidogens [58], but not for the methanogens, assuming low inhibition of methanogenic growth by excess amount of acetate.

$$\mu_a(S_g, S_a) = \hat{\mu}_a \cdot \frac{S_g}{(K_{sg} + S_g)} \cdot \frac{K_i}{(K_i + S_a)} \tag{8}$$

$$\mu_m(S_a) = \hat{\mu}_m \frac{S_a}{K_{sa} + S_a} \tag{9}$$

Acknowledgments
Thanks to Jan-Ulrich Kreft School of Biosciences, University of Birmingham for providing the original version of iDynoMiCS.

Funding
Research was funded by USU USTAR Grants Program, Huntsman Environmental Research Center and State of Utah Energy Research Triangle.

Authors' contributions
AD and HV equally contributed to the work by designing a model, performing the simulation, validating the results and drafting the manuscript. CM and NF interpreted part of the data and supervised the work. All authors read and approved the final manuscript.

Competing interests
The authors declare that they have no competing interests.

Author details
[1] Department of Biological Engineering, Utah State University, Old Main Hill 4105, 84322-4105 Logan, UT, USA. [2] Department of Computer Science, Utah State University, Old Main Hill 420, 84322-4205 Logan, UT, USA.

References
1. Hulshoff-Pol LW, Dolfing J, van Straten K, de Zeeuw WJ, Lettinga G. Pelletization of anaerobic sludge in up-flow anaerobic sludge bed reactors on sucrose-containing substrates. 3rd International Symposium on Microbial Ecology Proceedings. Washington; 1984, pp. 636–42.
2. Zeeuw WD. Acclimatization of anaerobic sludge for UASB-reactor start-up. PhD thesis, [SI: sn]. 1984.
3. Kosaric N, Blaszczyk R, Orphan L. Factors influencing formation and maintenance of granules in anaerobic sludge blanket reactors (UASBR). Water Sci Technol. 1990;22(9):275–82.
4. Tiwari MK, Guha S, Harendranath CS, Tripathi S. Influence of extrinsic factors on granulation in UASB reactor. Appl Microbiol Biotechnol. 2006;71(2):145–54.
5. Schmidt JEE, Ahring BK. Extracellular polymers in granular sludge from different upflow anaerobic sludge blanket (UASB) reactors. Appl Microbiol Biotechnol. 1994;42(2-3):457–62.
6. Liu Y, Xu HL, Yang SF, Tay JH. Mechanisms and models for anaerobic granulation in upflow anaerobic sludge blanket reactor. Water Res. 2003;37(3):661–73.
7. Kobayashi T, Xu KQ, Chiku H. Release of extracellular polymeric substance and disintegration of anaerobic granular sludge under reduced sulfur compounds-rich conditions. Energies. 2015;8(8): 7968–85.
8. Tay JH, Xu HL, Teo KC. Molecular mechanism of granulation. I: H+ translocation-dehydration theory. J Environ Eng. 2000;126(5):403–10.
9. Teo KC, Xu HL, Tay JH. Molecular mechanism of granulation. II: proton translocating activity. J Environ Eng. 2000;126(5):411–8.
10. Liu XW, Sheng GP, Yu HQ. Physicochemical characteristics of microbial granules. Biotechnol Adv. 2009;27(6):1061–70.
11. Batstone DJ, Picioreanu C, Van Loosdrecht MCM. Multidimensional modelling to investigate interspecies hydrogen transfer in anaerobic biofilms. Water Res. 2006;40(16):3099–108.
12. Lin Y, Yin J, Wang J, Tian W. Performance and microbial community in hybrid anaerobic baffled reactor-constructed wetland for nitrobenzene wastewater. Bioresour Technol. 2012;118:128–35.
13. Tartakovsky B, Guiot SR. Modeling and analysis of layered stationary anaerobic granular biofilms. Biotech Bioeng. 1997;54(2):122–30.
14. Arcand Y, Chavarie C, Guiot SR. Dynamic modelling of the population distribution in the anaerobic granular biofilm. Water Sci Technol. 1994;30(12):63–73.
15. Shayegan J, Ghavipanjeh F, Mehdizadeh1O H. Dynamic Modeling of Granular Sludge in UASB Reactors. Iranian J Chem Eng. 2005;2(1):53.
16. Skiadas IV, Ahring BK. A new model for anaerobic processes of up-flow anaerobic sludge blanket reactors based on cellular automata. Water Sci Technol. 2002;45(10):87–92.
17. Wimpenny JWT, Colasanti R. A unifying hypothesis for the structure of microbial biofilms based on cellular automaton models. FEMS Microbiol Ecol. 1997;22(1):1–16.
18. Rocheleau S, Greer CW, Lawrence JR, Cantin C, Laramée L, Guiot SR. Differentiation of Methanosaeta concilii andMethanosarcina barkeri in Anaerobic Mesophilic Granular Sludge by Fluorescent In Situ Hybridization and Confocal Scanning Laser Microscopy. Appl Environ Microbiol. 1999;65(5):2222–9.
19. Sekiguchi Y, Kamagata Y, Nakamura K, Ohashi A, Harada H. Fluorescence in situ hybridization using 16S rRNA-targeted oligonucleotides reveals localization of methanogens and selected uncultured bacteria in mesophilic and thermophilic sludge granules. Appl Environ Microbiol. 1999;65(3):1280–8.
20. Fang HH. Microbial distribution in UASB granules and its resulting effects. Water Sci Technol. 2000;42(12):201–8.
21. Batstone DJ, Keller J, Blackall LL. The influence of substrate kinetics on the microbial community structure in granular anaerobic biomass. Water Res. 2004;38(6):1390–404.
22. Picioreanu C, Batstone DJ, Van Loosdrecht MCM. Multidimensional modelling of anaerobic granules. Water Sci Technol. 2005;52(1-2):501–7.
23. Batstone DJ, Keller J, Angelidaki I, Kalyuzhnyi SV, Pavlostathis SG, Rozzi A, Sanders WTM, Siegrist H, Vavilin VA. The IWA anaerobic digestion model no 1 (ADM1). Water Sci Technol. 2002;45(10):65–73.
24. Batstone DJ, Puyol D, Flores-Alsina X, Rodr'iguez J. Mathematical modelling of anaerobic digestion processes: applications and future needs. Rev Environ Sci Bio/Technol. 2015;14(4):595–613.

25. Graner FMC, Glazier JA. Simulation of biological cell sorting using a two-dimensional extended potts model. Phys Rev Lett. 1992;69:2013–6. doi:10.1103/PhysRevLett.69.2013.

26. Mora Van Cauwelaert E, Del Angel A, Antonio J, Benítez M, Azpeitia EM. Development of cell differentiation in the transition to multicellularity: a dynamical modeling approach. Front Microbiol. 2015;6:603.

27. Marée AF, Hogeweg P. Modelling dictyostelium ciscoideum morphogenesis: the culmination. Bull Math Biol. 2002;64(2):327–53.

28. Durand M, Guesnet E. An efficient cellular potts model algorithm that forbids cell fragmentation. Comput Phys Commun. 2016;208:54–63.

29. Voss-Böhme A. Multi-scale modeling in morphogenesis: a critical analysis of the cellular potts model. PloS ONE. 2012 7(9):42852.

30. Lardon LA, Merkey BV, Martins S, Dötsch A, Picioreanu C, Kreft JU, Smets BF. iDynoMiCS: next-generation individual-based modelling of biofilms. Environ Microbiol. 2011;13(9):2416–34.

31. Baker QB, Podgorski GJ, Johnson CD, Varcis E, Flann NS. Bridging the multiscale gap: Identifying cellular parameters from multicellular data. In: IEEE Conference on Computational Intelligence in Bioinformatics and Computational Biology (CIBCB). 2015. p. 1–7.

32. Storck T, Picioreanu C, Virdis B, Batstone DJ. Variable cell morphology approach for individual-based modeling of microbial communities. Biophys J. 2014;106(9):2037–48.

33. Nishio N, Kuroda K, Nagai S, Others. Methanogenesis of glucose by defined thermophilic coculture of Clostridium thermoaceticum and Methanosarcina sp. J Ferment Bioeng. 1990;70(6):398–403.

34. Kalyuzhnyy SV, Gachok VP, Sklyar VI, Varfolomeyev SD. Kinetic investigation and mathematical modeling of methanogenesis of glucose. Appl Biochem Biotechnol. 1991;28(1):183–95.

35. Kalyuzhnyi SV, Davlyatshina MA. Batch anaerobic digestion of glucose and its mathematical modeling. I. Kinetic investigations. Bioresource Technol. 1997;59(1):73–80.

36. Fang C, Boe K, Angelidaki I. Anaerobic co-digestion of by-products from sugar production with cow manure. Water Res. 2011;45(11):3473–80. doi:10.1016/j.watres.2011.04.008.

37. Britz TJ, Van Schalkwyk C, Roos P. Development of a method to enhance granulation in a laboratory batch system. Water SA. 2002;28(1):49–54.

38. Kato MT, Field JA, Lettinga G. The anaerobic treatment of low strength wastewaters in UASB and EGSB reactors. Water Sci Technol. 1997;36(6-7):375–82.

39. Álvarez JA, Armstrong E, Gómez M, Soto M. Anaerobic treatment of low-strength municipal wastewater by a two-stage pilot plant under psychrophilic conditions. Bioresour Technol. 2008;99(15):7051–62. doi:10.1016/j.biortech.2008.01.013.

40. Kumar A, Yadav AK, Sreekrishnan TR, Satya S, Kaushik CP. Treatment of low strength industrial cluster wastewater by anaerobic hybrid reactor. Bioresource Technol. 2008;99(8):3123–9. doi:10.1016/j.biortech.2007.05.056.

41. Diya'uddeen BH, Daud WM, Aziz ARA. Treatment technologies for petroleum refinery effluents: A review. Process Saf Environ Protect. 2011;89(2):95–105. doi:10.1016/j.psep.2010.11.003.

42. Lens PNL, Gastesi R, Vergeldt F, van Aelst AC, Pisabarro AG, Van As H. Diffusional properties of methanogenic granular sludge: 1H NMR characterization. Appl Environ Microbiol. 2003;69(11):6644–9.

43. Pol LH, de Castro Lopes SI, Lettinga G, Lens PN. Anaerobic sludge granulation. Water Res. 2004;38(6):1376–89.

44. Alphenaar PA, Visser A, Lettinga G. The effect of liquid upward velocity and hydraulic retention time on granulation in UASB reactors treating wastewater with a high sulphate content. Bioresource Technol. 1993;43(3):249–58. doi:10.1016/0960-8524(93)90038-D.

45. MacLeod F, Guiot S, Costerton J. Layered structure of bacterial aggregates produced in an upflow anaerobic sludge bed and filter reactor. Appl Environ Microbiol. 1990;56(6):1598–607.

46. Gagliano MC, Ismail SB, Stams AJM, Plugge CM, Temmink H, Van Lier JB. Biofilm formation and granule properties in anaerobic digestion at high salinity. Water Res. 2017;121:61–71. doi:https://doi.org/10.1016/j.watres.2017.05.016, http://www.sciencedirect.com/science/article/pii/S0043135417303755.

47. Mahoney EM, Varangu LK, Cairns WL, Kosaric N, Murray RG. The effect of calcium on microbial aggregation during UASB reactor start-up. Water Sci Technol. 1987;19(1-2):249–60.

48. Show KY, Wang Y, Foong SF, Tay JH. Accelerated start-up and enhanced granulation in upflow anaerobic sludge blanket reactors. Water Res. 2004;38(9):2293–304. doi:10.1016/j.watres.2004.01.039.

49. Kang S, Kahan S, McDermott J, Flann N, Shmulevich I. Biocellion: accelerating computer simulation of multicellular biological system models. Bioinformatics. 2014;30(21):3101–8.

50. Macklin P, Frieboes HB, Sparks JL, Ghaffarizadeh A, Friedman SH, Juarez EF, Jonckheere E, Mumenthaler SM. Progress towards computational 3-d multicellular systems biology. In: Systems Biology of Tumor Microenvironment. Cham: Springer; 2016. p. 225–46.

51. Lardon LA, Merkey BV, Martins S, Dötsch A, Picioreanu C, Kreft J-UU, Smets BF. iDynoMiCS: next-generation individual-based modelling of biofilms. Environ Microbiol. 2011;13(9):2416–34. doi:10.1111/j.1462-2920.2011.02414.x.

52. Stewart PS. Diffusion in biofilms. J Bacteriol. 2003;185(5):1485–91.

53. SiÑEriz F, Pirt SJ. Methane production from glucose by a mixed culture of bacteria in the chemostat: the role of Citrobacter. Microbiology. 1977;101(1):57–64.

54. Moletta R, Verrier D, Albagnac G. Dynamic modelling of anaerobic digestion. Water Res. 1986;20(4):427–34.

55. Yang ST, Okos MR. Kinetic study and mathematical modeling of methanogenesis of acetate using pure cultures of methanogens. Biotech Bioeng. 1987;30(5):661–7.

56. Ibba M, Fynn GH. Two stage methanogenesis of glucose byAcetogenium kivui and acetoclastic methanogenic Sp. Biotechnol Lett. 1991;13(9):671–6.

57. Bhunia P, Ghangrekar MM. Analysis, evaluation, and optimization of kinetic parameters for performance appraisal and design of UASB reactors. Bioresour Technol. 2008;99(7):2132–40.

58. Gavala HN, Angelidaki I, Ahring BK. In: Ahring BK, Angelidaki I, de Macario EC, Gavala HN, Hofman-Bang J, Macario AJL, Elferink SJWHO, Raskin L, Stams AJM, Westermann P, Zheng D, editors. Kinetics and modeling of anaerobic digestion process. Berlin, Heidelberg: Springer; 2003, pp. 57–93. http://dx.doi.org/10.1007/3-540-45839-5_3.

59. Hobbie R, Roth BJ. Intermediate physics for medicine and biology. New York: Springer; 2007.

60. Haynes WM. the CRC Handbook of Chemistry and Physics 93RD Edition. Boca Raton: CRC Press; 2012.

61. Kubitschek HE. Cell volume increase in Escherichia coli after shifts to richer media. J Bacteriol. 1990;172(1):94–101.

62. Sowers KR, Baron SF, Ferry JG. Methanosarcina acetivorans sp. nov., an acetotrophic methane-producing bacterium isolated from marine sediments. Appl Environ Microbiol. 1984;47(5):971–8.

63. Gavala HN, Angelidaki I, Ahring BK. Kinetics and modeling of anaerobic digestion process In: Ahring BK, Angelidaki I, de Macario EC, Gavala HN, Hofman-Bang J, Macario AJL, Elferink SJWH, Raskin L, Stams AJM, Westermann P, Zheng D, editors. Biomethanation I. Berlin: Springer; 2003. p. 57–93. doi:10.1007/3-540-45839-5_3.

3

A mathematical model of mechanotransduction reveals how mechanical memory regulates mesenchymal stem cell fate decisions

Tao Peng[1†], Linan Liu[2†], Adam L MacLean[1], Chi Wut Wong[2], Weian Zhao[2] and Qing Nie[1*] (iD)

Abstract

Background: Mechanical and biophysical properties of the cellular microenvironment regulate cell fate decisions. Mesenchymal stem cell (MSC) fate is influenced by past mechanical dosing (memory), but the mechanisms underlying this process have not yet been well defined. We have yet to understand how memory affects specific cell fate decisions, such as the differentiation of MSCs into neurons, adipocytes, myocytes, and osteoblasts.

Results: We study a minimal gene regulatory network permissive of multi-lineage MSC differentiation into four cell fates. We present a continuous model that is able to describe the cell fate transitions that occur during differentiation, and analyze its dynamics with tools from multistability, bifurcation, and cell fate landscape analysis, and via stochastic simulation. Whereas experimentally, memory has only been observed during osteogenic differentiation, this model predicts that memory regions can exist for each of the four MSC-derived cell lineages. We can predict the substrate stiffness ranges over which memory drives differentiation; these are directly testable in an experimental setting. Furthermore, we quantitatively predict how substrate stiffness and culture duration co-regulate the fate of a stem cell, and we find that the feedbacks from the differentiating MSC onto its substrate are critical to preserve mechanical memory. Strikingly, we show that re-seeding MSCs onto a sufficiently soft substrate increases the number of cell fates accessible.

Conclusions: Control of MSC differentiation is crucial for the success of much-lauded regenerative therapies based on MSCs. We have predicted new memory regions that will directly impact this control, and have quantified the size of the memory region for osteoblasts, as well as the co-regulatory effects on cell fates of substrate stiffness and culture duration. Taken together, these results can be used to develop novel strategies to better control the fates of MSCs in vitro and following transplantation.

Keywords: Mesenchymal stem cell, ECM, *YAP/TAZ*, Cell fate decision, Stiffness sensing, Memory, Bistability, Nonlinear dynamics, Mathematical modeling

Background

Changes in cellular state can be regulated by mechanical signals from the cellular microenvironment, such as the local extracellular matrix (ECM) stiffness [1–4]. Recent studies into mechanotransduction have demonstrated that cells sense and integrate mechanical cues from the ECM, causing transcriptional changes to occur and influencing cell fate decisions [1–3, 5]. Mesenchymal stem cells (MSCs) are controlled by signals from the ECM and exhibit a wide range of differential gene expression patterns [1, 6]. The mechanisms governing how MSCs sense the surrounding ECM, and the myriad other factors affecting MSC fate, including interactions with proteins and ligands, tethering, and porosity, remain incompletely defined [3, 7]. Further understanding of

* Correspondence: qnie@uci.edu
†Equal contributors
[1]Department of Mathematics, Center for Complex Biological Systems, and Center for Mathematical and Computational Biology, University of California, Irvine, CA 92697, USA
Full list of author information is available at the end of the article

how differentiation cues are mediated by mechanical stimuli will help to facilitate new biomaterial design, cell-based therapeutics, and engineered tissue constructs for use in regenerative medicine.

The signals arising at the stem cell/substrate interface are complex and dynamic [7], however it has been shown that stiffness alone is enough to direct MSC differentiation [3, 4]. MSCs undergo neurogenic or adipogenic differentiation on soft substrates (<1 kPa), and myogenic or osteogenic differentiation on stiff substrates (>10 kPa) [1, 5] (Fig. 1). Upon further study, more complex differentiation patterns emerge. For example, it has been observed that cells cultured for a period of time on stiff substrates, such as standard tissue culture polystyrene (TCPS) plates, differentiate into osteogenic lineage cells even after being transferred from the stiff to a softer substrate [8]. Seeding MSCs on a phototunable substrate demonstrates that osteogenic patterns of gene expression persist even after

decreasing the stiffness of the substrate [8]. This "mechanical memory": the ability of MSCs to remember previous physical stimuli depends on both culture time and substrate stiffness (depicted in Fig. 1).

Due to mechanical memory, MSC differentiation in vitro can yield unpredictable (and undesirable) results. Mechanical memory also makes it very difficult to perform certain in vitro assays reliably, for example on extremely soft or stiff substrates, or assays with very long or short incubation periods. Such extreme culture conditions are nonetheless important to assess in order to fully elucidate the relationship between MSC fate and substrate stiffness [9]. In addition to the impracticality of performing short (i.e. seconds) or long (i.e. months) incubation experiments, experimental knock-downs of key genes involved in mechanotransduction, such as Yes-associated protein (*YAP*), can be lethal or highly toxic in vitro and in vivo [10, 11]. There is thus a need for *in*

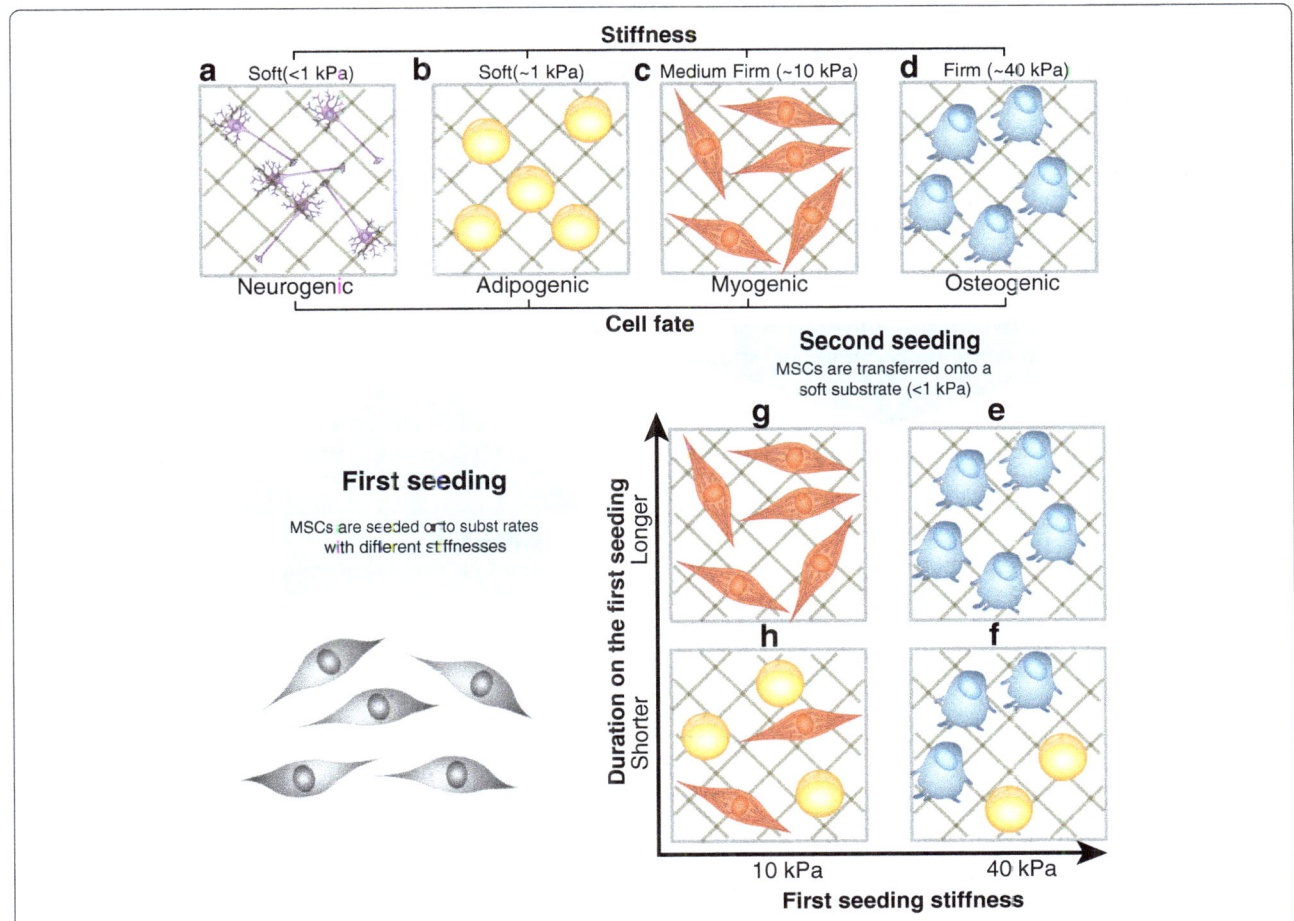

Fig. 1 Mesenchymal stem cells (MSCs) exhibit mechanical memory. **a**, **b**, **c**, **d**: MSCs differentiate into distinct lineages under different substrate stiffness conditions by upregulating lineage marker genes TUBB3 (<1 kPa stiffness, the neurogenic fate), PPARG (~1 kPa stiffness, the adipogenic fate), MYOD1 (~10 kPa stiffness, the myogenic fate), or RUNX2 (~40 kPa stiffness, the osteogenic fate). When re-seeded onto a soft substrate (~1 kPa), MSCs are expected to undergo adipogenic differentiation [1, 6, 64]. **e**, **f**: However, for higher first seeding stiffness values (>10 kPa), or for long first seeding durations (>10 days), mechanical memory leads to heterogeneous osteogenic differentiation [8]. **g**, **h**: The model predicts that for high first seeding stiffness values (~10 kPa), or for long first seeding durations, mechanical memory leads to heterogeneous myogenic differentiation

silico studies to simulate culture conditions and to map the MSC fate predictions to experimental results describing mechanically induced cell differentiation.

Several mathematical models of mechanotransduction have been built to describe cell differentiation directed by external mechanical stimuli [12, 13]. These include, for example, analysis of the role of *YAP/TAZ*, the transcriptional factors *YAP* and transcriptional co-activator with PDZ-binding motif (*TAZ*), in mechanosensing [14], and models that aim to predict cell differentiation during bone healing [12, 15, 16]. Mousavi et al. developed a 3D mechanosensing computational model to illustrate that matrix stiffness can regulate MSC fates. Their simulation results of MSC differentiation in response to substrate stiffness are in agreement with published experimental observations [13]. Burke et al. built a computational model to test whether substrate stiffness and oxygen tension regulate stem cell differentiation during fracture healing [12]. Their model predicted the presence of major processes involved with fracture healing, including cartilaginous bridging, endosteal and periosteal bony bridging, and bone remodeling, using parameters related to cell proliferation, oxygen tension, and substrate stiffness. However, these models are limited in that the effects of regulatory factors were not considered [12–16]. Furthermore, these studies used different models to represent different experimental observations. Hence it is difficult to describe the overall cell state space and to study the transitions between cell fates [12–16]. Thus, there is a need for a dynamic mathematical model, which can stimulate a continuous range of stiffness values and their associated cell fates.

Here we present a mathematical model of MSC differentiation controlled by the following set of core mechanisms

(Fig. 2 and Table 1) [1, 6, 9]. The MSCs sense the stiffness of their environment directly via their adhesion to the substrate. The transcriptional factors *YAP* and *TAZ* mediate the signal via their interaction with downstream genes involved in cell differentiation. *TUBB3*, a gene encoding Tubulin beta-3 chain tightly correlated with a neurogenic cell fate is expressed when MSCs receive stimuli from a soft stiffness environment (<1 kPa) [1]. *PPARG*, peroxisome proliferator-activated receptor gamma, encodes an adipogenic marker and has been shown to be turned on in soft stiffness environments (~1 kPa) [6]. *MYOD1*, myogenic differentiation 1, a myogenic gene turned on in medium-stiff environments (~10 kPa), encodes key factors regulating muscle differentiation [1]. *RUNX2*, runt-related transcription factor 2, an osteogenic gene which is upregulated in high stiffness environments (~40 kPa), is a key transcriptional factor involved in osteoblast differentiation [1] (Fig. 1). We use this set of four lineage-specific genes in our model to minimally describe the transcriptional changes observed during MSC differentiation into four distinct cell fates under the influence of mechanical stimuli mediated by *YAP/TAZ* signaling.

Based on the proposed regulatory network structure (Fig. 2), we simulate gene expression dynamics under different mechanical dosings. Each *in silico* experiment describes MSCs cultured in two passages: a first seeding and a second seeding. The substrate stiffness for the first seeding and the duration of the first seeding are particularly important in cell fate determination of MSCs. We also discover an important role for the second seeding stiffness through our simulation studies. Crucially, this two-seeding setup permits mechanical memory to be observed and studied. We assess when cell fates are determined not only by the current substrate stiffness but

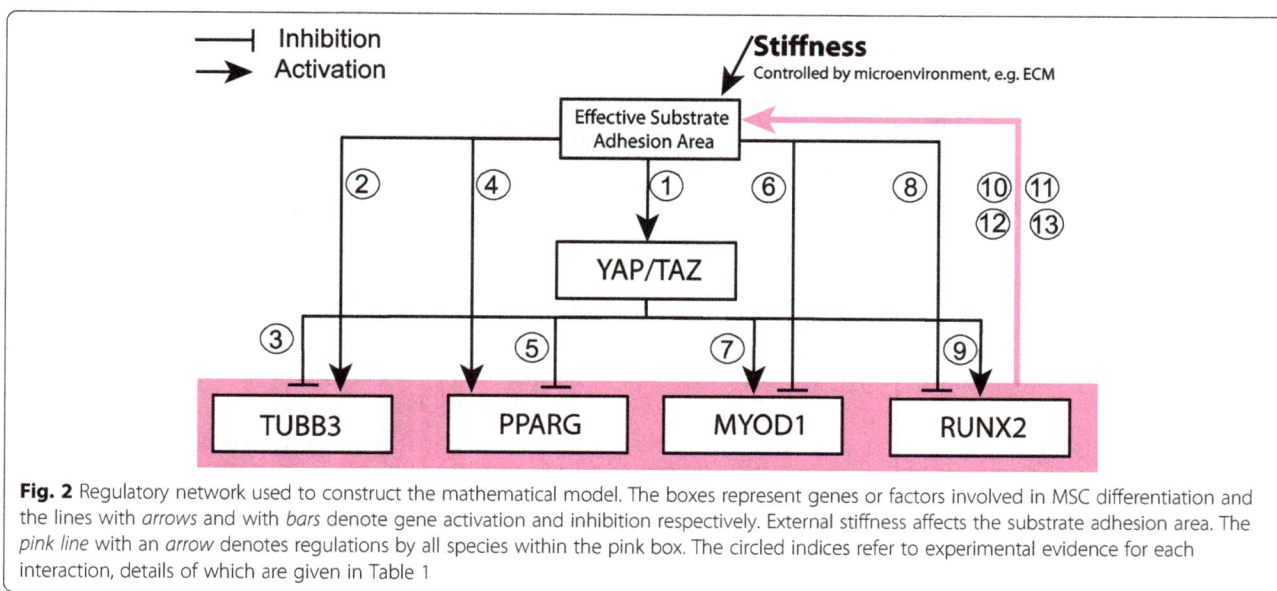

Fig. 2 Regulatory network used to construct the mathematical model. The boxes represent genes or factors involved in MSC differentiation and the lines with *arrows* and with *bars* denote gene activation and inhibition respectively. External stiffness affects the substrate adhesion area. The *pink line* with an *arrow* denotes regulations by all species within the pink box. The circled indices refer to experimental evidence for each interaction, details of which are given in Table 1

Table 1 The references of regulatory interactions in the network

Index of Arrows	Interactions	References
1	*YAP/TAZ* is identified as mechanical sensors and mediators.	Halder, G et al, 2012; Dupont S. et al. 2011. [6, 18]
3	The inhibition of *TUBB3* can be attenuated by *YAP* depletion.	Alarcon, C et al. 2009 [65]
5	*PPARG* can be bound to *TAZ*, which results in transcription inhibitions from the aP2 promoter.	Hong, J.H. et al, 2006.[21]
7	*TAZ* functions as an enhancer of *MYOD*-mediated myogenic differentiation.	Jeong, H. et al, 2010. [66]
9	*RUNX2* has binding domain to *TAZ* for osteocalcin expression.	Hong, J.H. et al, 2006. Hong, J.H. et al, 2005 [20, 21]
10,11,12,13	Increased cell spreading results in higher stiffness sensitivity via increased binding of integrins to the ECM.	Halder G et al, 2012. Sun Y et al, 2012. Bernabe B P et al, 2016. [6, 17, 67]
2,4,6,8	These arrows are necessary for the dynamics of *TUBB3*, *PPARG*, *MYOD1*, and *RUNX2* on all possible stiffness environment since *TUBB3*, *PPARG*, *MYOD1*, and *RUNX2* are expressed only on the super soft stiffness (< 1 kPa), the soft stiffness (~1 kPa), the medium stiffness (~10 kPa), and the high stiffness (~40 kPa) environment respectively.	Engler, A.J. et al,2006; Halder G et al, 2012 [1, 6]

also by past exposure and find that a memory region exists for each of the four MSC-derived cell lineages studied. Our model demonstrates that stiffness-based MSC differentiation results from non-cooperative regulation of representative genes. Moreover, we show that lowering the second seeding stiffness of MSCs leads to a more diverse palette of MSC fates.

Results

A mathematical model based on a mechanotransduction network

The following set of biological assumptions has been used to develop the mathematical model. MSCs differentiate according to their surrounding mechanical environment [2–4, 6, 17]. Directed differentiation towards a particular lineage can be guided if the cells are cultured in a microenvironment that mimics the tissue elasticity of the environment in vivo [2, 3, 17]. Stiff substrates promote cell-ECM adhesion interactions via integrins [6]. These adhesive interactions control the localization of downstream transcriptional factors *YAP* and *TAZ*, which have been identified as mechanical sensors and mediators of such signals [6, 18]. *YAP/TAZ* localizes in the cytoplasm on soft substrates (~1 kPa) and can re-localize to the nucleus on stiff substrates (~40 kPa), thus functioning as a mechanosensitive transcription factor [6, 18].

Additionally, *YAP/TAZ* has been reported to be an upstream factor of a number of genes associated with cell differentiation cues [6, 18, 19]. For example, the inhibition of *TUBB3* can be attenuated by *YAP* depletion, whereas that the factor *PPARG* binding to *TAZ* results in inhibition of transcription from the aP2 promoter [20, 21]. *TAZ* functions as an enhancer of *MYOD*-mediated myogenic differentiation. *RUNX2* can also bind to *TAZ* and cause osteocalcin to be expressed, thus promoting osteogenic differentiation [20, 21]. To describe these

interactions, we model *YAP/TAZ* as both a downstream factor of the mechanical stimulus from the ECM and an upstream factor of the selected cell lineage genes [1, 22] (Fig. 2 and Table 1). Previous references show an intriguing relationship between morphological changes to MSCs and their lineage differentiation potential, whereby morphological changes have been shown to be instrumental to the process of MSC differentiation [1, 17, 13, 23–25]. In particular, it was shown that MSC osteogenic differentiation is enhanced by the morphological change of MSCs and *MYOD1* induced the myogenic differentiation efficiency via the morphological change of MSCs [26, 27]. Other factors regulating cell spreading such as *NKX2.5* were integrated in the model implicitly [28]. Therefore, we model a feedback loop between the lineage-specific target genes and the cellular sensing of substrate stiffness.

In order to predict how mechanical dosing influences MSC differentiation, we use ordinary differential equations to model the MSC lineage regulatory network [29–32] (Fig. 2 and Table 1). We assume that changes in the stiffness of the substrate act as stimulus to the cell (mediated by stiffness receptors) [12, 33]. We use Hill functions to model the chemical activation/inhibition [31, 32, 34]. We model the feedback loop that controls mechanical memory via a non-cooperative regulation, i.e., any of the lineage-specific genes (*TUBB3*, *PPARG*, *MYOD1*, *RUNX2*) can increase the effective stiffness adhesion area (we use "OR-GATE" logic). The feedback loop controls the expression of *YAP/TAZ* and its downstream genes via the stimulus (i.e., the change in stiffness [8]). We also test a feedback model of cooperative regulations (where *TUBB3*, *PPARG*, *MYCD1* and *RUNX2* must act together to increase the effective stiffness adhesion area, i.e. "AND-GATE" logic) but find that it does not satisfy the dynamical requirements of the MSC differentiation system (see Methods for full details).

Model simulations predict mechanical memory regions for each lineage-specific gene

The non-cooperative regulation model displays multiple steady states over the behavioral regions that we have investigated (with first seeding stiffness values ranging from 0.1 kPa to greater than 100 kPa; Fig. 3). This range is sufficient to encompass all known in vitro studies [1, 6, 8]. In Fig. 3a and b the multiple steady states of *YAP/ TAZ* expression over the stiffness range studied are shown, and changes in the *YAP/TAZ* state can be visualized as the stiffness increases (blue lines) or decreases (red lines). The nonlinear relationship between *YAP/ TAZ* and the stiffness of the substrate along the blue lines is consistent with previous observations [9, 19].

Figure 3c demonstrates bistability in the relative gene expression of *TUBB3* (driver of neurogenic differentiation) downstream of *YAP/TAZ*. *TUBB3* is "OFF" when the stiffness is lower than 0.2 kPa. It will be turned "ON" as the stiffness increases to 0.25 kPa. It turns "OFF" again as the stiffness increases further. Meanwhile, *TUBB3* stays "ON" when the stiffness decreases below 0.2 kPa, thus highlighting the mechanical memory observed during neurogenic differentiation. Notably, *TUBB3* stays "OFF" as the stiffness decreases from 0.6 kPa. We define the region of stiffness from 0.25 to 0.55 kPa as a "differentiation memory region" for *TUBB3*. This means that if the first seeding stiffness is within this range, the cell will "remember" the stiffness of this first seeding substrate, and will differentiate according (towards a neurogenic fate) upon reseeding. Our model also predicts novel differentiation memory regions for *PPARG* (0.6 to 3 kPa; Fig. 3d) and

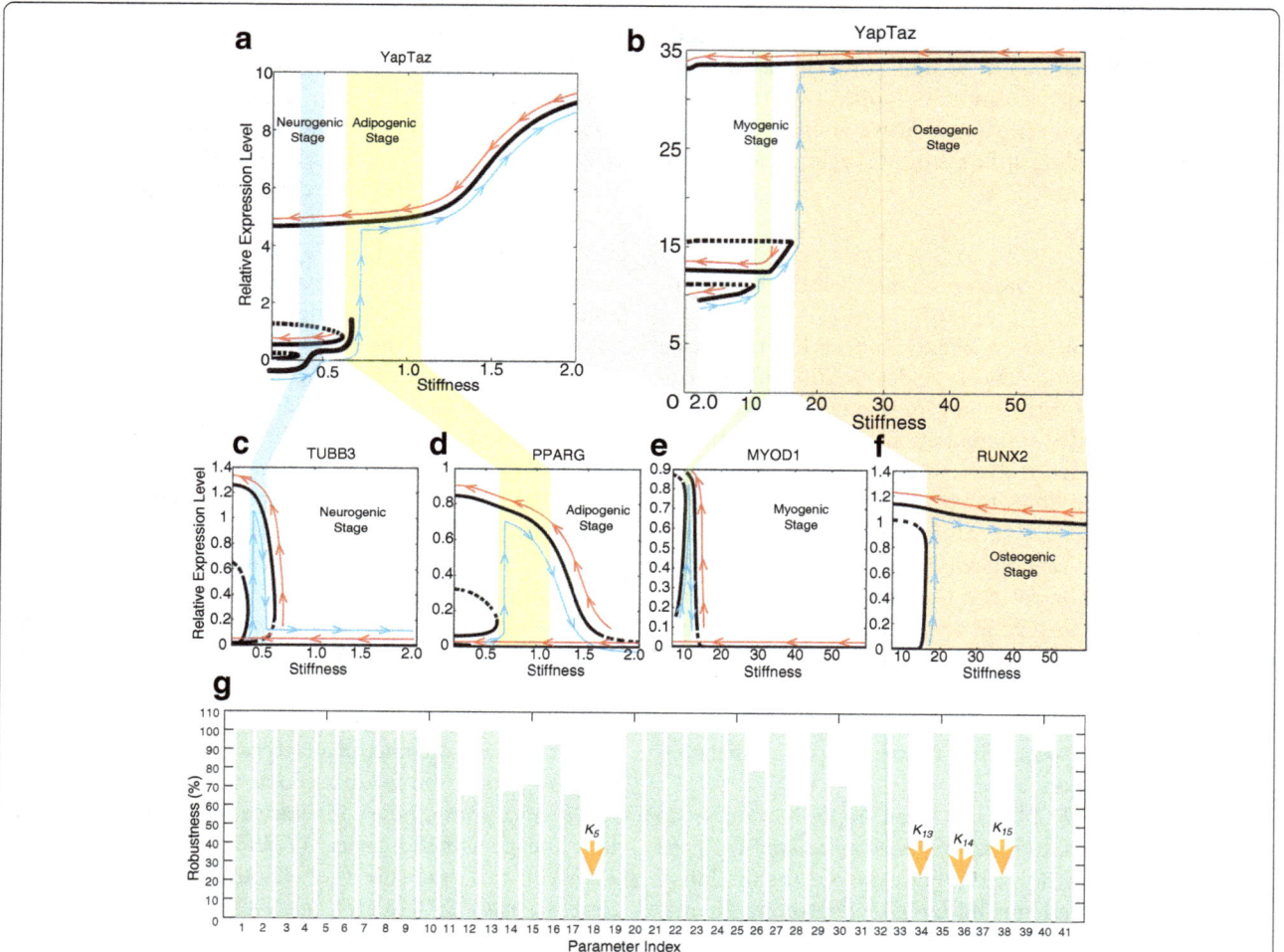

Fig. 3 Multistability in the MSC differentiation network. The relative expression level of YAP/TAZ in a stiffness range from 0.1 kPa to 60 kPa is shown (**b**), with inset (**a**). The relative expression levels of lineage-specific genes are shown in (**c-f**). On each plot the x-axis is the stiffness of the substrate and the y-axis is the relative gene expression level. *Blue lines* illustrate changes in the relative expression level as the stiffness increases; *red lines* illustrate changes in the relative expression level as the stiffness decreases. (**g**). The robustness of the parameters in the mathematical model. The x-axis is the parameter index, corresponding to the notation of Table 2. The y-axis is the robustness of the parameters (defined in Methods)

MYOD1 (10 to 15 kPa; Fig. 3e). *RUNX2* displays the largest differential memory region of the four lineage-specific marker genes studied.

Figure 3c-f collectively demonstrate a bistable region for each of the four lineage-specific genes studied. This is a startling prediction: that a region of mechanical memory exists for each of the cell fates, not just for osteogenic differentiation, as has been previously reported [8]. For neurogenic and adipogenic differentiation, the memory regions are smaller than that of osteoblasts yet may still be of great importance for stem cell fate regulation. The true contribution of each will require further study to elucidate, as a host of interacting factors contribute to the neurogenic and adipogenic cell fate decisions, including those which are not currently included in our model, such as the role of substrate-induced stemness and of epithelial to mesenchymal transition [35–37].

To test the robustness of the mathematical model we calculate the values of the robustness of each parameter in Eqs. (1,2,3,4,5 and 6) with respect to the memory and multistability of the system (full details of our methodology are in Methods). Out of the 41 parameters tested, 37 are robust to small changes for the majority of perturbations tested (and many of these 37 were robust

more than 80% of the time) (Fig. 3g). Four parameters are found to be sensitive to small perturbations. All of these four parameters are involved in myogenic or osteogenic differentiation. Both these processes involve relatively large memory regions, thus it is possible that following these perturbations memory is maintained over parts of – but not the entire – original memory regions. Overall, we find that the system displays robustness using the parameters given in Table 2, with regard to the memory effects and the multistability of the states.

A lower second seeding stiffness permits a greater number of MSC lineages

Potential energy landscape analysis is an appealing method with which we can investigate the system and study the MSC differentiation propensities under different conditions [38–40]. Since it is not possible to write down a complete expression for the potential energy of the system, we use an approximate method derived from mean field theory in order to calculate quasi-potential in terms of the six system variables [40, 41]. Explicitly, we calculate the potential of the system as $U(X) = -\ln(P_{ss}(X))$, where $P_{ss}(X)$ is the total probability of the state vector X, and X describes all the states of the system [40, 41].

Table 2 Parameter values of the mathematical model

Index	Parameter	Value	Estimated from references	Index	Parameter	Value	Estimated from references
1	k_1	0.2	[1,6]	2	k_2	2.2	[1,6]
3	k_3	5	[1,6]	4	k_4	9	[1,6,8]
5	k_5	4	[1,6]	6	k_6	2.9	[1,6]
7	k_7	3	[1,6]	8	k_8	5	[1,6,8]
9	k_9	2	[1,6,8]	10	K_1	600	[1,6]
11	n_1	4	[1,6]	12	K_2	1.1	[1,6]
13	n_2	2	[1,6]	14	K_3	1300	[1,6]
15	n_3	6	[1,6]	16	K_4	0.8	[1,6,8]
17	n_4	2	[1,6,8]	18	K_5	20,000	[1,6]
19	n_5	4	[1,6]	20	K_6	1	[1,6]
21	n_6	20	[1,6]	22	K_7	60,000	[1,6]
23	n_7	6	[1,6]	24	K_8	1.1	[1,6]
25	n_8	20	[1,6]	26	K_9	0.1	[1,6]
27	n_9	2	[1,6]	28	K_{10}	0.5	[1,6]
29	n_{10}	8	[1,6]	30	K_{11}	0.89	[1,6]
31	n_{11}	2	[1,6]	32	K_{12}	4	[1,6]
33	n_{12}	8	[1,6]	34	K_{13}	12	[1,6]
35	n_{13}	20	[1,6]	36	K_{14}	3	[1,6]
37	n_{14}	60	[1,6]	38	K_{15}	16	[1,6,8]
39	n_{15}	45	[1,6,8]	40	K_{16}	4.5	[1,6,8]
41	n_{16}	55	[1,6,8]	d_i $(i = 1,2,\cdots 6)$		1	Unconstrained

In order to visualize this potential function we project it onto a two-dimensional plane, defined by the species in our model: *YAP/TAZ*, and the effective stiffness adhesion area (SAA). In doing so we integrate out the four remaining system variables (*TUBB3, PPARG, MYOD1,* and *RUNX2*) [40, 41]. We are thus able to study how the potential depends on these variables for different stiffness values. In Fig. 4 we show the potential functions for four different conditions (we change the second-seeding stiffness values). Overall, we find that by reducing the second seeding stiffness, a greater number of steady states is permitted.

We simulate more than 10,000 initial conditions in order to avoid becoming trapped in local minima [40, 41]. We observe that across the entire landscape there are four stable states (or basins of attractions), representing neurogenic, adipogenic, myogenic, and osteogenic cell lineages. At a final stiffness of ~0.4 kPa, MSCs can differentiate into each of the four possible lineages (Fig. 4a). Only at such sufficiently small values for the second stiffness can MSCs differentiate into neurons: the basin of attraction for the neurogenic fate (i.e. the probability of differentiating into a neuron) is the smallest of the four fates. This means that mechanical memory is observed only over a small range of space. In comparison, a much greater mechanical memory effect is seen for the osteogenic lineage, corresponding to a larger basin of attraction. Figure 4b and c show the potential landscapes at second seeding stiffness values of ~0.8 kPa and ~12 kPa, respectively. The number of basins

decreases to three, and then two, as the second seeding stiffness increases. When the second seeding stiffness increases further to ~20 kPa, we have only one remaining basin of attraction, thus only one possible cell fate: in this region the largest mechanical memory effect is seen, and osteogenic differentiation dominates. These data intriguingly suggest that simply by controlling the substrate stiffness upon re-seeding we can control the number of cell fates that are accessible to MSCs.

The duration of the initial seeding determines the fate of an MSC

In addition to studying the effect of the second seeding stiffness on the fate of MSCs, we perform tests to assess the agreement between our model and in vitro observations regarding MSC differentiation [1, 18]. Specifically, we manipulate the stiffness of the second seeding substrate and the duration of the first seeding, and find, consistent with previous studies [5, 42], that both of these variables play an important role in the fate determination of an MSC upon differentiation. In addition these simulation results highlight several new phenomena.

In order to examine how the first seeding duration affects MSC fates, we use a non-dimensionalized version of the model, that is, we express time in relative units. In Fig. 5a, the first and second seeding stiffness values are 30 kPa and 0.4 kPa, respectively. When the duration of the first seeding time is 50 (blue line), MSCs differentiate into osteoblasts (consistent with [5]): *RUNX2* is the

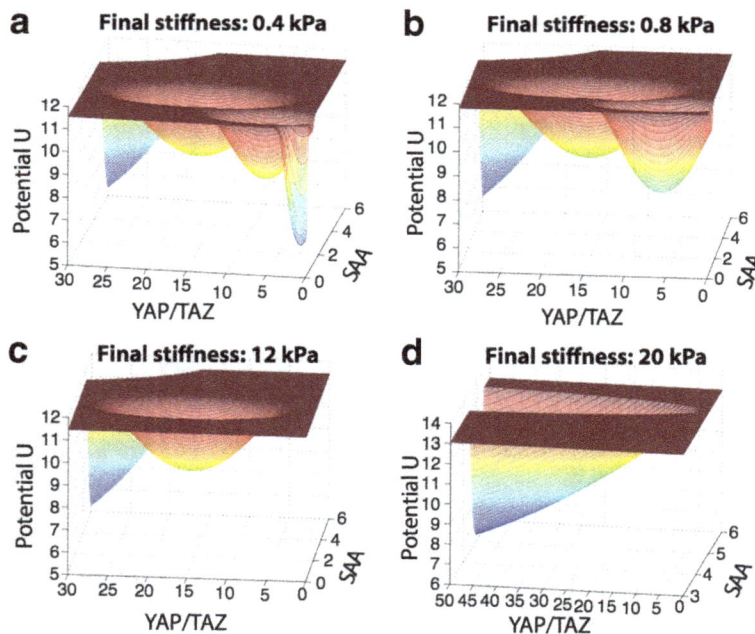

Fig. 4 Potential landscapes of the regulatory network under different stiffness conditions. In each figure the relative stiffness level (input to the system) is plotted on the x-axis, the relative expression level of YAP/TAZ is plotted on the y-axis, the energy potential function U is plotted on the z-axis. Potential energy landscapes are shown with stiffness values of ~0.4 kPa (**a**), ~0.8 kPa (**b**), ~12 kPa (**c**) and ~20 kPa (**d**)

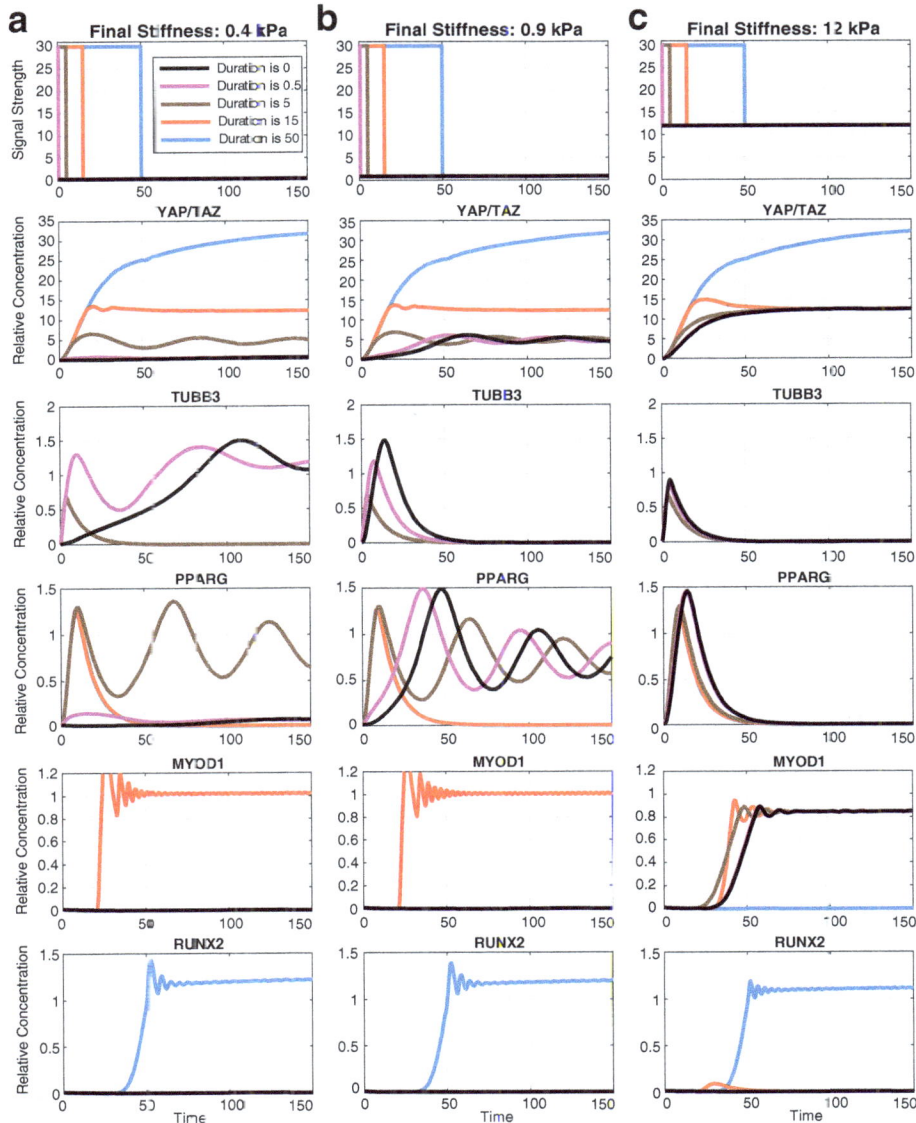

Fig. 5 The duration of the first seeding regulates MSC fates via mechanical memory. The first seeding stiffness in this figure is 30 kPa. The second seeding stiffness is 0.4 kPa (**a**), 0.9 kPa (**b**) or 12 kPa (**c**). When the duration of the first seeding is 50 (*blue lines*), MSCs undergo osteogenic differentiation according to memory. When the duration of the first seeding is 15 (*red lines*), MSCs undergo myogenic differentiation. When the duration of the first seeding is 5 (*brown lines* in columns A and B), MSCs differentiate into adipocytes or myogenic cells. When the duration of the first seeding is 0.5 (*pink lines* in column A), MSCs are able to undergo adipogenic, myogenic, or neurogenic differentiation. Finally, when the duration of the first seeding is 0 (*black lines*), MSCs are able to undergo adipogenic, myogenic, or neurogenic differentiation

only gene that is highly expressed under this condition. When the first seeding duration is 15 (red line), MSCs differentiate into skeletal muscle cells (*MYOD1* high); when the first seeding duration is five (brown line), MSCs differentiate into adipocytes (*PPARG* high). Finally when the first seeding duration is 0.5 or 0 (pink and black lines), MSCs differentiate into neurogenic cells (*TUBB3* high). These results are consistent with previous studies and highlight the breadth of control that mechanical memory enables: MSCs can be directed to four different fates by changing only the duration of the

first seeding, keeping both of the first and the second substrate stiffness values constant. Although mechanical memory is not observed when the first seeding duration is less than 0.5, for the first seeding durations greater than five, we predict that mechanical memory will influence MSC fates, directing MSCs towards myogenic or adipogenic lineages.

Mechanical memory persists when the second seeding stiffness increases, but the number of fates accessible to an MSC decreases, as described in previous sections. In Fig. 5b the second seeding stiffness is 0.9 kPa. When the relative duration of the first seeding is 50 (blue line), MSCs differentiate into osteoblasts according to mechanical memory. When the relative duration of the first seeding is 15 (red line), MSCs differentiate into myocytes (again, influenced by memory). When the relative duration of the first seeding is 5, 0.5 or 0, however (brown, pink or black lines), MSCs differentiate into adipocytes: mechanical memory is not present when the second seeding duration is less than 15.

Figure 5c shows the dynamics of the system when the second seeding stiffness is 12 kPa. For the longest first seeding duration (blue line), MSCs differentiate into osteoblasts, as above, but when the duration is 15 or lower (red, brown, pink or black lines), MSCs differentiate into myocytes. These data illustrate that as the second seeding stiffness increases, the range of first seeding durations over which mechanical memory is observed decreases, which is consistent with the observation from Yang et al [8]. At a second seeding stiffness of 12 kPa, the memory effect is observed only for osteogenic differentiation, and not for any other lineages. Intriguingly, higher first seeding stiffness values for shorter periods of time might accelerate an MSC towards lineage commitment. *TUBB3* expression approaches the steady state quickly following stimulation on a 30 kPa substrate for a relative time of 0.5 (Fig. 5a, pink line). Compare this to the differentiation characteristics of an MSC seeded only on a 0.3 kPa substrate (Fig. 5a, black line); the latter takes a longer time to differentiate.

Feedback signaling onto the effective substrate adhesion area

Mechanotransduction pathways may contain positive feedback loops in which integrin engagement activates actomyosin cytoskeleton contractility, resulting in morphological changes affecting the adhesion area of the substrate [1, 17, 18, 23–27]. Here we assess the importance of such feedback. Figure 6 shows the relative expression levels of the lineage-specific genes at steady states for a range of substrate stiffness values. In Fig. 6a, we block the feedback from *TUBB3* onto the effective substrate adhesion area. We see that the bistability that was observed in Fig. 3 is no longer present: no hysteresis effect can be seen when the substrate stiffness is increased or decreased (illustrated by the blue and red lines). Thus, no mechanical memory effect remains for *TUBB3* during MSCs differentiation. Similar results are obtained for *PPARG* (Fig. 6b), *MYOD1* (Fig. 6c) and *RUNX2* (Fig. 6d) when the final seeding stiffness is 0.9 kPa, 10 kPa and 16 kPa, respectively. The mechanical memory of the genes disappeares when the feedback loops are removed. Collectively our simulation results illustrate that the feedback loops downstream of the stiffness of substrates are necessary for the mechanical memory.

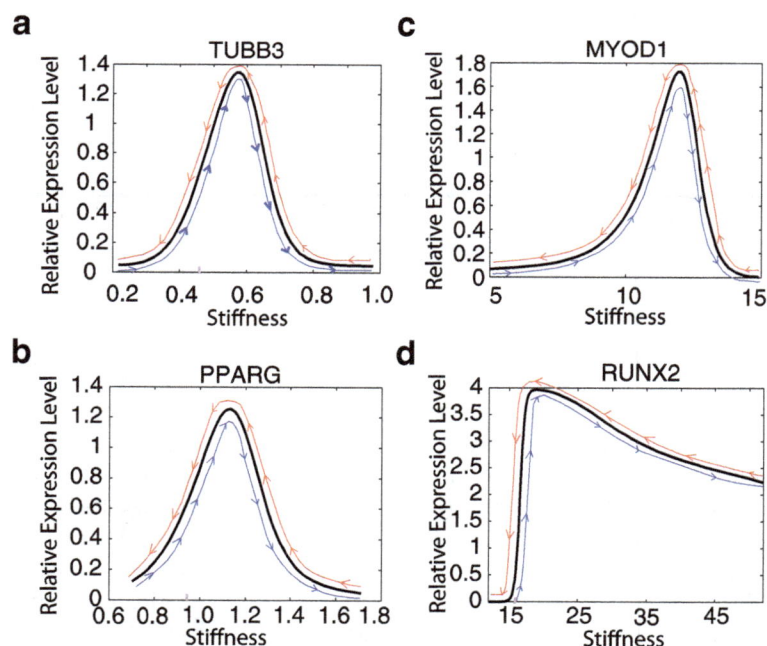

Fig. 6 The MSC network precludes multistability when feedback loops are blocked. Shown are the steady states of TUBB3 (**a**), PPARG (**b**), MYOD1 (**c**), and RUNX2 (**d**) under different stiffness values. In each figure the x-axis denotes the stiffness and the y-axis denotes the relative expression levels of specific lineage genes at steady states (*black lines*). The blue lines illustrate how the relative gene expression at the steady state changes as the stiffness increases. The red lines illustrate how the relative gene expression level at the steady state changes as the stiffness decreases

Noise can induce fate switching during MSC differentiation

There is inherent noise in gene expression dynamics [43, 44]. We employ a stochastic differential equation (SDE) model (described in Methods) to study the effects of gene expression noise on MSC differentiation [45, 46]. We find that SDE simulations broadly recapitulate the results obtained in the deterministic case, however under certain conditions fate switching is observed. In Fig. 7 we simulate a system of SDEs based on the deterministic model with multiplicative noise added to the expression level of each gene; blue and dark green lines describe the relative gene expression under the deterministic model, while pink and black lines describe analogous results under the SDE model. We vary the initial seeding stiffness while keeping the second seeding stiffness constant at 12 kPa. In the deterministic case, we see that *MYOD1* is expressed when the value of the initial stiffness is 12 kPa, and not when the value is 34 kPa. Conversely, *RUNX2* is not expressed at an initial stiffness of 12 kPa, but is expressed when the initial stiffness is 34 kPa: here stem cells are differentiating according to mechanical memory.

In the stochastic case, a different picture emerges. First we note that the memory effect observed for osteogenic differentiation in the deterministic case (driven by *RUNX2* expression) is preserved under the stochastic model (Fig. 7 black line). However, in the stochastic case, at 12 kPa, *MYOD1* is expressed transiently: as its expression declines to zero, *RUNX2* is turned on. Thus noise has induced a fate transition between myogenic and osteogenic lineages. At 34 kPa no such transitions are observed: *RUNX2* is expressed constitutively.

Discussion

Mesenchymal stem cell fate can be controlled by mechanical dosing [1]. Mechanical memory (past mechanical dosing) also affects stem cell fate, particularly when the initial substrate is stiff [8], it is difficult however to experimentally test the effects of mechanical memory over a wide range of culture conditions. Here we have presented a mathematical model that allows such tests to be performed, producing several striking predictions. We first assessed whether the model is able to recapitulate experimental studies, and find that it does agree with evidence showing MSC differentiation into neurons or adipocytes on softer substrates, and myocytes or osteoblasts on stiffer substrates. We then analyzed model behavior over longer timescales, and found that a mechanical memory region exists for each of these MSC-derived cell lineages, with substantial variation in the memory stiffness range for each cell fate. Previously, a memory region has only been observed during osteogenic differentiation, and even then, only qualitative assessment of its behavior was made. We are able to provide bounds on the substrate stiffness ranges permissive of memory effects for all four lineages.

Upon re-seeding MSCs onto a second substrate, the stem cells differentiate according to mechanical memory under certain conditions. We predict that (in addition to the stiffness of the first substrate) the duration of the first seeding also directly influences stem cell memory. By changing only the duration of the initial seeding we can directly influence cell fate. The number of fates accessible to the MSC can also be controlled by the final seeding stiffness. Landscape analysis demonstrates that,

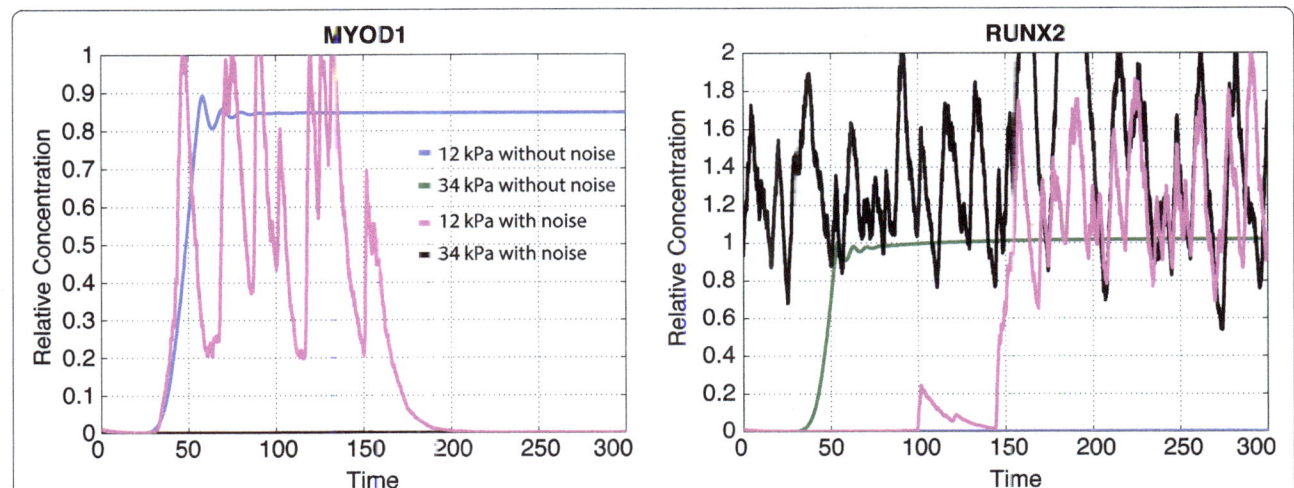

Fig. 7 Stochastic gene expression dynamics under different stiffness conditions. The green and blue lines depict the relative expression levels of genes from the deterministic model. The magenta and black lines depict the relative expression levels of genes from the stochastic differential equation model with noise term ~ $N(0,0.05)$. Blue and magenta lines represent a first-seeding stiffness of 12 kPa, green and black lines represent a first-seeding stiffness of 34 kPa. The final seeding stiffness is 12 kPa in all cases

for a constant first seeding stiffness and duration, a higher second seeding stiffness limits the number of MSC fates accessible, and that a sufficiently low final seeding stiffness is permissive of differentiation into all four cell fates. We also found that the feedback loop connecting lineage-specific genes to the effective surface adhesion area is critical for the mechanical memory of MSC differentiation. This might be due to integrin—substrate binding, or morphological changes that occur upon differentiation [1, 3, 7, 17].

As well as their direct relevance for in vitro studies, our model predictions also have important implications for the design of regenerative therapeutics. A major challenge here is lack of precision in cell fate control following transplantation. A better understanding of the relationship between mechanical conditions, culture duration, and stem cell fates is needed. By defining the substrate stiffness limits that regulate MSC fates, this study provides means to design experimental protocols that constrain cells to be confined within fate boundaries, thus avoiding differentiation towards an undesirable fate [47–50]. Mechanical memory could be employed advantageously here, e.g. by preconditioning MSCs via mechanical dosing. An improved understanding of the MSC mechanotransduction pathway will also affect our ability to control multipotency, and should enable us to better culture undifferentiated MSCs in vitro.

In order to study additional effects of the mechanotransduction pathway on stem cell fate, a model that describes a larger regulatory network is needed. Cell-cell interactions have not yet been incorporated into our model, although there is a large body of work detailing the importance of the microenvironment (i.e. the effects of cell-cell interactions and of the niche) on stem cell differentiation [30, 51]. In addition, we have chosen a small set of four lineage-specific genes in order to minimize the size of the model's parameter space. Clearly a greater number of genes are involved in the regulation of MSC fate; without a description of this larger transcriptional network we will not be able to describe nuances of mechanically-induced MSC fate dynamics. However, we believe that the dynamics – and the attractors corresponding to differentiated cell states observed here constitute core pathway mechanisms that would still underlie cell fate decisions in a larger network.

Conclusions

In this study we sought to investigate the mechanisms of control exerted via mechanical forces upon mesenchymal stem cells during culture and differentiation. Simulations of the gene expression dynamics under different mechanical dosing conditions have led to several predictions. We found that non-cooperative gene regulation is the most plausible mechanism to describe MSC differentiation and we predict that mechanical memory is a general mechanism affecting

all of the MSC-derived lineages in this model. We found that the duration of the initial culture and the substrate stiffness during this initial culture are particularly crucial in determining the MSC fates. In addition, we were able to show that a lower final-seeding substrate stiffness permitted a greater number of MSC fates.

Through careful analysis, the ever-expanding body of high-throughput transcriptomic data will enable the study of ever-more complex gene networks. Both the MSC fate transcriptional network structure and the dynamics of the network need to be inferred from data. Spatial interactions, e.g. arising from niche-mediated effects on MSCs, may necessitate a move towards a suitable model framework such as partial differential equations or cell-based (e.g. Cellular Potts) models. Once a clearer picture emerges, it will be possible to extend our model with the incorporation of relevant new signaling interactions. In doing so, we hope to provide further insight into the complex networks of regulation underpinning mesenchymal stem cell fate.

Methods
A dynamical model of mesenchymal stem cell fate
We model a simplified gene regulatory network that underpins MSC fate with ordinary differential equations (ODEs) [31, 32].

$$\frac{d[SAA]}{dt} = k_1 \underbrace{\frac{(S/K_1)^{n_1} + ([TUBB3]/K_2)^{n_2}}{1 + (S/K_1)^{n_1} + ([TUBB3]/K_2)^{n_2}}}_{\boxed{10}}$$

$$+ k_2 \underbrace{\frac{(S/K_3)^{n_3} + ([PPARG]/K_4)^{n_4}}{1 + (S/K_3)^{n_3} + ([PPARG]/K_4)^{n_4}}}_{\boxed{11}}$$

$$+ k_3 \underbrace{\frac{(S/K_5)^{n_5} + ([MYOD1]/K_6)^{n_6}}{1 + (S/K_5)^{n_5} + ([MYOD1]/K_6)^{n_6}}}_{\boxed{12}}$$

$$+ k_4 \underbrace{\frac{(S/K_7)^{n_7} + ([RUNX2]/K_8)^{n_8}}{1 + (S/K_7)^{n_7} + ([RUNX2]/K_8)^{n_8}}}_{\boxed{13}} - d_1[SAA] \tag{1}$$

$$\frac{d[YAPTAZ]}{dt} = \underbrace{k_5[SAA]}_{\boxed{1}} - d_2[YAPTAZ] \tag{2}$$

$$\frac{d[TUBB3]}{dt} = k_6 \underbrace{\frac{([SAA]/K_9)^{n_9}}{1 + ([SAA]/K_9)^{n_9} + ([YAPTAZ]/K_{10})^{n_{10}}}}_{\boxed{2,3}} - d_3[TUBB3] \tag{3}$$

$$\frac{d[PPARG]}{dt} = k_7 \underbrace{\frac{([SAA]/K_{11})^{n_{11}}}{1 + ([SAA]/K_{11})^{n_{11}} + ([YAPTAZ]/K_{12})^{n_{12}}}}_{\boxed{4,5}} - d_4[PPARG]$$

$$(4)$$

$$\frac{d[MYOD1]}{dt} = k_8 \underbrace{\frac{([YAPTAZ]/K_{13})^{n_{13}}}{1 + ([SAA]/K_{14})^{n_{14}} + ([YAPTAZ]/K_{13})^{n_{13}}}}_{\boxed{6,7}} - d_5[MYOD1]$$

$$(5)$$

$$\frac{d[RUNX2]}{dt} = k_9 \underbrace{\frac{([YAPTAZ]/K_{15})^{n_{15}}}{1 + ([SAA]/K_{16})^{n_{16}} + ([YAPTAZ]/K_{15})^{n_{15}}}}_{\boxed{8,9}} - d_6[RUNX2]$$

$$(6)$$

Where S and $[SAA]$, are the relative levels of the stiffness (input to the system) and of the effective stiffness adhesion area, respectively. $[YAPTAZ]$, $[TUBB3]$, $[PPARG]$, $[MYOD1]$, and $[RUNX2]$ denote the relative concentrations of YAP/TAZ, TUBB3, PPARG, MYOD1, and RUNX2. Since concentration and time in the model are given in relative units, i.e. are dimensionless, then all parameters in the above equations are also dimensionless. d_i ($i = 1, 2, ..., 6$) in Eqs. (1,2,3,4,5 and 6) are the degradation rates of the corresponding genes/factors. The terms denoted by the label (1, 2, ..., 9) under the brackets in Eqs. (1,2,3,4,5 and 6) are the active/inhibitive regulations acting on $[SAA]$, $[YAPTAZ]$, $[TUBB3]$, $[PPARG]$, $[MYOD1]$, and $[RUNX2]$, where the numbers in rectangle boxes are consistent with the circled indices shown in Fig. 2 [52]. All values of parameters in Eqs. (1,2,3,4,5 and 6) shown in Table 2 are estimated or approximated according to the behaviours that we sought to describe. Parameters values are fit to qualitative features of the biological system [1, 6, 8, 9, 19] (See Additional file 1). The data required performing full inference of the parameters are as-yet unavailable, however the results of our sensitivity analysis show that the models results do not depend crucially on specific values of parameters of the model.

Cooperative regulation model

The terms (10, 11, 12, 13) in Eq. (1) are based on the non-cooperative regulations of MSCs stiffness sensing. Meanwhile, we model the regulations as the cooperative one and Eq. (1) is rewritten below [53].

Rehfeldt et al showed the "switch-like" nonlinear relationship between S and SAA expanding from 0.5 kPa to much large stiffness (>60 kPa) and TUBB3, PPARG, MYOD1, and RUNX2 are turned on in their specific ranges of stiffness, which are relatively disjoint [52, 53]. In particular, the stiffness range for the myogenic differentiation is far away from the one for adipogenic differentiation. Based on the properties of the system, we can rewrite our model into four different submodels under the corresponding stiffness ranges. They are shown as follows.

$$\frac{d[SAA]}{dt} = k_1 \frac{(S/K_1)^{n_1} + ([TUBE3]/K_2)^{n_2}}{1 + (S/K_1)^{n_1} + ([TUBB3]/K_2)^{n_2}} - d_1[SAA]$$

$$(8)$$

$$\frac{d[SAA]}{dt} = k_1 \frac{(S/K_1)^{n_1} + ([PPARG]/K_3)^{n_3}}{1 + (S/K_1)^{n_1} + ([PPARG]/K_3)^{n_3}} - d_1[SAA]$$

$$(9)$$

$$\frac{d[SAA]}{dt} = k_1 \frac{(S/K_1)^{n_1} + ([MYOD1]/K_4)^{n_4}}{1 + (S/K_1)^{n_1} + ([MYOD1]/K_4)^{n_4}} - d_1[SAA]$$

$$(10)$$

$$\frac{d[SAA]}{dt} = k_1 \frac{(S/K_1)^{n_1} + ([RUNX2]/K_5)^{n_5}}{1 + (S/K_1)^{n_1} + ([RUNX2]/K_5)^{n_5}} - d_1[SAA]$$

$$(11)$$

The difficulty is to determine the values of K_1. If K_1 is less than 1000, the hill function in Equation (7) is saturated for high stiffness levels (> 10,000) and it means that the models cannot distinguish the myogenic differentiation and osteogenic differentiation since Eqs. (10 and 11) both approach the limit $\frac{d[SAA]}{dt} = k_1 - d_1[SAA]$. If K_1 is greater than 10,000, then the model cannot describe the system for low stiffness levels (< 1000) with that TUBB3 and PPARG cannot express under the low stiffness levels since Eqs. (8 and 9) will respectively approach the limit:

$$\frac{d[SAA]}{dt} = k_1 \frac{([TUBB3]/K_2)^{n_2}}{1 + ([TUBB3]/K_2)^{n_2}} - d_1[SAA]$$

$$\frac{d[SAA]}{dt} = k_1 \frac{([PPARG]/K_3)^{n_3}}{1 + ([PPARG]/K_3)^{n_3}} - d_1[SAA]$$

Thus the cooperative regulation model is unable to accurately describe the MSC differentiation system over the range of stiffness values considered.

$$\frac{d[SAA]}{dt} = k_1 \underbrace{\frac{(S/K_1)^{n_1} + ([TUBB3]/K_2)^{n_2} + ([PPARG]/K_3)^{n_3} + ([MYOD1]/K_4)^{n_4} + ([RUNX2]/K_5)^{n_5}}{1 + (S/K_1)^{n_1} + ([TUBB3]/K_2)^{n_2} + ([PPARG]/K_3)^{n_3} + ([MYOD1]/K_4)^{n_4} + ([RUNX2]/K_5)^{n_5}}}_{\boxed{10,11,12,13}} - d_1[SAA] \quad (7)$$

Sensitivity analysis

In order to calculate the sensitivities of the parameters shown in Table 2 with respect to the memory and multi-stability of the system, we sample 1000 values between 0.2 kPa and 42 kPa; they are taken as the stiffness of the system and they are vectorized as the stiffness vector S_b. We then calculate the steady states, Q_b^{Upper} and Q_b^{Lower}, corresponding to the steady states on the lower bifurcation branch (indicated by blue arrowhead lines in Fig. 3c-f, and to the steady states on the upper bifurcation branch (indicated by red arrowhead lines in Fig. 3c-f) for each of the genes: *TUBB3*, *PPARG*, *MYOD1*, and *RUNX2*, using the parameters in Table 2. In order to calculate the sensitivity of each parameter, we perturbe it 1000 times under the constraint of CV (coefficient of variance) = 0.05, and calculate the steady states Q_P^{Upper} (with the same initial conditions as Q_b^{Upper}), and Q_p^{Lower} (with the same initial conditions as Q_b^{Lower}). We perform such comparisons – for each of the four genes – for a total of 41 parameters and 1000 perturbations, thus for the parameter vector P_i^j ($i = 1, 2, ..., 41; j = 1, 2, ..., 1000$), i.e. the j-th perturbation of the i-th parameter. We count the number (N_i) of P_i^j that satisfies $||Q_P^{Upper} - Q_b^{Upper}||_2 + ||Q_P^{Lower} - Q_b^{Lower}||_2 <$ *TOL*. The tolerance, *TOL*, is set such that the perturbed parameter vector gave rise to the same number of steady states as for the unperturbed case (i.e. multi-stability and the memory effect is maintained; we set *TOL* = 4). The robustness R_i of the i-th parameter is defined as $\frac{N_i}{10}\%$ and the sensitivity S_i of the i-th parameter is $1 - \frac{N_i}{10}\%$. The robustness values for each of the 41 parameters are shown in the bar graph (See Fig. 3g) and the index of the parameters in the graph is consistent with the one in Table 2. Four of them are sensitive than the rest and they are marked by yellow arrows in the following bar graph.

Steady state analysis

We compute the steady states of the dynamical system under different S in Eqs. (1, 2, 3, 4, 5 and 6). Here we use the continuation method to compute the steady states and their branches [54, 55].

Landscape potential using a mean field self-consistent approximation and Gaussian approximation

Here we derive an approximation for the potential energy of the system. Starting from the Fokker-Planck equation, we calculate the steady state probability distributions using a self-consistent mean field method [56–58]. The probability function $P(X,t)$ satisfies the following diffusion equation:

$$\frac{\partial P(X,t)}{\partial t} = -\frac{\partial}{\partial X}[F(X,S)P(X,t)] + D\frac{\partial^2}{\partial X^2}[d(X)P(X,t)] \quad (12)$$

where $F(X,S)$ and $d(X)$ are the drift and diffusion part respectively and the noise is weak, i.e. $D << 1$. Note that X is a vector of species ($[SAA],[YAPTAZ],[TUBB3],[PPARG],[MYOD1],[RUNX2]$) but we have dropped the arrow notation for convenience below. We factor the original probability function using the self-consistent mean field approach [59], $P(X,t) = \prod_{i=1}^{n} P(X_i,t)$ to reduce the computational complexity of solving the original equation on the probability, similar to a previous study [57]. We use the Gaussian distribution to approximate the true distribution [57], leading to a description for the mean and variance of the gene expression:

$$\overline{X}'(t) = F(\overline{X}(t), S) \quad (13)$$

$$\sigma'(t) = \sigma(t)A^T(t) + A(t)\sigma(t) + 2D[\overline{X}(t)] \quad (14)$$

where $\overline{X}$ is the mean value of $X(t)$, $\sigma(t)$ is the variance matrix, the matrix element $\alpha_{ij}(t)$ of $A(t)$ is $\frac{\partial F_i(\overline{X}(t))}{\partial \overline{X}_j(t)}$, i.e. A is the Jacobian matrix.

Since we consider the steady states, then we need to compute $\overline{X}^{(j)}(\infty)$ and $\sigma^{(j)}(\infty)$ from $\overline{X}'(t) = 0$ and $\sigma'(t) = 0$, for $j = 1,2,...,m$ respectively, where m is the number of basins of attraction. We consider only diagonal elements of $\sigma^{(j)}(\infty)$ from mean field splitting approximation. For each variable $\overline{X}_i^{(j)}(\infty)$, the probability distribution can be estimated using the mean and variance and based on Gaussian approximation [57, 60].

$$P^{(j)}(X_j, \infty) = \frac{1}{\sqrt{2\pi\sigma^{(j)}(\infty)}}\exp\left[-\frac{\left[X_i\overline{X}_i^{(j)}(\infty)\right]^2}{2\sigma^{(j)}(\infty)}\right] \quad (15)$$

If $m = 1$, we can use Eq. (6) to compute the probability distribution of the single basin of attraction. If $m > 1$, then the system permits multistability, and for each basin of attraction we compute its probability distribution. The probability function thus becomes a weighted sum of the probabilities given for each basin of attraction,

$$P(X_i, \infty) = \sum_{j=1}^{m} \omega_j P^{(j)}(X_i, \infty)$$

where ω_j is the weighting coefficient of the j-th basin. Assume m attractors, then the number of simulations that end up in each attractor is $N_1, N_2, ..., N_m$. The weighting coefficient for the j-th basin is then calculated as $\omega_j = N_j / \sum_{i=1}^{m} N_i$. Finally, we calculate the potential landscapes based on $U(X) = -\ln P(X, \infty)$ [61, 62].

A stochastic differential equation model

A stochastic differential equation (SDE) model for the regulatory network can be constructed via the addition of a noise term [43, 45, 46, 63]:

$$dX(t) = F(X(t), S)dt + \eta X(t)dW(t) \qquad (16)$$

where $W(t)$ denotes the scalar white noise (or Wiener process), and η is the noise coefficient.

Abbreviations

ECM: Extracellular matrix; MSC: Mesenchymal stem cell; MYOD1: Myogenic Differentiation 1; ODE: Ordinary differential equation; PPARG: Peroxisome proliferator-activated receptor gamma; RUNX2 Runt-related transcription factor 2; SAA: Effective stiffness adhesion area; SDE: Stochastic differential equation; TAZ: Transcriptional coactivator with PDZ-binding motif; TCPS: Tissue culture polystyrene; TUBB3: The gene encode Tubulin beta-3 chain; YAP: Yes-associated protein

Acknowledgements

Not applicable.

Funding

QN was partially supported by NSF grants DMS1161621, and DMS1562176, and NIH grants P50GM076516, R01GM107264, and RC1NS095355. WZ was partially supported by the NIH grant 1DP2CA195763.

Authors' contributions

The overall work was conceived and designed by TP, __, WZ and QN. Simulations and implementations of the model were carried out by TP. Data analysis was performed by TP, LL, ALM, CWW, WZ, and QN. TP and LL wrote the first draft of the manuscript, and TP, LL, ALM, CWW, WZ, and QN together finalized the manuscript. All authors have read and approved the final version of the manuscript.

Competing interests

The authors declare that they have no competing interests.

Author details

[1]Department of Mathematics, Center for Complex Biological Systems, and Center for Mathematical and Computational Biology, University of California, Irvine, CA 92697, USA. [2]Department of Pharmaceutical Sciences, Department of Biomedical Engineering, Department of Biological Chemistry, Sue and Bill Gross Stem Cell Research Center, Chao Family Comprehensive Cancer Center & Edwards Life sciences Center for Advanced Cardiovascular Technology, University of California, 845 Health Sciences Road, Irvine, CA 92697, USA.

References

1. Engler AJ, Sen S, Sweeney HL, Discher DE. Matrix elasticity directs stem cell lineage specification. Cell. 2006;126:677–89.
2. Guilak F, Cohen DM, Estes BT, Gimble JM, Liedtke W, Chen CS. Control of stem cell fate by physical interactions with the extracellular matrix. Cell Stem Cell. 2009;5:17–26.
3. Trappmann B, Gautrot JE, Connelly JT, Strange DGT, Li Y, Oyen ML, et al. Extracellular-matrix tethering regulates stem-cell fate. Nat Mater. 2012;11:642–9.
4. Khetan S, Guvendiren M, Legant WR, Cohen DM, Chen CS, Burdick JA. Degradation-mediated cellular traction directs stem cell fate in covalently crosslinked three-dimensional hydrogels. Nat Mater. 2013;12:458–65.
5. Ivanovska IL, Shin J-W, Swift J, Discher DE. Stem cell mechanobiology: diverse lessons from bone marrow. Trends Cell Biol. 2015;25(9):523–32.
6. Halder G, Dupont S, Piccolo S. Transduction of mechanical and cytoskeletal cues by YAP and TAZ. Nat Rev Mol Cell Biol. 2012;13:591–600.
7. Wen JH, Vincent LG, Fuhrmann A, Choi YS, Hribar KC, Taylor-Weiner H, et al. Interplay of matrix stiffness and protein tethering in stem cell differentiation. Nat Mater. 2014;13:979–87.
8. Yang C, Tibbitt MW, Basta L, Anseth KS. Mechanical memory and dosing influence stem cell fate. Nat Mater. 2014:13(6):645–52.
9. Rehfeldt F, Brown AEX, Raab M, Cai S, Zajac AL, Zemel A, et al. Hyaluronic acid matrices show matrix stiffness in 2D and 3D dictates cytoskeletal order and myosin-II phosphorylation within stem cells. Integr Biol. 2012;4:422–30.
10. Raghunathan VK, Morgan JT, Dreier B, Reilly CM, Thomasy SM, Wood JA, et al. Role of substratum stiffness in modulating genes associated with extracellular matrix and mechanotransducers YAP and TAZ. Invest Ophthalmol Vis Sci. 2013;54:378–86.
11. Azzolin L, Panciera T, Soligo S, Enzo E, Bicciato S, Dupont S, et al. YAP/TAZ incorporation in the β-Catenin destruction complex orchestrates the Wnt response. Cell. 2014;158:157–70.
12. Burke DP, Kelly DJ. Substrate stiffness and oxygen as regulators of stem cell differentiation during skeletal tissue regeneration: a mechanobiological model. PLoS ONE. 2012;7(7):e40737.
13. Mousavi SJ, Doweidar MH. Role of mechanical cues in cell differentiation and proliferation: a 3D numerical model. PLoS ONE. 2015;10:e0124529.
14. Sun M, Spill F, Zaman MH. A computational model of YAP/TAZ mechanosensing. Biophys J. 2016;110:2540–50.
15. Kang K-T, Park J-H, Kim H-J, Lee H-M, Lee K-I, Jung H-H, et al. Study on differentiation of mesenchymal stem cells by mechanical stimuli and an algorithm for bone fracture healing. Tissue Eng Regen Med. 2011;8:359–70.
16. Stops AJF, Heraty KB, Browne M, O'Brien FJ, McHugh PE. A prediction of cell differentiation and proliferation within a collagen–glycosaminoglycan scaffold subjected to mechanical strain and perfusive fluid flow. J Biomech. 2010;43:618–26.
17. Sun Y, Chen CS, Fu J. Forcing stem cells to behave: a biophysical perspective of the cellular microenvironment. Annu Rev Biophys. 2012; 41:519–42.
18. Dupont S, Morsut L, Aragona M, Enzo E, Giulitti S, Cordenonsi M, et al. Role of YAP/TAZ in mechanotransduction. Nature. 2011;474:179–83.
19. Swift J, Ivanovska IL, Buxboim A, Harada T, Dingal PCDP, Pinter J, et al. Nuclear lamin-A scales with tissue stiffness and enhances matrix-directed differentiation. Science. 2013;341:1240104–4.
20. Hong JH, Hwang ES, McManus MT, Amsterdam A, Tian Y, Kalmukova R, et al. TAZ, a transcriptional modulator of mesenchymal stem cell differentiation. Science. 2005;309:1074–8.
21. Hong JH, Yaffe MB. TAZ - A beta-catenin-like molecule that regulates mesenchymal stem cell differentiation. Cell Cycle. 2006;5:176–9.
22. Harada S, Rodan GA. Control of osteoblast function and regulation of bone mass. Nature. 2003;423:349–55.
23. Yourek G, Hussain MA, Mao JJ. Cytoskeletal changes of mesenchymal stem cells during differentiation. ASAIO J. 2007;53:219–28.
24. Kilian KA, Bugarija B, Lahn BT, Mrksich M. Geometric cues for directing the differentiation of mesenchymal stem cells. Proc Natl Acad Sci. 2010;107:4872–7.
25. Maharam E, Yaport M, Villanueva NL, Akinyibi T, Laudier D, He Z, et al. Rho/Rock signal transduction pathway is required for MSC tenogenic differentiation. Bone Res. 2015;3:15015.
26. Matsuoka F, Takeuchi I, Agata H, Kagami H, Shiono H, Kiyota Y, et al. Morphology-based prediction of osteogenic differentiation potential of human mesenchymal stem cells. PLoS ONE. 2013;8:e55082.

27. Rogov IA, Volkova IM, Kuleshov KV, Savchenkova IP. in vitro myogenic differentiation of bovine multipotent mesenchymal stem cells taken from bone marrow and adipose tissue. Agricult Biol. 2012;6:66–72.

28. Dingal PCDP, Bradshaw AM, Cho S, Raab M, Buxboim A, Swift J, et al. Fractal heterogeneity in minimal matrix models of scars modulates stiff-niche stem-cell responses via nuclear exit of a mechanorepressor. Nat Mater. 2015;14:951–60.

29. Adler M, Mayo A, Alon U. Logarithmic and power law input-output relations in sensory systems with fold-change detection. PLoS Comput Biol. 2014; 10(8):e1003781.

30. Peng T, Peng H, Choi D, Su J, Chang C-C, Zhou X. Modeling cell-cell interactions in regulating multiple myeloma initiating cell fate. IEEE J Biomed Health Inform. 2013;18:484–91.

31. Smolen P. Modeling transcriptional control in gene networks—methods, recent results, and future directions. Bull Math Biol. 2000;62:247–92.

32. Ben-Tabou de-Leon S, Davidson EH. Modeling the dynamics of transcriptional gene regulatory networks for animal development. Dev Biol. 2009;325:317–28.

33. Lv H, Li L, Sun M, Zhang Y, Chen L, Rong Y, et al. Mechanism of regulation of stem cell differentiation by matrix stiffness. Stem Cell Res Ther. 2015;6:1.

34. Gesztelyi R, Zsuga J, Kemeny-Beke A, Varga B, Juhasz B, Tosaki A. The Hill equation and the origin of quantitative pharmacology. Arch Hist Exact Sci. 2012;66:427–38.

35. Gilbert PM, Havenstrite KL, Magnusson KEG, Sacco A, Leonardi NA, Kraft P, et al. Substrate elasticity regulates skeletal muscle stem cell self-renewal in culture. Science. 2010;329:1078–81.

36. Lu D, Luo C, Zhang C, Li Z, Long M. Differential regulation of morphology and stemness of mouse embryonic stem cells by substrate stiffness and topography. Biomaterials. 2014;35:3945–55.

37. Choi B, Park KS, Kim JH, Ko KW, Kim JS, Han DK, et al. Stiffness of hydrogels regulates cellular reprogramming efficiency through mesenchymal-to-epithelial transition and stemness markers. Macromol Biosci. 2016;16:199–206.

38. Wang J, Zhang K, Xu L, Wang E. Quantifying the waddington landscape and biological paths for development and differentiation. Proc Natl Acad Sci. 2011;108:8257–62.

39. Baker WL. A review of models of landscape change. Landscape Ecol. Kluwer Academic Publishers; 1989;2:111-33.

40. Wang J, Xu L, Wang E. Potential landscape and flux framework of nonequilibrium networks: robustness, dissipation, and coherence of biochemical oscillations. Proc Natl Acad Sci. 2008;105:12271–6.

41. Wang J, Li C, Wang E. Potential and flux landscapes quantify the stability and robustness of budding yeast cell cycle network. Proc Natl Acad Sci. 2010;107:8195–200.

42. Murphy WL, McDevitt TC, Engler AJ. Materials as stem cell regulators. Nat Mater. 2014;13:547–57.

43. Elowitz MB, Levine AJ, Siggia ED, Swain PS. Stochastic gene expression in a single cell. Science. 2002;297:1183–6.

44. Chalancon G, Ravarani CNJ, Balaji S, Martinez-Arias A, Aravind L, Jothi R, et al. Interplay between gene expression noise and regulatory network architecture. Trends Genet. 2012;28:221–32.

45. Chen M, Wang L, Liu CC, Nie Q. Noise attenuation in the ON and OFF states of biological switches. ACS Synth Biol. 2013;2:587–93.

46. Holmes WR, Nie Q. Interactions and tradeoffs between cell recruitment, proliferation, and differentiation affect CNS regeneration. Biophys J. 2014; 106:1528–36.

47. Cox TR, Erler JT. Remodeling and homeostasis of the extracellular matrix: implications for fibrotic diseases and cancer. Dis Model Mech. 2011;4:165–78.

48. Mendez MG, Janmey PA. Transcription factor regulation by mechanical stress. Int J Biochem Cell Biol. 2012;44:728–32.

49. Paszek MJ, Zahir N, Johnson KR, Lakins JN, Rozenberg GI, Gefen A, et al. Tensional homeostasis and the malignant phenotype. Cancer Cell. 2005;8:241–54.

50. Levental KR, Yu H, Kass L, Lakins JN, Egeblad M, Erler JT, et al. Matrix crosslinking forces tumor progression by enhancing integrin signaling. Cell. 2009;139:891–906.

51. Gattazzo F, Urciuolo A, Bonaldo P. Extracellular matrix: a dynamic microenvironment for stem cell niche. Biochim Biophys Acta. 1840;2014: 2506–19.

52. Prinz H. Hill coefficients, dose–response curves and allosteric mechanisms. J Chem Biol. 2010;3:37–44.

53. Ingalls B. Mathematical modelling in systems biology: an introduction. 2013.

54. Allgower EL, Georg K. Numerical continuation methods: an introduction. 2012.

55. Dhooge A, Govaerts W, Kuznetsov YA, Meijer HGE, Sautois B. New features of the software MatCont for bifurcation analysis of dynamical systems. Math Comput Model Dyn Syst. 2008;14:147–75.

56. Li Q, Wennborg A, Aurell E, Dekel E, Zou J-Z, Xu Y, et al. Dynamics inside the cancer cell attractor reveal cell heterogeneity, limits of stability, and escape. Proc Natl Acad Sci. 2016;113:2672–7.

57. van Kampen NG. Stochastic processes in physics and chemistry. 1985.

58. Li C, Wang J. Quantifying cell fate decisions for differentiation and reprogramming of a human stem cell network: landscape and biological paths. PLoS Comput Biol. 2013;9:e1003165.

59. Sasai M, Wolynes PG. Stochastic gene expression as a many-body problem. Proc Natl Acad Sci. 2003;100:2374–9.

60. Hu G. Stochastic forces and nonlinear systems. 1994.

61. Li C, Wang J. Landscape and flux reveal a new global view and physical quantification of mammalian cell cycle. Proc Natl Acad Sci. 2014;111:14130–5.

62. Li C, Wang E, Wang J. Landscape topography determines global stability and robustness of a metabolic network. ACS Synth Biol. 2012;1(6):229–39.

63. Allen E. Modeling with Itô stochastic differential equations. 2007.

64. McBeath R, Pirone DM, Nelson CM, Bhadriraju K, Chen CS. Cell shape, cytoskeletal tension, and RhoA regulate stem cell lineage commitment. Dev Cell. 2004;6:483–95.

65. Alarcón C, Zaromytidou A-I, Xi Q, Gao S, Yu J, Fujisawa S, et al. Nuclear CDKs drive smad transcriptional activation and turnover in BMP and TGF-β pathways. Cell. 2009;139:757–69.

66. Jeong H, Bae S, An SY, Byun MR, Hwang J-H, Yaffe MB, et al. TAZ as a novel enhancer of MyoD-mediated myogenic differentiation. FASEB J. 2010;24:3310–20.

67. Bernabé BP, Shin S, Rios PD, Broadbelt LJ, Shea LD, Seidlits SK. Dynamic transcription factor activity networks in response to independently altered mechanical and adhesive microenvironmental cues. Integr Biol. 2016;8:844–60.

4

Clostridium butyricum maximizes growth while minimizing enzyme usage and ATP production: metabolic flux distribution of a strain cultured in glycerol

Luis Miguel Serrano-Bermúdez[1], Andrés Fernando González Barrios[2], Costas D. Maranas[3] and Dolly Montoya[1]* ⓘ

Abstract

Background: The increase in glycerol obtained as a byproduct of biodiesel has encouraged the production of new industrial products, such as 1,3-propanediol (PDO), using biotechnological transformation via bacteria like *Clostridium butyricum*. However, despite the increasing role of *Clostridium butyricum* as a bio-production platform, its metabolism remains poorly modeled.

Results: We reconstructed *i*Cbu641, the first genome-scale metabolic (GSM) model of a PDO producer *Clostridium* strain, which included 641 genes, 365 enzymes, 891 reactions, and 701 metabolites. We found an enzyme expression prediction of nearly 84% after comparison of proteomic data with flux distribution estimation using flux balance analysis (FBA). The remaining 16% corresponded to enzymes directionally coupled to growth, according to flux coupling findings (FCF). The fermentation data validation also revealed different phenotype states that depended on culture media conditions; for example, *Clostridium* maximizes its biomass yield per enzyme usage under glycerol limitation. By contrast, under glycerol excess conditions, *Clostridium* grows sub-optimally, maximizing biomass yield while minimizing both enzyme usage and ATP production. We further evaluated perturbations in the GSM model through enzyme deletions and variations in biomass composition. The GSM predictions showed no significant increase in PDO production, suggesting a robustness to perturbations in the GSM model. We used the experimental results to predict that co-fermentation was a better alternative than *i*Cbu641 perturbations for improving PDO yields.

Conclusions: The agreement between the predicted and experimental values allows the use of the GSM model constructed for the PDO-producing *Clostridium* strain to propose new scenarios for PDO production, such as dynamic simulations, thereby reducing the time and costs associated with experimentation.

Keywords: *Clostridium butyricum*, 1,3-propanediol, Genome-scale metabolic model, Objective function

Background

The rising biodiesel industry has resulted in a major overproduction of glycerol as a byproduct, which now threatens the economic viability of this industry [1, 2]. This situation has spurred research into glycerol utilization as a carbon source [3–5] and for the generation of products such as 1,3-propanediol (PDO), a precursor of important commercial polymers, such as polyester and polyurethane [6, 7]. PDO can be biosynthesized from glycerol by bacteria such as *Clostridium butyricum* or *Klebsiella* spp. [3, 7]. *Clostridium* species are the more attractive alternative because they are safer and achieve higher yields than *Klebsiella* [8]. However, industrial PDO production using bacteria is still limited by insufficient yields, which presents a serious obstacle to the competitiveness of this process [9–11]. Therefore, strategies such as fed-batch cultures and random mutagenesis have been developed, resulting in improvements in PDO production

* Correspondence: dmontoyac@unal.edu.co
[1]Bioprocesses and Bioprospecting Group, Universidad Nacional de Colombia. Ciudad Universitaria, Carrera 30 No. 45-03, Bogotá, D-C, Colombia
Full list of author information is available at the end of the article

of up to 137% and 78%, respectively [11–13]. A more detailed understanding of the metabolic pathways in species such as *Clostridium butyricum* could therefore shed light on a better ways to promote glycerol transformation to PDO in this organism.

Metabolism studies of glycerol by the anaerobic bacterium *Clostridium butyricum* have generally focused on its central metabolism, which is composed of oxidative and reductive branches [14]. The oxidative branch is mainly related to the production of ATP and reducing equivalents (NADH), with the formation of acetic and butyric acids as byproducts. By contrast, the reductive branch produces PDO while simultaneously regenerating reducing equivalents by conversion of NADH to NAD [7, 9, 15]. Bizukojc et al. [16] reported the most detailed metabolic model for a PDO producer *Clostridium* strain, indicating the functioning of 77 reactions and 69 metabolites. The model, in addition to the oxidative and reductive branches, also included simplified synthesis reactions for amino acids, macromolecules, and biomass. However, at present, metabolic models based on genome annotation information, also known as genome-scale metabolic (GSM) models [17, 18], are lacking for *Clostridium butyricum*.

A proteomics study of the native Colombian strain *Clostridium* sp. IBUN 158B cultured in glycerol [19] has provided experimental validation of the enzyme expression involved in PDO metabolic networks in this specie. The proteome contained 21 enzymes classified as follows: one from the reductive branch (PDO dehydrogenase), three from the oxidative branch, eleven from carbohydrate synthesis, four from amino acid synthesis, and two from nucleotide synthesis. Gungormusler et al. [20] also used proteomics for the experimental detection of 262 different enzymes expressed by *Clostridium butyricum* 5521 cultured in glycerol. Nevertheless, despite this experimental information and the computational tools available, the prediction of PDO production by *Clostridium* based on its metabolic behavior is still limited.

One computational tool commonly employed for metabolic modeling is flux balance analysis (FBA). FBA allows the use of a steady state assumption of defined culture conditions to predict the phenotype of one microorganism based on its GSM model [21–25]. However, a GSM model expressed as stoichiometric matrix is an undetermined system, that is, it has more reactions than metabolites. This creates a situation with infinite solutions, so an objective function is required to predict the microorganism phenotype. FBA then becomes an optimization process in which the constraints are the culture conditions, mass balances, and thermodynamic feasibilities [22, 25–28].

In general, predictions using GSM models assume biomass yield maximization as the objective function, based on the assumption that cells have evolved to select the most efficient pathways that achieve the best yields [29]. Nevertheless, predictions with biomass maximization do not always capture the cellular physiology, and alternative objective functions have been developed [28, 30–33]. Studies have included error minimization by bi-level optimization [30, 34, 35], objective function selection by Bayesian inference [31] or by Euclidian distance minimization [32], and linear combination of objective functions [28, 36]. The results, overall, highlight that a cell does not maximize biomass yield under scenarios like substrate excess, so that one single function is unable to predict all the evaluated scenarios [28, 32, 33, 37–39].

For these reasons, the initial purpose of the present research was to construct the first GSM model of a PDO producer *Clostridium* strain. The biological model selected was the Colombian strain *Clostridium* sp. IBUN 13A, a strain isolated by our Bioprocesses and Bioprospecting Group. This strain is a natural PDO producer and has been employed over the last 20 years in several studies aimed at understanding PDO production, including the annotation of its genome [40–42]. Additionally, as second objective, our intent was to predict the phenotypic states of this bacterium during culture in glycerol and in other substrates using the GSM model and FBA with the adequate objective function. Our overall aim was to evaluate the effect of perturbations in the constructed GSM model on PDO yield improvements.

Results and discussion
Genome-scale metabolic model *i*Cbu641 reconstruction and curation

A draft metabolic model for the *Clostridium* sp. IBUN 13A strain was constructed based on RAST annotation [41]. The draft was composed of 641 genes, 365 enzymes, 671 reactions, and 606 metabolites. GapFind [43] analysis of the draft model identified 303 blocked metabolites, which were reduced to 63 by adding 59 reactions based on experimental fermentation evidence from *Clostridium butyricum* cultured in glycerol [15, 19, 44] and on curated GSM models from other solventogenic clostridia [16, 45–52]. The biomass reaction was adapted from *C. beijerinckii* GSM [45], which does not account for the proton formation associated with ATP hydrolysis during the growth-associated maintenance (GAM), as is also observed in *C. acetobutylicum* [49]. This excluded proton would accumulate in the biomass, thereby preventing stabilization of the biomass charge. By contrast, the GSM models of *C. thermocellum* [50], *C. ljungdahlii* [52], and *C. cellulolyticum* [51] include this proton production in their biomass reactions. Therefore, the proton formation was included in the present biomass reaction to resolve the inconsistency in the elemental composition of the biomass, as well as the charge balance. The elemental composition per C atom, calculated based only on stoichiometric consumption of precursors, was therefore $CH_{1.624}O_{0.456}N_{0.216}P_{0.033}S_{0.0047}$.

After curation, elemental balancing, and loop deletion, the constructed iCbu641 GSM model included 641 genes, 891 reactions, and 701 metabolites. Table 1 summarizes the main features of the curated metabolic network, and Fig. 1 shows the pathway distribution of the cytosolic reactions. Comparison with other GSM models of solventogenic Clostridium strains showed 11 unique enzymes, including as PDO dehydrogenase (EC.1.1.1.202) and glycerol dehydratase (EC.4.2.1.30). These two enzymes function in lipid metabolism but are associated with the reductive branch of glycerol metabolism [14].

The iCbu641 metabolic network is the first GSM model curated for a PDO producing Clostridium strain [17, 18, 49] (See Additional files 1 and 2 for complete metabolic model at excel format and SBML format, respectively). The iCbu641 model also includes all the enzymes associated with glycolysis and the pentose phosphate pathway and most of the enzymes involved in the TCA cycle. The enzymes from the TCA cycle that were not included are malate dehydrogenase (EC.1.1.5.4), succinate-CoA ligase (EC.6.2.1.4 - EC.6.2.1.5) and fumarate reductase (EC.1.3.5.4), which were not detected in the genome. Therefore, additional experimentation is required to verify the presence or absence of genes encoding these three enzymes in the genome of Clostridium sp. IBUN 13A. The model is able to synthesize de novo all the precursors involved in the biomass reaction (e.g., amino acids, nucleotides, fatty acids, teichoic acid, and cofactors).

Flux distribution prediction using flux balance analysis with glycerol as substrate

The constructed iCbu641 GSM model and FBA were employed to predict the flux distribution of Clostridium butyricum cultured in glycerol. FBA was solved using linear programming (LP) with biomass maximization as

an objective function. When compared with the experimental data, FBA predicted a biomass overestimation and no PDO production (Fig. 2 – Blue lines), giving biomass yield ($Y_{X/S}$) and PDO yield ($Y_{PDO/S}$) errors of 300% and 100%, respectively [44]. Therefore, taking into account only biomass yield, we could infer that, experimentally, Clostridium butyricum does not grow optimally in glycerol. This can be explained mathematically, because the objective function and all constraints (mass balances and thermodynamic feasibilities) used were linear. This means that the optimum found is indeed a vertex of the feasible solution space [38], where no PDO could be produced.

Contrary to FBA predictions, PDO production in Clostridium sp. IBUN 13A is related to dha operon expression and the enzyme activity of both PDO dehydrogenase (EC.1.1.1.202) and glycerol dehydratase (EC.4.2.1.30), which have been experimentally detected in the presence of glycerol as main carbon source [40, 53, 54]. The lack in predictions can be interpreted biologically based on redox balance together with the capability of C. butyricum to produce formic acid and hydrogen (H_2) [44]. Clostridium butyricum is an anaerobic bacterium, so according to the redox balance, the substrate must act simultaneously as an acceptor and donor of electrons [55]. However, the aforementioned capability allows to LP optimization predicts that the substrate could be used mostly as an electron donor, thereby generating more ATP and biomass. This prediction is achieved due to the regeneration of reducing equivalents through formic acid and H_2 formation, which would require no substrate as an electron acceptor to produce reduced compounds such as PDO. In other words, in the vertex predicted by LP optimization, the substrate is used mainly as an electron donor due to the formation of formic acid and H_2, and no PDO is produced.

Similar results were observed using glucose as a substrate, where formic acid and H_2 were overproduced instead of reduced products like butanol or ethanol, which have been detected experimentally (data not shown) [56]. Consequently, LP optimization could be considered as unsuitable for predicting the experimental yields of Clostridium butyricum, which is capable of producing formic acid or H_2. Nevertheless, LP optimization is commonly employed in solventogenic Clostridium strains, although all of them are able to produce at least hydrogen [45, 47, 48, 50, 51]. This lack of prediction could be solved using experimental constraints of formic acid and H_2 secretion in LP optimization; however, no linear trend was observed, especially in H_2 secretion [44].

A new objective function of maximizing biomass yield per flux unit (Equation 1) was therefore employed to improve the FBA predictions. This objective function is based on the hypothesis that cells operate to maximize biomass yield while minimizing enzyme usage [32]. This

Table 1 Main features of the iCbu641 metabolic network

Feature		Number
Genes		641
Enzymes		365
Total Reactions		891
	Cytosolic reactions[a]	727
	Transport reactions	86
	Exchange reactions	78
Total Metabolites		701
	Blocked metabolites	63

[a]Includes 17 simplified biomass and macromolecule synthesis reactions [45] and 59 reactions added in the curation

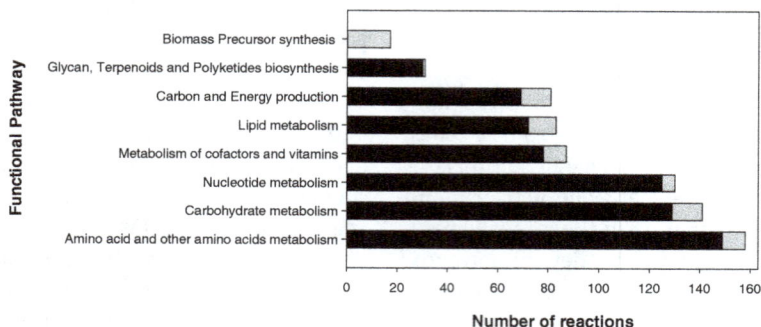

Fig. 1 Distribution of cytosolic reactions in the *i*Cbu641 GSM model by functional pathway. Notation (■) Gene-Associated reactions (▪) Non Gene-Associated reactions

non-linear programming (NLP) optimization is non-convex, but Schuetz et al. suggested that the predicted local optimum is indeed the global optimum [32]. In addition to the new objective function, a non-linear constraint was employed, which corresponds to the maximum acetic acid secretion flux and has an allosteric trend in the function of glycerol uptake flux [44]. Therefore the constraint was not used during simulations with other carbon sources, as glucose. This allosteric trend of acetate production has been previously reported for *Clostridium butyricum* cultured in glycerol as a mechanism to control acetyl-CoA/CoA and ATP/ADP ratios [57]. The new simulations (Fig. 2 – Red lines) predicted $Y_{X/S}$ and $Y_{PDO/S}$ errors of nearly 4.5%, and 1.5%, respectively, when compared with the experimental values obtained under limiting conditions of glycerol (<15 g/L) [44]. Therefore, the error reduction using NLP predictions suggests the resolution of the redox balance problems observed in LP optimization caused by the capability of producing formic acid and H_2. It also suggests that *Clostridium butyricum* indeed minimizes enzyme usage and prefers short pathways (PDO

production) to maximize its growth under limiting nutrient conditions.

$$\text{Max } \frac{\mu}{\sum_{i=1}^{n} v^2} \tag{1}$$

Identical phenotypic predictions were also observed using both LP and NLP optimizations when the production of hydrogen and formic acid were blocked as additional constraints (data not shown). This validates, on the one hand, the effect of these products in the lack of prediction using LP optimization. On the other hand, NLP optimization indeed predicts global optima, as suggested Schuetz et al. [32].

However, the new objective function overestimated $Y_{X/S}$ under glycerol-excess conditions (without reaching inhibition conditions) [44], suggesting that *Clostridium butyricum* grows under sub-optimal conditions when the substrate is present in excess, as is also observed in *E. coli* [32]. This behavior is understandable if thermodynamics is considered. Growth, as with any reaction, is

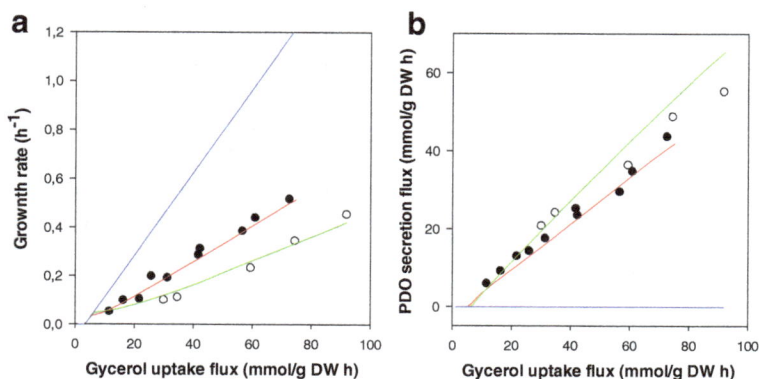

Fig. 2 Robustness analyses in function of glycerol uptake flux. **a** Specific growth rate μ and **b** PDO secretion flux v_{PDO}. Notation: Experimental data from Solomon et al. [44] at (●) glycerol limitation and (○) glycerol excess. FBA predictions using: biomass maximization (Blue Line), biomass maximization per enzyme usage (Red Line) and biomass maximization while minimizing both enzyme usage and ATP production (Green Line)

thermodynamically feasible if its driving force, expressed as Gibbs free energy, is negative ($\Delta G_{growth} < 0$). Growth Gibbs free energy can be expressed as a function of biomass yield and the Gibbs free energies for catabolism and anabolism, as shown Eq. 2 [55]. Therefore, an improvement in thermodynamic feasibility will be coupled to biomass yield reduction, as occurs in scenarios such as substrate competition with other microorganisms [55]. This suggests that a substrate excess condition could induce the above scenario in order for the organism to prevail in this environment.

$$\Delta G_{growth} = \frac{1}{Y_{X/S}} \Delta G_{catabolism} + \Delta G_{anabolism}$$

$$(2)$$

The sub-optimal behavior was accounted for in the objective function of Eq. 1 by the use of a tunable weighting factor (w) that minimizes both the enzyme usage and ATP production of the network (Equation 3). The ATP incorporation in the objective function is based on experimental results [15, 44] and because a reduction in ATP production is related to the respective biomass yield reduction. Therefore, a weight w equal to 1 corresponds to the optimal conditions of Eq. 1. The use of weight factors has been described by Torres et al., who added ATP and NADH/NADPH minimization or maximization to the objective function, which improved *S. cerevisiae* growth predictions up to 98% [36, 39]. On this basis, simulations under excess conditions were made using a weight factor equal to 0.04 ($w = 0.04$), where the average experimental errors were minimal [44]. The average $Y_{X/S}$ and $Y_{PDO/S}$ errors were 5.3% and 2.5%, respectively, as shown Fig. 2 – Green Lines, confirming the ability of FBA to provide accurate predictions of these results through NLP optimization. However, this weight factor only applies to *Clostridium butyricum* cultured anaerobically in glycerol; additional experimental data would validate its usage.

$$\text{Max} \quad \frac{\mu}{w\left[\sum_{i=1}^n v^2\right] + (1-w)\left[v_{ATP\ prod}^2\right]} \quad w \in (0,1) \quad (3)$$

We further validated the necessity of employing the weight factor at sub-optimal conditions of glycerol by comparing both predicted phenotypes with the results reported by Zeng [15]. Zeng found that the directionality of the reaction catalyzed by ferredoxin reductase (EC.1.18.1.2 or EC.1.18.1.3) changes if glycerol is in limitation or in excess. Under the limiting condition, ferredoxin reductase consumes reducing equivalents and produces more H_2 and less PDO than is observed under the excess condition, where ferredoxin reductase

produces reducing equivalents. Consequently, Zeng quantified ferredoxin reductase directionality by calculating the ratio between hydrogen and reduced ferredoxin formed ($\alpha_{H_2/Fd}$). The experimental ratios are 1.1 and 0.4 under limiting and excess glycerol conditions, respectively. Similarly, the ratios predicted by FBA are 1.46 and 0.40 at optimal and sub-optimal conditions, respectively. Therefore, the predictions for both conditions agree with the reported values and validate the utility of the weight factor in predicting glycerol cultures.

We also compared the flux prediction for the 365 enzymes present in *i*Cbu641 with the experimental expression of the 286 enzymes detected by Gungormusler et al. in the proteome of the *Clostridium butyricum* DSM 10702 strain cultured under limiting glycerol conditions [20]. An analysis of the 174 common enzymes using flux couple finding (FCF) revealed 80 enzymes that were partially or fully coupled to growth and 67 enzymes that were directionally coupled to growth [58]. The remaining 27 enzymes were blocked and were therefore excluded from the comparison. These 27 enzymes could be blocked due to a lack of gene annotation or exclusion from biomass synthesis (e.g., terpenoid synthesis). Terpenoids have been detected in the cell walls of *Clostridium* strains and may arise as a stress response to the acids formed during culture [59]. The terpenoid pathway could be unblocked if these metabolites are added to biomass synthesis, but their concentration levels first need accurate quantification [59].

Assuming a qualitative correlation between the expression and flux for the 147 enzymes included in the comparison, FBA was able to predict the expression of 123 of them (83.7%). The remaining 24 non-predicted enzymes correspond only to directionally coupled growth and most of them (21 enzymes) are involved in carbohydrate metabolism and the synthesis of nitrogenous compounds (amino acids and nucleotides). The absence of a prediction for these enzymes could be due the presence of alternate pathways and isozymes, as is the case for asparagine synthase (EC.6.3.5.4), isocitrate dehydrogenase (NADP) (EC.1.1.1.42), and glycerol-3-phosphate dehydrogenase (EC.1.1.1.94), where the alternative enzymes are asparagine synthetase (EC.6.3.1.1), isocitrate dehydrogenase (NAD) (EC.1.1.1.41), and glycerol kinase (EC.2.7.1.30), respectively. Another possible cause is the inability of FBA to predict regulation mechanisms [60], as is the case for the enzyme pyruvate phosphate dikinase (EC.2.7.9.1), which is involved in the gluconeogenesis pathway but appears to act in the place of pyruvate kinase, consuming AMP instead ADP [61, 62]. Other enzymes, such as nicotinate phosphoribosyltransferase (EC.6.3.4.21) or pyrimidine nucleoside phosphorylase (EC.2.4.2.2), show a lack of prediction because they are involved in RNA or DNA fragment recycling [63].

Finally, 1,4-alpha-glucan branching enzyme (EC.2.4.1.18) or starch synthase (EC.2.4.1.21) were not predicted as they are part of granulose synthesis, a process that is not included in the biomass reaction. Granulose is a polysaccharide employed as carbon source during sporulation; therefore, it is produced during exponential growth [64]. (See Additional file 3 for complete proteomic comparison results).

Flux distribution prediction using FBA and other carbon sources

The robustness of iCbu641 was further tested by comparing the experimental and simulated data using other substrates at optimal conditions, Eq. 1. We first compared the FBA predictions and experimental yields of Clostridium butyricum W5 cultured in glucose [56], as shown in Table 2. We observed an accuracy of nearly 97% for predicting the biomass yield using this substrate, confirming the ability to predict not only glycerol cultures but also glucose cultures. All the reported experimental yields also agreed with their respective predicted feasible ranges calculated using flux variability analysis (FVA).

Junghare et al. [65] also evaluated biomass and hydrogen production of Clostridium butyricum TM-9A using different carbohydrates. Our comparison of the experimental yields with the predicted yields obtained through FBA is shown in Table 3. The trends in the predicted $Y_{X/S}$ agree with the experimentally obtained values, showing smaller yields for pentoses and the highest yield for the trisaccharide raffinose, while the yields using monosaccharides were lower than those obtained using disaccharides. However, some differences are observed between the experimental and predicted values. First, ribose and xylose had considerably higher experimental than predicted yields. Second, the experimental yield for cellobiose was much lower than the predicted yield. Finally, although the simulations predicted the same yields for arabinose and ribose, their experimental yields differed. The first case could be a result of an incomplete curation of iCbu641 related to pentose consumption; therefore, more experimental information is needed. The

Table 2 Comparison of the experimental and simulated yields (mol/mol) of Clostridium butyricum W5 cultured in glucose [56]

Product	Experimental Yield	Predicted feasible range
Biomass	0.0270	0.0279 (0.0228–0.0330)
Acetate	0.172	0.574 (0–1.053)
Lactate	0.566	0.179 (0–0.773)
Butyrate	0.295	0.114 (0–0.454)
H_2	1.325	0.661 (0.046–1.345)
Ethanol	0.043	0.215 (0–0.714)

Table 3 Comparison experimental and simulated yields (mol/mol) of Clostridium butyricum TM-9A cultured in different carbohydrates [65]

Carbohydrate	Experimental Yields		Predicted Yields	
	$Y_{X/S}^a$	$Y_{H2/S}$	$Y_{X/S}^a$	Feasible range $Y_{H2/S}$
Arabinose	2.5%	0.067	6.7%	0.204–1.101
Ribose	25.3%	0.843	6.7%	0.236–1.100
Xylose	32.3%	0.589	11.1%	0.327–1.406
Mannose	30.8%	0.668	36.2%	0.046–1.346
Fructose	32.2%	0.848	37.2%	0.092–1.374
Galactose	35.9%	0.864	29.1%	0.478–1.711
Cellobiose	35.9%	0.945	59.9%	0.141–2.468
Trehalose	65.7%	1.612	72.3%	0.030–2.485
Sucrose	74.9%	1.494	70.1%	0–2.323
Raffinose	100.0%	2.716	100.0%	0–3.141

[a]Biomass yields were normalized based on the raffinose value

second and third cases could be due to miscalculation of the experimental yields of arabinose and cellobiose, since these substrates were not consumed completely, as can be inferred from their pH reports [65]. The last can point to the possibility that Clostridium was not well adapted to these substrates and may have needed to undergo more generations to reach its optimal growth [33]. The hydrogen yields ($Y_{H2/S}$) showed experimental values that were mostly within their respective predicted feasible ranges. The only unpredicted $Y_{H2/S}$ corresponded to arabinose, which supports the necessity of complementing iCbu641 for consumption of pentoses. Consequently, iCbu641 has the capacity to be employed to predict Clostridium butyricum growth using different carbohydrates as substrates.

We also assumed a qualitative correlation between the enzyme flux and mRNA expression and compared FBA predictions and experimental transcriptomics results for the C. butyricum strain CWBI 1009 cultured in glucose [66]. Of the 288 enzymes shared by both systems, 51 were blocked according FCF analysis and excluded from comparison. Among the remaining 237 enzymes, FBA predicted the activity of the 123 enzymes as partially and fully coupled to growth and 56 enzymes directionally coupled to growth, for a total of 179 predicted enzymes (75.5%). Similar to the proteomics comparison, the last 58 non-predicted enzymes were also directionally coupled to growth, and their lack of prediction is consistent with the reasons mentioned in the proteomics comparison. However, the lack of prediction of enzymes involved in thiamine production is highlighted, which is because this cofactor is not included as a biomass precursor. A similar situation happens with holo-ACP synthetase (EC.2.7.8.7), an enzyme involved in the CoA hydrolysis required to synthetize acyl carrier proteins

(ACPs) [67]. (See Additional file 3 for complete transcriptomic comparison results).

Qualitative comparisons between proteomic and transcriptomic data require the assumption that enzymes are active, but this depends on different factors, like post-translational modifications, allosteric control, etc. [68]. Such factors are also a problem even when quantitative omics data are used in mathematical approaches that are employed to predict phenotype states [69]. However, qualitative proteomic and transcriptomic comparisons with FBA predictions have been previously reported using E. coli K12 by Lewis et al. [70], who found predictions for up to 82% of the evaluated enzymes, and most of the unpredicted ones were isozymes, in agreement with our results.

Finally, the knockout mutant of Clostridium butyricum W5 obtained by ClosTron technology [71] was employed to evaluate the ability of iCbu641 to predict yields after perturbations in GSM. This mutant has butyric acid production blocked; its experimental yields are shown in Table 4. The wild type strain yields were predicted using FBA, while the mutant strain yields were obtained using regulatory on/off minimization (ROOM) [72]. ROOM was employed because it is better at predicting mutant phenotypic states when compared to other approaches, like minimization of metabolic adjustments (MOMA) [39]. This is because the ROOM approach seeks to maintain both the metabolic network and gene expression stabilities, as determined experimentally [72]. Simulations predicted an increased yield of ethanol using the mutant strain, and this was experimentally detected (Table 4). A biomass yield ($Y_{X/S}$) reduction was also predicted for the mutant strain. By contrast, the experimental reduction of hydrogen yield in the mutant was not predicted; however, this is not conclusive since the FVA ranges of the wild type and mutant strains did agree with their experimental values

Modeling scenarios with PDO yield increment

Three scenarios were evaluated to predict an increase in $Y_{PDO/S}$ using iCbu641. The first strategy was to use ROOM to predict single and double mutants through in silico enzyme deletion. The 145 enzymes associated with reactions directionally coupled to growth at culture

conditions, as reported by Comba et al. [19], were considered in the analysis. Enzymes fully or partially coupled to growth were not considered, since their deletions would affect the culture time and therefore increase the fermentation costs. A total of 18 enzymes catalyzed at least two reactions that could block growth if simultaneously deleted, including dihydrodipicolinate reductase (EC.1.17.1.8) or proline oxidase (EC.1.5.1.2). Similar results were obtained for the 22 double enzyme deletions, such prephenate dehydrogenase (EC.1.3.1.12) and prephenate dehydrogenase (NADP) (EC.1.3.1.13) or shikimate dehydrogenase (EC.1.1.1.25) and quinate/shikimate dehydrogenase (EC.1.1.1.282). Most of the single and double deletions enhanced the $Y_{PDO/S}$ by up to 1% in relation to the wild type strain. One of the best mutant predicted had simultaneous deletion of lactate dehydrogenases (EC.1.1.1.27 and EC.1.1.1.28), which increased $Y_{PDO/S}$ only to nearly 1.2% (See Additional file 4 for complete mutant prediction results).

Single knockout mutants obtained from Colombian strain Clostridium sp. IBUN 158B [73] to improve the PDO production were used for validation. This is a different PDO producer strain, but research shows that the native strains currently sequenced share at least 99% of the genome (Article in preparation). Therefore, no significant differences were expected between the metabolic models of 13A and 158B, supporting the use of these mutants in validation. The inactivated enzymes were hydrogenase (ΔhydA-420 s), lactate dehydrogenase (ΔldhA-508 s), and 3-hydroxybutyryl-CoA dehydrogenase (Δhbd-414 s), corresponding the lack of production of hydrogen, lactic acid, and butyric acid, respectively. Montoya [73] reported that these three mutants were viable, as shown Table 5; however, he cultured only two of them due lack of time during his doctoral research. The biomass yield predictions for the mutants were overestimated, although a trend was predicted. This overestimation could be due to smaller glycerol uptake fluxes for the mutants, which would result in inadequate biomass yield normalization; however, the lack of measurements limits the elaboration of better comparisons.

The experimental values for PDO yields agreed with the predicted range, validating the ROOM simulations. Moreover, the FVA of the wild type strain indicated a

Table 4 Comparison experimental and simulated yields of wild type and mutant strains of Clostridium butyricum W5 cultured in glucose [71]

Product	Experimental yields		Simulated yields (FVA range)	
	Wild strain	Mutant strain	Wild strain (FBA)	Mutant strain (ROOM)
H₂	1.25[a]	0.69	0.661 (0.046–1.345)	0.694 (0–1.787)
Ethanol	0.18	3.31 [a]	0.215 (0–0.714)	0.680 (0–1.069)
Biomass[b]	100%	99.2%[a]	100%	95.0%

[a]Values calculated from information reported by Cai et al. [71]
[b]Yields reported as percentages based on the wild type strain

Table 5 Comparison experimental and simulated yields of wild type and mutant strains of *Clostridium* sp. IBUN 158B cultured in glycerol [73]

Strain	Experimental yields		Simulated yields (FVA range)	
	$Y_{X/S}^a$	$Y_{PDO/S}$	$Y_{X/S}^a$	$Y_{PDO/S}$
Wild strain	100.0 ± 8%	0.538 ± 0.047	100%	0.588 (0.479–0.656)
Δ*hydA*-420 s	46.7 ± 9%	0.465 ± 0.031	75.8%	0.610 (0.478–0.721)
Δ*ldhA*-508 s	61.6 ± 3%	0.579 ± 0.021	98.9%	0.595 (0.489–0.659)
Δ*hbd*-414 s	Not available [b]		95.2%	0.572 (0.448–0.656)

[a]Biomass Yields were normalized based on the wild type strain value
[b]Data not measured experimentally by Montoya [73]

PDO maximum flux that was 11.6% higher than the flux predicted by FBA; therefore, none of the mutants would show an increase in the PDO yield greater than this value without affecting biomass yield. This validates the single and double mutant predictions and suggests that mutant elaboration by *knockout* is an inadequate strategy for improving PDO yields. Similar results were obtained using *Optknock*, with up to 3 deletions [35, 74], where the maximum PDO yield predicted was 0.712 deleting hydrogen and butanol production. This agreed with the ROOM predictions shown in Table 5 and indicated a biomass yield reduction of nearly 28%. The *Optknock* maximum PDO yield value was 17.4% higher than the predicted value under glycerol limitation, but it was only 0.4% higher than the predicted value under glycerol excess, which reinforces that blocking reactions are useless. This can be understood by considering that PDO is a primary metabolite and its production is associated with growth [10, 75].

The second strategy evaluated was perturbation in the biomass composition by simultaneous random variation of stoichiometric coefficients of 44 precursors and 8 macromolecules. This strategy was evaluated since modifications in the culture media can affect biomass composition, such as accumulation of lipids during nitrogen starvation or protein accumulation with excess of nitrogen in the culture medium [76]. Normal distributions with relative standard deviations of 30% were employed for all stoichiometric coefficients considered in the perturbation. Figure 3 shows the $Y_{X/S}$ and $Y_{PDO/S}$ correlation values, indicating low correlation with the precursors (fatty acids, amino acids, nucleotides, polar lipids and cofactors) but a higher correlation with macromolecules, especially proteins (the main biomass component, accounting for 86.7% on a molar basis). This could suggest that a low protein content in the cell could improve $Y_{X/S}$ and reduce the $Y_{PDO/S}$, due their negative and positive correlations, respectively. However, the relative standard deviations obtained for $Y_{X/S}$ and $Y_{PDO/S}$ were 3.2% and 0.45%, respectively, meaning that the model is sufficiently stable to perturbations in biomass composition. These predictions suggest that changes in culture media aimed at modifying

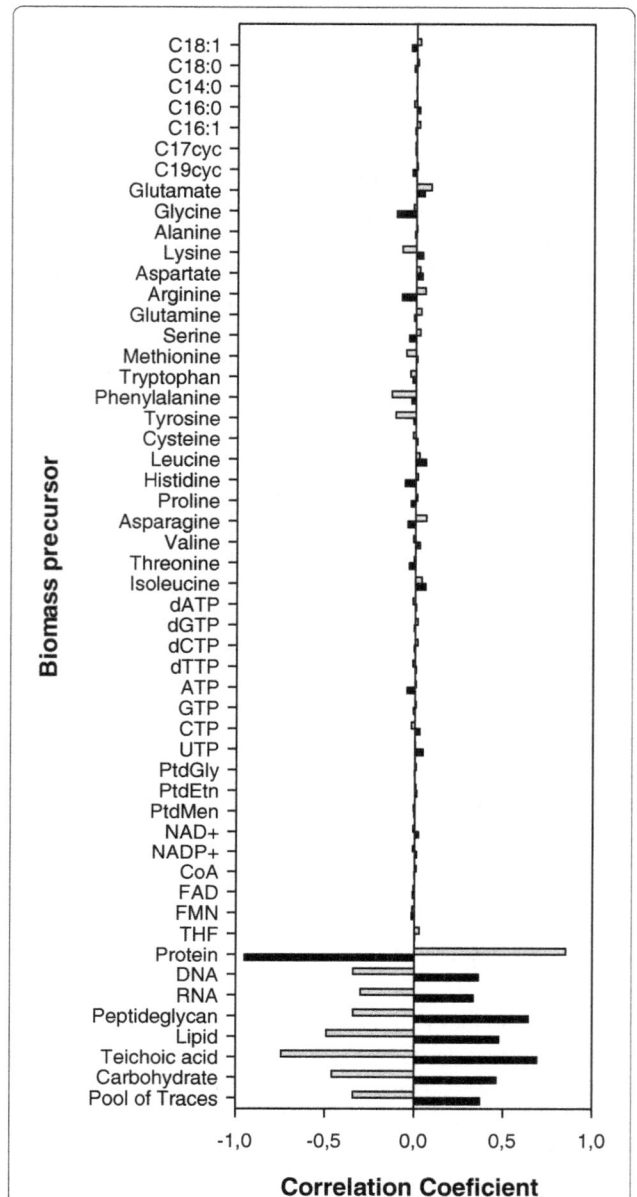

Fig. 3 Pearson correlation coefficients of biomass and PDO yields to biomass precursors. Notation: (■) $Y_{X/S}$ correlation. (■) $Y_{PDO/S}$ correlation. Pearson values were calculated using a normalized random distribution of biomass precursors; a relative standard distribution of 30% was employed in all precursors. The covariance analyses were made with 9035 random combinations of precursor compositions

biomass composition and enhancing PDO yields would be unnecessary. However, modifications in the culture media could reduce PDO yield, as Dabrock et al. found using iron in excess [77].

The third strategy evaluated was to use two substrates simultaneously: glucose and glycerol. Glucose is used as carbon source, while glycerol is used to maintain redox balance, therefore generating higher $Y_{PDO/S}$ values than obtained using glycerol as single substrate [78]. Figure 4a shows the experimental values and FVA ranges, which suggest that the cell operates at optimal conditions to produce PDO when glucose is present in the medium. This validation allowed an evaluation of the $Y_{PDO/S}$ at different glucose and glycerol uptake fluxes, as shown Fig. 4b, which shows no PDO production ($Y_{PDO/S} = 0$) in the absence of a glycerol uptake flux. The results also show the complete transformation of glycerol to PDO ($Y_{PDO/S} = 1$), using ratios of at least 0.375 between uptake fluxes of glucose and glycerol. Therefore, these predictions permit the proposal that co-fermentation is the best alternative for improving biomass and PDO yields. However, a priori prediction of the molar ratio between these substrates that could allow these flux ratios is difficult.

Conclusions

We generated iCbu641, the first curated genome-scale metabolic model for a PDO producer Clostridium strain. During iCbu641 validation, we solved flux balance analysis using LP optimization; however, according to the experimental data, the model predicted errors of nearly 300% for biomass yield and failed to predict PDO production. Therefore, NLP optimization was employed in FBA simulations, and the new objective function maximized biomass yield per flux unit [32]. The validation allowed prediction of appropriate growth and PDO production of cultures under glycerol limitation, but it still overestimated the experimental yields of cultures under glycerol excess. Thus, sub-optimal growth predictions under glycerol excess were achieved through a second NLP optimization, where ATP minimization was added to objective function. Therefore, both objective functions were able to predict Clostridium butyricum growth and PDO production under limiting and excess glycerol

conditions. Additional validations were developed using proteomics and transcriptomics data, as well data from knockout mutants, which allowed verification of the accuracy of predicting perturbations of iCbu641. All validations were completed using experimental data from different Clostridium butyricum strains and suggested that iCbu641 is an agnostic GSM model at the state steady, but the differences may be observed during dynamic predictions.

Subsequently, perturbations in the metabolic network and biomass composition were proposed to increase the PDO yield predictions. However, these perturbations predicted no significant increments. We also evaluated glucose-glycerol co-fermentation as a strategy to improve PDO yields. We found that a ratio of glucose and glycerol uptake fluxes greater than or equal to 0.375 would allow the complete conversion of glycerol to PDO; however, experimental analysis is needed to find the molar ratio that allows the achievement of this flux ratio. Finally, predictions of PDO production in state steady cultures using iCbu641 allows the proposal of this GSM model for predicting dynamic cultures (i.e. batch and fed-batch fermentations) capable of increasing PDO production, thereby minimizing the need for direct experimental efforts.

Methods

Genomic scale metabolic model iCbu641 reconstruction
The draft genome was obtained for the Colombian-native strain Clostridium sp. IBUN 13A, isolated and stored by the Bioprocesses and Bioprospecting Research Group from the Institute of Biotechnology of the Universidad Nacional de Colombia. This draft genome was previously sequenced and annotated and is available in GenBank with the accession no NZ_JZWG00000000.1 [41]. The strategy used for initial manual curation was

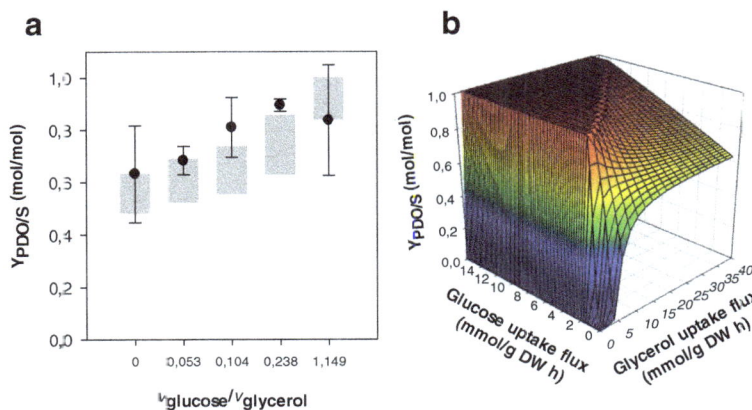

Fig. 4 PDO yields using glucose and glycerol co-fermentation. **a** Comparison of experimental (scatter dots with standard deviation as error bars [78]) and FVA range prediction (vertical boxes) of $Y_{PDO/S}$ in the function of the glucose/glycerol uptake flux ratio. **b** Maximum $Y_{PDO/S}$ predicted using FVA at different glucose and glycerol uptake fluxes. A ratio of glucose/glycerol uptake fluxes greater than or equal to 0.375 allows a complete conversion of glycerol to PDO ($Y_{PDO/S} = 1$). By contrast, PDO production is 0 without glycerol uptake flux ($Y_{PDO/S} = 0$)

reverse engineering proposed by Senger and Papoutsakis [47]; automated curation was also employed using GapFind and GapFill [43]. The GSM models of the solventogenic *Clostridium* strains *C. acetobutylicum* [46–49], *C. thermocellum* ATCC 27405 [50], *C. beijerinckii* NCIMB 8052 [45], *C. ljungdahlii* ATCC 55383 [52], and *C. cellulolyticum H10* [51] were used as a database for the curation. The resulting network is based on KEGG nomenclature, whereas the SEED database [79] was used in mass and charge balances at pH 7. Finally, thermodynamically infeasible loops were eliminated according to the methodology of Schellenberger et al. [80].

The values for growth-associated maintenance (GAM) and non-growth–associated maintenance (NGAM) were reported for *Clostridium acetobutylicum* ATCC 824 by Lee et al. and are 40 mmol·ATP·g^{-1} and 5 mmol·ATP·g^{-1}·h^{-1}, respectively [46]. An allosteric model was also included as an upper bound constraint for the acetic acid secretion flux in the function of glycerol uptake flux. This trend was obtained using the experimental data reported by Solomon et al. [44] and Papanikolaou et al. [81]. The kinetic model was expressed as a logistic function, with 0.158 and 11.5·mmol·g^{-1} h^{-1} as initial and maximum values, respectively, and –0.0879·g·h·mmol^{-1} as the accumulation rate, as shown in Eq. 4.

$$v_{acetic\ acid} \le \frac{11.5^*0.158^* e^{\left(-0.0859^* v_{glycerol}\right)}}{11.5 + 0.158^*\left(e^{\left(-0.0859^* v_{glycerol}\right)} - 1\right)}$$

(4)

Flux balance analysis

The dynamics of the mass balance of metabolite x_i involved in N reactions is described in Eq. 5, where S_{ij} is the stoichiometric coefficient of metabolite i in reaction j, and v_j is the flux value in which this reaction occurs [25, 80]. Now, assuming a steady state, Eq. 5 can be expressed for M metabolites; however, since $N > M$, the prediction of fluxes v_j can be achieved using FBA, which maximizes or minimizes an objective function Z (Equation 6). The constraints of this function are the mass balances for the M metabolites and the upper v_j^{max}, and lower v_j^{min} bounds of the N fluxes v_j [28, 31]. Additionally, feasible ranges of fluxes predicted by FBA are calculated using FVA. Since the objective functions employed are non-linear, the objective function value Z calculated with FBA has to be relaxed by 5%, as suggested Mahadevan et al., Eq. 7 [82]. The mutant phenotypes were predicted using the ROOM approach, Eq. 8, where b_j is the binary number of reaction j [72]. Also, $v_{j,wild}^l$ and $v_{j,wild}^u$ are the lower and upper confidence limits of wild type flux j. δ and ε are relative and absolute tolerance ranges, respectively.

$$\frac{dx_i}{dt} = \sum_{j=1}^{N} S_{ij} v_i$$

(5)

$$\text{Max/Min} \quad Z = f\left(v_j\right)$$
Subject to
$$\sum_{j=1}^{N} S_{ij} v_j = 0, \quad \forall i \in 1,...,M$$
$$v_j^{min} \le v_j \le v_j^{max}, \quad \forall j \in 1,...,N$$

(6)

$$\text{Max/Min} \quad v_j$$
Subject to
$$\sum_{j=1}^{N} S_{ij} v_j = 0, \quad \forall i \in 1,...,M$$
$$v_j^{min} \le v_j \le v_j^{max}, \quad \forall j \in 1,...,N$$
$$Z_{relaxed} = \begin{cases} 1.05Z & \text{if } Z \text{ was minimized} \\ 0.95Z & \text{if } Z \text{ was maximized} \end{cases}$$

(7)

$$\text{Min} \sum_{j=1}^{N} b_j$$
Subject to
$$\sum_{j=1}^{N} S_{ij} v_{j,mutant} = 0, \quad \forall i \in 1,...,M$$
$$v_{j,mutant}^{max} \wedge v_{j,mutant}^{min} \begin{cases} 0 & \forall\ v_j \text{ catalyzed only by enzyme ec} \\ v_{j,wild}^{max} \wedge v_{j,wild}^{min} & \forall\ v_j \begin{cases} \text{not cat by enzyme ec} \\ \text{cat by isozyme ofec} \end{cases} \end{cases}$$
$$v_{j,mutant} - b_j\left(v_{j,mutant}^{max} - v_{j,wild}^u\right) \le v_{j,wild}^u \quad \forall\ j \in 1,...,N$$
$$v_{j,mutant} - b_j\left(v_{j,mutant}^{min} - v_{j,wild}^l\right) \ge v_{j,wild}^l \quad \forall\ j \in 1,...,N$$
$$v_{j,wild}^u = v_{j,wild} + \delta|v_{j,wild}| + \varepsilon \quad \forall\ j \in 1,...,N$$
$$v_{j,wild}^l = v_{j,wild} - \delta|v_{j,wild}| - \varepsilon \quad \forall\ j \in 1,...,N$$
$$b_j = [0,1] \quad \forall\ j \in 1,...,N$$
$$\delta \approx 0.05 \quad \varepsilon \approx 0.001$$
Subject to
FBA of wild type strain

(8)

Experimental validation

The experimental validation used two robustness analyses, as reported by Price et al. [83]: the former for the growth rate (μ) and the latter for PDO secretion flux (v_{PDO}), both according to the glycerol uptake flux ($v_{Glycerol}$). The robustness analyses were made using different objective functions for both glycerol limited and glycerol excess conditions. The objective function employed under glycerol limitation was biomass maximization per enzyme usage. Under glycerol excess, the biomass was maximized, while both enzyme usage and ATP production were minimized, where ATP production corresponds only to thermodynamically feasible reactions able to produce ATP. According to KEGG nomenclature, these reactions are R00156, R00158, R00200, R00315, R00332, R00512, R00570, R00722, R01512, R01547, R01665, R01688, R01724, R02090, R02093, R02094, R02098, R02326, R02331, R03005, R03035, R03530, and R03920. The predicted values were compared with the experimental values reported for the *Clostridium butyricum* DSM 5431 strain cultured in glycerol limited and glycerol excess conditions [44].

The prediction capability of enzyme expression was also evaluated by comparing the enzymes present in both metabolic model *i*Cbu641 and the experimental proteome of the strain *Clostridium butyricum* DSM 10702 cultured in glycerol [20]. The enzymes in the model were classified as blocked or directionally, partially or fully coupled to growth using FCF, according to the methodology reported by Burgard et al. [58]. Blocked enzymes were excluded from the comparison; this comparison assumed that all the expressed enzymes were active and catalyzed some reaction. Therefore, the enzyme expression could be: a) predicted when the flux of some of the reactions catalyzed by such enzyme is different to zero; b) not predicted when all the fluxes of reactions catalyzed by the enzyme expressed in the proteome are equal to zero.

Validation was also obtained using data from experimental cultures of *Clostridium butyricum* strains in substrates other than glycerol, such as glucose [56] and other carbohydrates [65]. The transcriptome reported by Calusinska et al. [66] for strain *C. butyricum* CWBI 1009, was also used for validation; this organism had been cultured in glucose using batch fermentation with uncontrolled pH. The mRNA detected in the transcriptome coded a total of 913 enzymes (532 unique and 381 redundant), where values at the exponential growth phase were used in a similar way to those from the proteomic data.

In silico perturbations of the metabolic model

Regulatory on/off minimization (ROOM) [72] was employed as a strategy for prediction of the phenotypic states of mutants by *knockout* of enzymes directionally coupled to growth. In total, 145 single mutants were evaluated. Double deletion was also studied by simultaneously blocking two enzymes, excluding enzymes previously detected as essential in the single deletion; this led to evaluation of almost 7900 double mutants.

We also made perturbations in the biomass composition expressed by the variation of stoichiometric coefficients of 8 macromolecules, 7 fatty acids, 20 amino acids, 8 nucleotides, 3 polar lipids, and 6 cofactors. FBA was performed using different compositions randomly generated using normal distributions with standard deviations equivalent to 30% the values used in GSM model *i*Cbu641. A total number of 10,000 simulations were made, where 965 of them were excluded because at least one of the concentrations was negative. Coefficients of correlation were calculated for biomass and PDO yields using the remaining 9035 biomass precursor combinations.

Technical implementation

FBA, FVA, and ROOM were computer simulated using GAMS (General Algebraic Modeling System, GAMS Development Corp., Washington, DC) software V.24.2.2 r44857 for Linux. Linear and Nonlinear Programming (LP and NLP) were developed with solver CONOPT v3.15 N, and Mixed Integer Programming (MIP) was developed with solver CPLEX 12.6.0.0. Data were analyzed using Microsoft Excel® 2010.

Additional files

Additional file 1: Excel format of *i*Cbu641 Genome Scale Metabolic Model.

Additional file 2: SBML format of *i*Cbu641 Genome Scale Metabolic Model.

Additional file 3: Proteomic and transcriptomic comparison results.

Additional file 4: Results of modeling scenarios with PDO yield increment.

Abbreviations

ACP: Acyl carrier proteins; FBA: Flux balance analysis; FCF: Flux couple finding; FVA: Flux variability analysis; GAM: Growth-associated maintenance; GSM: Genome-scale metabolic; LP: Linear programming; MIP: Mixed integer programming; MOMA: Minimization of metabolic adjustments; NGAM: Non-growth-associated maintenance; NLP: Non-linear programing; PDO: 1,3-Propanediol; ROOM: Regulatory On/Off Minimization

Acknowledgements

The authors would also like to thank Satyakam Dash and Juan Rosas for helpful discussions and suggestions. Finally, iCbu641 GSM model was deposited in BioModels [84] and assigned the identifier MODEL1704210000.

Funding

This work was supported by COLCIENCIAS, Colombia.

Authors' contributions

LMSB performed simulations, analyzed data, and wrote the paper. LMSB, AFGB, and DMC conceived the study and participated in its design and coordination. CDM helped with key analyses of the data. All authors read and approved the final manuscript.

Competing interests

The authors declare that they have no competing interest.

Author details

[1]Bioprocesses and Bioprospecting Group, Universidad Nacional de Colombia. Ciudad Universitaria, Carrera 30 No. 45-03, Bogotá, D.C, Colombia. [2]Grupo de Diseño de Productos y Procesos (GDPP), Departamento de Ingeniería Química, Universidad de los Andes, Carrera 1 N.° 18A – 12, Bogotá, Colombia. [3]Department of Chemical Engineering, The Pennsylvania State University, University Park, PA 16802, USA.

References

1. Ayoub M, Abdullah AZ. Critical review on the current scenario and significance of crude glycerol resulting from biodiesel industry towards more sustainable renewable energy industry. Renew Sust Energ Rev. 2012;16(5):2671–86.
2. EIA. International Energy Statistics: Total Biofuels Production. 2014. [http://www.eia.gov/cfapps/ipdbproject/IEDIndex3.cfm?tid=79&pid=79&aid=1]. Accessed 19 May 2016.
3. Almeida JRM, Fávaro LCL, Quirino BF. Biodiesel biorefinery: opportunities and challenges for microbial production of fuels and chemicals from glycerol waste. Biotechnol Biofuels. 2012;5 art 48:1–16.
4. Yazdani SS, Gonzalez R. Anaerobic fermentation of glycerol: a path to economic viability for the biofuels industry. Curr Opin Biotechnol. 2007;18(3):213–9.
5. Amaral PFF, Ferreira TF, Fontes GC, Coelho MAZ. Glycerol valorization: new biotechnological routes. Food Bioprod Process. 2009;87(3):179–86.
6. Saxena RK, Anand P, Saran S, Isar J. Microbial production of 1,3-propanediol: recent developments and emerging opportunities. Biotechnol Adv. 2009;27(6):895–913.
7. Kaur G, Srivastava AK, Chand S. Advances in biotechnological production of 1,3-propanediol. Biochem Eng J. 2012;64:106–18.
8. Drozdzyńska A, Leja K, Czaczyk K. Biotechnological production of 1,3-propanediol from crude glycerol. Biotechnologia. 2011;92(1):92–100.
9. Dobson R, Gray V, Rumbold K. Microbial utilization of crude glycerol for the production of value-added products. J Ind Microbiol Biotechnol. 2012;39(2):217–26.
10. Kubiak P, Leja K, Myszka K, Celińska E, Spychała M, Szymanowska PD. Czaczyk K, Grajek W: physiological predisposition of various clostridium species to synthetize 1,3-propanediol from glycerol. Process Biochem. 2012;47(9):1308–19.
11. Wilkens E, Ringel AK, Hortig D, Willke T, Vorlop KD. High-level production of 1,3-propanediol from crude glycerol by clostridium butyricum AKR102a. Appl Microbiol Biotechnol. 2012;93(3):1057–63.
12. González-Pajuelo M, Meynial-Salles I, Mendes F, Andrade JC, Vasconcelos I, Soucaille P. Metabolic engineering of clostridium acetobutylicum for the industrial production of 1,3-propanediol from glycerol. Metab Eng. 2005;7(5–6):329–36.
13. Otte B, Grunwaldt E, Mahmoud O, Jennewein S. Genome shuffling in clostridium diolis DSM 15410 for improved 1,3-propanediol production. Appl Environ Microbiol. 2009;75(24):7610–6.
14. Celińska E. Debottlenecking the 1,3-propanediol pathway by metabolic engineering. Biotechnol Adv. 2010;28(4):519–30.
15. Zeng AP. Pathway and kinetic analysis of 1,3-propanediol production from glycerol fermentation by clostridium butyricum. Bioprocess Eng. 1996;14(4):169–75.
16. Bizukojc M, Dietz D, Sun J, Zeng AP. Metabolic modelling of syntrophic-like growth of a 1,3-propanediol producer, clostridium butyricum, and a methanogenic archeon, Methanosarcina mazei, under anaerobic conditions. Bioprocess Biosyst Eng. 2010;33(4):507–23.
17. Senger RS, Yen JY, Fong SS. A review of genome-scale metabolic flux modeling of anaerobiosis in biotechnology. Curr Opin Chem Eng. 2014;6:33–42.
18. Dash S, Ng CY, Maranas CD. Metabolic modeling of clostridia: current developments and applications. FEMS Microbiol Lett. 2016;363(4):fnw004.
19. Comba González N, Vallejo AF, Sánchez-Gómez M, Montoya D. Protein identification in two phases of 1,3-propanediol production by proteomic analysis. J Proteome. 2013;89:255–64.
20. Gungormusler-Yilmaz M, Shamshurin D, Grigoryan M, Taillefer M, Spicer V, Krokhin OV, et al. Reduced catabolic protein expression in clostridium butyricum DSM 10702 correlate with reduced 1,3-propanediol synthesis at high glycerol loading. AMB Express. 2014;4:63.
21. Kauffman KJ, Prakash P, Edwards JS. Advances in flux balance analysis. Curr Opin Biotechnol. 2003;14(5):491–6.
22. Orth JD, Thiele I, Palsson BO. What is flux balance analysis? Nat Biotechnol. 2010;28(3):245–8.
23. Chen Q, Wang Z, Wei DQ. Progress in the applications of flux analysis of metabolic networks. Chin Sci Bull. 2010;55(22):2315–22.
24. Mahadevan R, Edwards JS, Doyle Iii FJ. Dynamic flux balance analysis of diauxic growth in Escherichia coli. Biophys J. 2002;83(3):1331–40.
25. Raman K, Chandra N. Flux balance analysis of biological systems: applications and challenges. Brief Bioinform. 2009;10(4):435–49.
26. Min Lee J, Gianchandani EP, Eddy JA, Papin JA. Dynamic analysis of integrated signaling, metabolic, and regulatory networks. PLoS Comput Biol. 2008;4(5):e1000086.
27. Haggart CR, Bartell JA, Saucerman JJ, Papin JA. Whole-genome metabolic network reconstruction and constraint-based modeling. Methods Enzymol. 2011;500:411–33.
28. García Sánchez CE, Vargas García CA, Torres Sáez RG. Predictive potential of flux balance analysis of Saccharomyces cerevisiae using as optimization function combinations of cell compartmental objectives. PLoS One. 2012;7(8):e43006.
29. Westerhoff HV, Winder C, Messiha H, Simeonidis E, Adamczyk M, Verma M, et al. Systems biology: the elements and principles of life. FEBS Lett. 2009;583(24):3882–90.
30. Gianchandani EP, Oberhardt MA, Burgard AP, Maranas CD, Papin JA. Predicting biological system objectives de novo from internal state measurements. BMC Bioinformatics. 2008;9:43.
31. Knorr AL, Jain R, Srivastava R. Bayesian-based selection of metabolic objective functions. Bioinformatics. 2007;23(3):351–7.
32. Schuetz R, Kuepfer L, Sauer U. Systematic evaluation of objective functions for predicting intracellular fluxes in Escherichia coli. Mol Syst Biol. 2007;3:119.
33. Ibarra RU, Edwards JS, Palsson BO. Escherichia coli K-12 undergoes adaptive evolution to achieve in silico predicted optimal growth. Nature. 2002; 420(6912):186–9.
34. Burgard AP, Maranas CD. Optimization-based framework for inferring and testing hypothesized metabolic objective functions. Biotechnol Bioeng. 2003;82(6):670–7.
35. Chowdhury A, Zomorrodi AR, Maranas CD. Bilevel optimization techniques in computational strain design. Comput Chem Eng. 2015;72:363–72.
36. Vargas García CA, García Sánchez C, Arguello Fuentes H, Torres Sáez RG. Balance de Flujos Metabólicos en Saccharomyces cerevisiae basado en Compartimentalización Intracelular. Rev Colomb Biotecnol. 2013;15(2):18-28.
37. Feist AM, Palsson BO. The biomass objective function. Curr Opin Microbiol. 2010;13(3):344–9.
38. Gianchandani EP, Chavali AK, Papin JA. The application of flux balance analysis in systems biology. WIREs Syst Biol Med. 2010;2(3):372–82.
39. García Sánchez CE, Torres Sáez RG. Comparison and analysis of objective functions in flux balance analysis. Biotechnol Prog. 2014;30(5):985–91.
40. Barragán CE, Gutiérrez-Escobar AJ, Castaño DM. Computational analysis of 1,3-propanediol operon transcriptional regulators: insights into clostridium sp. glycerol metabolism regulation. Univ Sci. 2015;20(1):129–40.
41. Rosas-Morales JP, Perez-Mancilla X, Lopez-Kleine L, Montoya-Castano D, Riano-Pachon DM. Draft genome sequences of clostridium strains native to Colombia with the potential to produce solvents. Genome Announc. 2015;3(3):e00486-15.
42. Montoya D, Spitia S, Silva E, Schwarz WH. Isolation of mesophilic solvent-producing clostridia from Colombian sources: physiological characterization,

solvent production and polysaccharide hydrolysis. J Biotechnol. 2000;79(2):117–26.

43. Satish Kumar V, Dasika MS, Maranas CD. Optimization based automated curation of metabolic reconstructions. BMC Bioinformatics. 2007;8:212.

44. Solomon BO, Zeng AP, Biebl H, Schlieker H, Posten C, Deckwer WD. Comparison of the energetic efficiencies of hydrogen and oxychemicals formation in Klebsiella pneumoniae and clostridium butyricum during anaerobic growth on glycerol. J Biotechnol. 1995;39(2):107–17.

45. Milne CB, Eddy JA, Raju R, Ardekani S, Kim PJ, Senger RS, et al. Metabolic network reconstruction and genome-scale model of butanol-producing strain Clostridium Beijerinckii NCIMB 8052. BMC Syst Biol. 2011;5:130.

46. Lee J, Yun H, Feist AM, Palsson BØ, Lee SY. Genome-scale reconstruction and in silico analysis of the clostridium acetobutylicum ATCC 824 metabolic network. Appl Microbiol Biotechnol. 2008;80(5):849–62.

47. Senger RS, Papoutsakis ET. Genome-scale model for clostridium acetobutylicum: part I. Metabolic network resolution and analysis. Biotechnol Bioeng. 2008;101(5):1036–52.

48. McAnulty MJ, Yen JY, Freedman BG, Senger RS. Genome-scale modeling using flux ratio constraints to enable metabolic engineering of clostridial metabolism in silico. BMC Syst Biol. 2012;6:42.

49. Dash S, Mueller TJ, Venkataramanan KP, Papoutsakis ET, Maranas CD. Capturing the response of clostridium acetobutylicum to chemical stressors using a regulated genome-scale metabolic model. Biotechnol Biofuels. 2014;7:144.

50. Roberts SB, Gowen CM, Brooks JP, Fong SS. Genome-scale metabolic analysis of clostridium thermocellum for bioethanol production. BMC Syst Biol. 2010;4:31.

51. Salimi F, Zhuang K, Mahadevan R. Genome-scale metabolic modeling of a clostridial co-culture for consolidated bioprocessing. Biotechnol J. 2010;5(7): 726–38.

52. Nagarajan H, Sahin M, Nogales J, Latif H, Lovley DR, Ebrahim A, Zengler K: Characterizing acetogenic metabolism using a genome-scale metabolic reconstruction of Clostridium ljungdahlii. Microbial Cell Factories. 2013;12(1).

53. Quilaguy-Ayure DM, Montoya-Solano JD, Suárez-Moreno ZR, Bernal-Morales JM, Montoya-Castaño D. Analysing the dhaT gene in Colombian clostridium sp. (clostridia) 1,3-propanediol-producing strains. Univ Sci. 2010;15(1):17–26.

54. Cárdenas DP, Pulido C, Aragón OL, Aristizábal FA, Suárez ZR, Montoya D. Evaluación de la producción de 1,3-propanodiol por cepas nativas de Clostridium sp. mediante fermentación a partir de glicerol USP y glicerol industrial subproducto de la producción de biodiesel. Revista Colombiana De Ciencias Químico Farmacéuticas. 2006;35(1):120–37.

55. Von Stockar U, Maskow T, Liu J, Marison IW, Patiño R. Thermodynamics of microbial growth and metabolism: an analysis of the current situation. J Biotechnol. 2006;121(4):517–33.

56. Cai G, Jin B, Saint C, Monis P. Metabolic flux analysis of hydrogen production network by clostridium butyricum W5 effect of pH and glucose concentrations. Int J Hydrog Energy. 2010;35(13):6681–90.

57. Saint-Amans S, Girbal L, Andrade J, Ahrens K, Soucaille P. Regulation of carbon and electron flow in clostridium butyricum VPI 3266 grown on glucose-glycerol mixtures. J Bacteriol. 2001;183(5):1748–54.

58. Burgard AP, Nikolaev EV, Schilling CH, Maranas CD. Flux coupling analysis of genome-scale metabolic network reconstructions. Genome Res. 2004;14(2):301–12.

59. Ladygina N, Dedyukhina EG, Vainshtein ME. A review on microbial synthesis of hydrocarbons. Process Biochem. 2006;41(5):1001–14.

60. Covert MW, Schilling CH, Palsson B. Regulation of Gene expression in flux balance models of metabolism. J Theor Biol. 2001;213(1):73–88.

61. Feng X-M, Cao L-J, Adam RD, Zhang X-C, Lu S-Q. The catalyzing role of PPDK in Giardia lamblia. Biochem Biophys Res Commun. 2008;367(2):394–8.

62. Wood HG, O'Brien WE, Michaels G. Properties of carboxytransphosphorylase; pyruvate, phosphate dikinase; pyrophosphate-phosphofructikinase and pyrophosphate-acetate kinase and their roles in the metabolism of inorganic pyrophosphate. Adv Enzymol Relat Areas Mol Biol. 1977;45:85–155.

63. Vinitsky A, Grubmeyer C. A new paradigm for biochemical energy coupling. Salmonella typhimurium nicotinate phosphoribosyltransferase. J Biol Chem. 1993;268(34):26004–10.

64. Al-Hinai MA, Jones SW, Papoutsakis ET. The clostridium sporulation programs: diversity and preservation of endospore differentiation. Microbiol Mol Biol Rev. 2015;79(1):19–37.

65. Junghare M, Subudhi S, Lal B. Improvement of hydrogen production under decreased partial pressure by newly isolated alkaline tolerant anaerobe,

66. Calusinska M, Hamilton C, Monsieurs P, Mathy G, Leys N, Franck F, et al. Genome-wide transcriptional analysis suggests hydrogenase- and nitrogenase-mediated hydrogen production in clostridium butyricum CWBI 1009. Biotechnol Biofuels. 2015;8:27.

67. Mootz HD, Finking R, Marahiel MA. 4'-Phosphopantetheine transfer in primary and secondary metabolism of Bacillus Subtilis. J Biol Chem. 2001;276(40):37289–98.

68. Winter G, Krömer JO. Fluxomics - connecting 'omics analysis and phenotypes. Environ Microbiol. 2013;15(7):1901–16.

69. Machado D, Herrgård M. Systematic evaluation of methods for integration of Transcriptomic data into constraint-based models of metabolism. PLoS Comput Biol. 2014;10(4):e1003580.

70. Lewis NE, Hixson KK, Conrad TM, Lerman JA, Charusanti P, Polpitiya AD, et al. Omic data from evolved E. coli are consistent with computed optimal growth from genome-scale models. Mol Syst Biol. 2010;6:390.

71. Cai G, Jin B, Saint C, Monis P. Genetic manipulation of butyrate formation pathways in clostridium butyricum. J Biotechnol. 2011;155(3):269–74.

72. Shlomi T, Berkman O, Ruppin E. Regulatory on/off minimization of metabolic flux changes after genetic perturbations. Proc Natl Acad Sci U S A. 2005;102(21):7695–700.

73. Montoya Solano JD: Metabolic engineering of the Colombian strain Clostridium sp. IBUN 158B in order to improve the bioconversion of glycerol into 1,3-propanediol. Germany: University of Ulm; 2012.

74. Burgard AP, Pharkya P, Maranas CD. OptKnock: a Bilevel programming framework for identifying Gene knockout strategies for microbial strain optimization. Biotechnol Bioeng. 2003;84(6):647–57.

75. Chatzifragkou A, Aggelis G, Gardeli C, Galiotou-Panayotou M, Komaitis M, Papanikolaou S. Adaptation dynamics of clostridium butyricum in high 1,3-propanediol content media. Appl Microbiol Biotechnol. 2012;95(6):1541–52.

76. Lari Z, Moradi-kheibari N, Ahmadzadeh H, Abrishamchi P, Moheimani NR, Murry MA. Bioprocess engineering of microalgae to optimize lipid production through nutrient management. J Appl Phycol. 2016;28(6):3235–50.

77. Dabrock B, Bahl H, Gottschalk G. Parameters affecting solvent production by clostridium pasteurianum. Appl Environ Microbiol. 1992;58(4):1233–9.

78. Malaoui H, Marczak R. Influence of glucose on glycerol metabolism by wild-type and mutant strains of clostridium butyricum E5 grown in chemostat culture. Appl Microbiol Biotechnol. 2001;55(2):226–33.

79. Henry CS, Dejongh M, Best AA, Frybarger PM, Linsay B, Stevens RL. High-throughput generation, optimization and analysis of genome-scale metabolic models. Nat Biotechnol. 2010;28(9):977–82.

80. Schellenberger J, Lewis NE, Palsson BØ. Elimination of thermodynamically infeasible loops in steady-state metabolic models. Biophys J. 2011;100(3):544–53.

81. Papanikolaou S, Ruiz-Sanchez P, Pariset B, Blanchard F, Fick M. High production of 1,3-propanediol from industrial glycerol by a newly isolated clostridium butyricum strain. J Biotechnol. 2000;77(2–3):191–208.

82. Mahadevan R, Schilling CH. The effects of alternate optimal solutions in constraint-based genome-scale metabolic models. Metab Eng. 2003;5(4):264–76.

83. Price ND, Reed JL, Palsson BØ. Genome-scale models of microbial cells: evaluating the consequences of constraints. Nat Rev Microbiol. 2004;2(11):886–97.

84. Chelliah V, Juty N, Ajmera I, Ali R, Dumousseau M, Glont M, et al. BioModels: ten-year anniversary. Nucleic Acids Res. 2015;43(D1):D542–8.

5

Evaluation and improvement of the regulatory inference for large co-expression networks with limited sample size

Wenbin Guo[1,2], Cristiane P. G. Calixto[2], Nikoleta Tzioutziou[2], Ping Lin[3], Robbie Waugh[2,4], John W. S. Brown[2,4] and Runxuan Zhang[1*] (ID)

Abstract

Background: Co-expression has been widely used to identify novel regulatory relationships using high throuput measurements, such as microarray and RNA-seq data. Evaluation studies on co-expression network analysis methods mostly focus on networks of small or medium size of up to a few hundred nodes. For large networks, simulated expression data usually consist of hundreds or thousands of profiles with different perturbations or knock-outs, which is uncommon in real experiments due to their cost and the amount of work required. Thus, the performances of co-expression network analysis methods on large co-expression networks consisting of a few thousand nodes, with only a small number of profiles with a single perturbation, which more accurately reflect normal experimental conditions, are generally uncharacterized and unknown.

Methods: We proposed a novel network inference methods based on Relevance Low order Partial Correlation (RLowPC). RLowPC method uses a two-step approach to select on the high-confidence edges first by reducing the search space by only picking the top ranked genes from an intial partial correlation analysis and, then computes the partial correlations in the confined search space by only removing the linear dependencies from the shared neighbours, largely ignoring the genes showing lower association.

Results: We selected six co-expression-based methods with good performance in evaluation studies from the literature: Partial correlation, PCIT, ARACNE, MRNET, MRNETB and CLR. The evaluation of these methods was carried out on simulated time-series data with various network sizes ranging from 100 to 3000 nodes. Simulation results show low precision and recall for all of the above methods for large networks with a small number of expression profiles. We improved the inference significantly by refinement of the top weighted edges in the pre-inferred partial correlation networks using RLowPC. We found improved performance by partitioning large networks into smaller co-expressed modules when assessing the method performance within these modules.

Conclusions: The evaluation results show that current methods suffer from low precision and recall for large co-expression networks where only a small number of profiles are available. The proposed RLowPC method effectively reduces the indirect edges predicted as regulatory relationships and increases the precision of top ranked predictions. Partitioning large networks into smaller highly co-expressed modules also helps to improve the performance of network inference methods.

The RLowPC R package for network construction, refinement and evaluation is available at GitHub: https://github.com/wyguo/RLowPC.

Keywords: Gene co-expression networks, Gene regulatory networks, Network method evaluation, Partial correlation, Synthetic data

* Correspondence: Runxuan.zhang@hutton.ac.uk
[1]Information and Computational Sciences, The James Hutton Institute, Invergowrie, Dundee, Scotland DD2 5DA, UK
Full list of author information is available at the end of the article

Background

Over the last fifteen years, there has been a growing interest in reverse engineering of Gene Regulatory Networks (GRNs) that aim to infer complex graphs representing transcriptional regulatory relationships, directly from gene expression profiles [1–15]. Due to its low computational complexity as well as lower requirements for the number of samples, co-expression network analysis has been widely used to infer gene regulatory networks from high throughput expression data, such as microarray or RNA-seq data [10, 16–19]. Typically thousands of genes/transcripts of special interest (e.g. differentially expressed) are utilized to construct the co-expression network in an experiment. Top candidates whose expression correlates with the gene of interest are usually further examined to identify novel regulators/targets. Despite this approach being widely used, there is a general lack of studies on the precision (the fraction of inferred regulatory relationships that are correct) and recall (the fraction of regulatory relationships that are inferred) expected.

Considerable effort has been made to evaluate the performance and robustness of GRN inference methods. The majority of evaluations were implemented on in silico datasets simulated from reference networks with sizes up to a few hundred or 1–2000 genes. Numerous studies using a range of network sizes, time-series data and perturbations have compared different analysis methods. Results are variable in terms of the top-performing method (Summaries in Additional file 1: Table S1). A series of studies have been carried out by the Dialogue for Reverse Engineering Assessments and Methods (DREAM) project, which generates challenges and organizes contests annually. The DREAM3 challenge presents gene network inference problems based on in silico networks of sizes ranging from 10, 50 and 100 genes [20–24]. Gene expression data was simulated using these networks for the following scenarios: 1) the steady state of the unperturbed networks, as well as steady state of the network where every gene is knocked out or down; and 2) 4, 23 and 46 different time series for the size 10, 50 and 100 networks respectively, with 21 time points for each time series. For example, for the network of size 100, there are a total of 1067 gene expression profiles with different perturbations and knockout/knockdown experiments available to make the inference. The inference methods: Scan Bayesian Model Averaging (ScanBMA), Gene Network Inference with Ensemble of trees (GENIE3) and Minimum Redundancy NETworks using Backward elimination (MRNETB) were the top performers in three different studies using the DREAM4 challenge time-series data,

which is composed of five perturbation experiments for size 10 networks and ten perturbation experiments for size 100 networks, each with 21 time points [24–27] (Additional file 1: Table S1). Besides the DREAM benchmark datasets, the Bayesian Network (BN), Graphical Gaussian models (GGMs) and Relevance Network (RN) methods were compared using expression simulations of 100 sample points for a size 11 network with BN and GGM performing best [12]. The Algorithm for the Reconstruction of Accurate Cellular Networks (ARACNE) method had a much better performance than BN and RN on expression data with 1000 samples simulated from size 100 networks [28] while MRNET was the top ranked method when compared to the RN, ARACNE and Context likelihood or relatedness (CLR) methods on 30 datasets with different network sizes (from 100 to 1000) and sample sizes (from 100 to 1000) [29] (Additional file 1: Table S1).

A few studies aimed to evaluate network methods on larger networks of a few thousand genes. In the DREAM5 challenge, Least Absolute Shrinkage and Selection Operator (LASSO), CLR and GENIE3 are top performers among more than 30 network inference methods on a size 1643 network with 805 simulated gene expression profiles, where a list of regulators (potential transcriptional factors) are given [30]. Ten network inference methods on size 1000 network from S. Rogers [31], size 300 and 1000 networks from SynTReN [32] and size 1565 and 2000 networks from GeneNetWeaver (GNW) [24] were assessed using simulated datasets of 1000, 800, 1000, 1565 and 2000 experiments individually. CLR, GENIE3 and MRNET were the top performers in this study [33]. Similarly, ARACNE, GeneNet, Weighted Correlation Network Analysis (WGCNA) and Sparse PArtial Correlation Estimation (SPACE) were compared using size 17, 44, 83, 231, 612 and 1344 networks over datasets with 20, 50, 100, 200, 500 and 1000 sample points simulated from Gaussian distribution [34]. GeneNet ranked in the first place followed by ARACNE (Additional file 1: Table S1).

Despite the large number of evaluation studies, none have explored the normal experimental situation where a regulatory network is generated which involves hundreds and thousands of genes with only a small number of profiles being available. The assessments in the literature were based on either small and medium sized networks or datasets with a large number of samples. The evaluation conclusions were also based on a large amount of simulated expression profiles which would be difficult to validate experimentally due to the prohibitive cost or the amount of work in real experiments [35, 36].

Distinguishing direct regulatory interactions from indirect associations has been one of the major challenges in gene regulatory network constructions [2, 21] (see Fig. 1a). Partial Correlation (PC) is one of the methods used as a solution to distinguish direct from indirect edges of each pair of

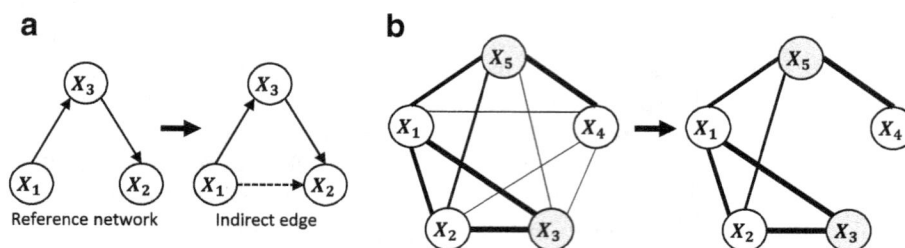

Fig. 1 Indirect edge and RLowPC network. **a** An indirect association from $X_1 \rightarrow X_2$ could arise from a regulatory structure of $X_1 \rightarrow X_3 \rightarrow X_2$. **b** RLowPC network inference. In an RLowPC network, firstly only the top ranked edges are kept in a pre-inferred PC network and then for each pair of genes, only the immediate neighbours will be regressed for PC calculation. In this example only the top 6 of 10 edges with highest correlations are kept and PC between X_1 and X_2 is re-calculated by removing the effects from two immediate neighbouring nodes (X_3, X_5). The correlation values are represented by the thickness of the edges

candidates by calculating the correlations after removing the linear dependencies from the remaining genes (see Fig. 1b). Other methods dealing with indirect connections include Partial Correlation coefficient with Information Theory (PCIT), ARACNE, MRNET and MRNETB. PCIT and ARACNE use the Information Theory of Data Processing Inequality method to remove the weakest gene association in each possible triplet structure in a network [37]. PCIT uses first order PC (removing the linear dependencies from the third gene in each possible triplet) to measure the significance of edge associations [38], whilst ARACNE uses Mutual Information (MI) to measure the associations between any two edges in each possible triplet [28]. MRNET uses a minimum redundancy feature selection method [39], where for each candidate gene in a MI network, it selects a subset of its highly relevant genes while minimising the MI between the selected genes [29]. MRNETB is an improved version of MRNET using a backward selection strategy starting from assuming that all genes are connected to the candidates. Less relevant genes are eliminated until the difference between the MI between a candidate and its neighbours and the MI within the neighbours are optimised [27].

Given that the search space for regulatory relationships expands factorially with the number of genes included in the network, the precision and recall of regulatory inference decrease with the increase of the network size. As gene clusters with highly cohesive patterns give rise to high correlations between all pairs of the genes in that cluster, the top ranked highly co-expressed genes may also be prone to errors of indirect associations. Here, we have developed a new method named Relevance Low order Partial Correlation (RLowPC), which is a refinement of top inferred edges by Partial Correlation methods. RLowPC selects top ranked edges from an inferred PC network as a reduced search space for indirect edges. We evaluated RLowPC alongside PC, PCIT, ARACNE, MRNET, MRNETB, and CLR on simulated time-series data and the summaries of the evaluated network inference methods is shown in Table 1. Precision and Area Under Precision-Recall curves (AUPR) were used as metrics to show that RLowPC outperforms the other methods.

Methods
Relevance low order partial correlation (RLowPC)
The conventional pair-wise PC measures correlations after linear dependencies on all the remaining genes are removed, the majority of which may not connect to the candidates, especially in large networks where the majority of the genes only have few linked neighbours [40, 41]. Low order partial correlation methods have been proposed and

Table 1 Summaries of the evaluated network inference methods

Category	Methods	Cor-based	MI-based	Ref.
Deal with indirect edges explicitly	RLowPC	Yes		
	PC	Yes		[2, 45]
	PCIT	Yes	Yes	[33, 38, 50, 51]
	MRNET		Yes	[29, 33, 39, 50]
	MRNETB		Yes	[27, 29, 33, 50]
	ARACNE		Yes	[28, 33, 50]
Not deal with indirect edges	Cor	Yes		
	CLR		Yes	[33, 48, 50]
	Random			

Nine correlation-based, MI-based and random network inference methods have been compared and evaluated in this study. The methods are classified into two main groups: Deal with indirect edges explicitly and Not deal with indirect edges

utilized in the past to reduce computational complexity without much sacrifice in prediction accuracy. For example, de la Fuente et al. [42] proposed to calculate up to second order partial correlations regressing against all the remaining genes. This method was improved by confining the second order partial correlation calculation only in cases where both zero and one order PC are non-zero [43]. Our proposed RLowPC method, firstly, reduces the search space by only picking the top ranked genes from partial correlation analysis and, secondly, computes the PC by only removing the linear dependencies from the shared neighbours in the confined search space, largely ignoring the genes showing lower association and which are less relevant in the pair-wise PC calculation. The implementation details are shown below:

Algorithm: RLowPC

//**Step 1**: Pre-inferring a PC network
Input gene expression matrix with N genes;
for each pair of genes (X_i, X_j), where $i, j \in N$ **do**

    Construct pair-wise Pearson correlation matrix $\Omega = (\rho_{X_i X_j})$;

    if $\Omega = (\rho_{X_i X_j})$ is positive definite and invertible **then**

        Inverse of correlation matrix $P = (p_{X_i X_j}) = \Omega^{-1}$;

    else

        Inverse of correlation matrix $P = (p_{X_i X_j}) = \widehat{P}$, where $\widehat{P}$ is from shrinkage

        estimation;

    end if

    Construct PC matrix $\widehat{\Omega} = (\hat{\rho}_{X_i X_j})$, where $\hat{\rho}_{X_i X_j} = -p_{X_i X_j} / \sqrt{p_{X_i X_i} p_{X_j X_j}}$.
end for

//**Step 2**: Extracting top ranked edges
Select top t edges with the highest PC values in $\widehat{\Omega}$ and form a subnetwork $\widehat{\Omega}_{sub}$ as the new search space.

//**Step 3**: Calculating RLowPC values
for each pair of candidates $(X_i, X_j) \in \widehat{\Omega}_{sub}$ connected by an edge **do**

    Find n neighbours shared by X_i and X_j in $\widehat{\Omega}_{sub}$

    if $n == 0$ **then**

        RLowPC between X_i and X_j: $\hat{\rho}'_{X_i X_j}$=Pearson correlation $\rho_{X_i X_j}$;

    else

        Construct a $(n + 2) \times (n + 2)$ Pearson correlation matrix Ω' using candidates X_i, X_j

        and all their n neighbours.

        if Ω' is positive definite and invertible **then**

            Inverse of correlation matrix $P' = \Omega'^{-1}$;

        else

            Inverse of correlation matrix $P' = \widehat{P}'$, where $\widehat{P}'$ is from shrinkage

            estimation;

        end if

        RLowPC between X_i and X_j: $\hat{\rho}'_{X_i X_j} = -p'_{X_i X_j} / \sqrt{p'_{X_i X_i} p'_{X_j X_j}}$, where

        $p'_{X_i X_j}, p'_{X_i X_i}$ and $p'_{X_j X_j} \in P'$;

    end if

end for
Output RLowPC values $\hat{\rho}'_{X_i X_j}$

For PC and shrinkage PC calculation we have used ppcor R package [44] and corpcor R package [45], respectively.

Gene expression data simulation

The main purpose of this study is to evaluate the performance of different network inference methods on datasets that reflect real experimental setup: large number of genes in the network with limited sample sizes and perturbations. Here, to evaluate the proposed methods comprehensively, large scale gene expression datasets were generated based on a variety of network structures using GNW version 3.1 [22, 24]. We used in silico size 100 networks in DREAM4, extracted size 500 and 1000 networks from a source E.coli network with 1565 nodes and 3758 edges and size 2000 and 3000 networks from a Yeast source network with 4441 nodes and 12,873 edges as reference networks. The source networks were provided by GNW [22, 24]. The networks were denoted as GNW100, GNW500, GNW1000,

GNW2000 and GNW3000. Summaries for data generation can be found in Table 2. For each size, network extraction was repeated five times yielding five networks with different structures and kinetics for statistical analysis of the results. To generate the time-series, transcription kinetic models of reference networks were firstly generated in GNW by removing self-regulatory interactions and randomly assigning transcription factor (TF) genes to groups to produce protein binding complexes. In the time-series simulation procedure, Stochastic Differential Equations (SDEs) were used to model the transcription kinetics, gene activation by protein complexes, gene perturbations, mRNA and protein production and degradation. One-third of the genes in each time-series were randomly selected and perturbed from steady state at the initial time-point. Perturbations were implemented by varying the activation strengths in the protein binding simulations to enhance or inhibit the downstream expression of target genes. The perturbations were sustained until the

Table 2 Source network structures and synthetic datasets

Network name		TF-gene networks	Gene No.	Edge No.	Network density	Data generator	Data type	Ref.
GNW100	GNW100_1	DREAM4 in Silico size 100	100	176	0.0356	The TF-gene reference networks were subsets of source networks in GNW. In each dataset, 1/3 genes were randomly selected and perturbed. Each experiment was sampled at 21 time points. 3 replicates were generated by adding different amount of noises. The noises are simulated by GNW. All the parameter settings were defaults in GNW.	Time-series data with multifactorial perturbation	[22, 24]
	GNW100_2		100	249	0.0503			
	GNW100_3		100	195	0.0394			
	GNW100_4		100	211	0.0426			
	GNW100_5		100	193	0.0390			
GNW500	GNW500_1	E.coli	500	1365	0.0109			
	GNW500_2		500	867	0.0069			
	GNW500_3		500	1107	0.0089			
	GNW500_4		500	947	0.0076			
	GNW500_5		500	1272	0.0102			
GNW1000	GNW1000_1	E.coli	1000	2337	0.0047			
	GNW1000_2		1000	2455	0.0049			
	GNW1000_3		1000	2089	0.0042			
	GNW1000_4		1000	2171	0.0043			
	GNW1000_5		1000	2249	0.0045			
GNW2000	GNW2000_1	Yeast	2000	4738	0.0024			
	GNW2000_2		2000	4467	0.0022			
	GNW2000_3		2000	5055	0.0025			
	GNW2000_4		2000	5283	0.0026			
	GNW2000_5		2000	4817	0.0024			
GNW3000	GNW3000_1	Yeast	3000	7515	0.0017			
	GNW3000_2		3000	7998	0.0018			
	GNW3000_3		3000	7626	0.0017			
	GNW3000_4		3000	8075	0.0018			
	GNW3000_5		3000	7333	0.0016			

A number of directed network structures were generated from source networks provided by GNW. The network names, gene and edge numbers for each structure are listed in the table. Network density is defined as the true edges divided by all possible edges. The network structures were used to simulate the time-series datasets using GNW

middle of the time-series at time point 11 when the activation strengths were changed back to initial levels. A random noise term proportional to production and degradation was introduced in the SDE model, inducing high noise for activated genes and low noise for inactivated genes. The coefficient to control the noise amplitude was set to 0.05. Another random noise, which was independent to the noise in SDEs, was added at the final step to the expression data to simulate technical variations [46]. The parameters for activation strengths, production, degradation and noises were set as defaults in GNW. The time-series generation were repeated five times yielding five different time-series with different initial conditions and perturbations. Average results obtained from these time series as well as five different network structures are reported in this study. Parameter setting details are shown in Additional file 1: Figure S3 and Additional file 2: Configuration file for GeneNetWeaver. Three biological replicates were generated for each time-series. By using the replicates, analysis of variance was carried out to select genes with significant expression changes across all 21 time-points with p-value cut-off of 0.001. In each experiment, there are only 63 gene expression profiles generated from one perturbation used for the network construction. The repeated generation of time series data as well as the network extraction are only used for statistical purposes to take the average and calculate the variations.

Evaluation of the network inference methods

Besides the methods mentioned earlier, we also included Pearson correlation, which has been the most commonly used method to identify correlated gene pairs, as well as random guessed network, which serves a baseline for network inference performances. We also included the CLR method, which although not partial correlation-based, has been shown to perform well in several studies [30, 33, 47–49]. We divided the methods under investigation into two groups. Group one includes all the methods that deal with indirect edges explicitly, which are RLowPC, PC, PCIT, ARACNE, MRNET and MRNETB. Group two are the methods which do not deal with indirect edges explicitly and they are CLR, Pearson correlation and random guessed networks. For MI-based methods, such as ARACNE, MRNET, MRNETB and CLR networks, we have used the minet R package with default parameters [50]. The MI matrices of the methods were approximated using Pearson correlation directly from continuous time-series data [27, 49]. The PC matrices were calculated by a shrinkage approach using corpcor R package [45]. The Boolean PCIT adjacency matrices were calculated using PCIT R package [38, 51], which was used as a weight to Pearson correlation networks [33]. For the RLowPC

method, the top (1500, 2000, 3000, 5000, 8000) weighted edges of inferred PC networks in GNW100, GNW500, GNW1000, GNW2000 and GNW3000 datasets were selected as search space for indirect edges. Details for tools used in the network inference analyses can be found in Table S2 in Additional file 1. In each inferred network, the top 1000 edge predictions was used to calculate True Positive (TP), False Positive (FP), True Negative (TN) and False Negative (FN) by comparing to the reference networks. The precision (TP/(TP + FP)) and pAUPR (partial plot of Area Under Precision against Recall = TP/(TP + FN)) values were calculated by picking the top ranked edges. pAUROC (partial Area Under the Receiver-Operating curve) was also calculated and the results were shown in the Supplementary material. All the evaluation of network inference methods was based on undirected network structures and the self-regulation edges were removed.

Results

RLowPC significantly improves the precision and recall in top predictions

Figure 2 illustrates the average pAUPR values, which are the partial Area Under Precision against Recall of the top 1000 predictions, for the different methods for different network sizes. Firstly, all methods except one case for ARACNE, outperformed the random guessed network, which proves the utility of such co-expression network analysis methods. Secondly, the performances of all methods are quite consistent across different network sizes. Within Group One, RLowPC consistently performs better than all of the other methods, with MRNET/MRNETB being the next best. Within Group two, CLR clearly outperforms the most commonly employed Pearson correlation method. The differences of pAUPR values between different methods were determined using a Student t-test in pairs between RLowPC and the other eight methods (Fig. 2). Results show that the RLowPC method is able to improve the pAUPR among the top edges significantly compared to other methods except for a few cases. The pAUROC show similar results (Additional file 1: Figure S1).

We further divided the top 1000 predictions into groups of top 1–100, 101–500 and 501–1000 (Fig. 3). The plots indicate that, once again, the precision of RLowPC method outperformed all others, regardless of which group within the top 1000 genes were selected for investigation. MRNET, MRNETB and CLR again showed slightly better performance than PC, PCIT and ARACNE and correlation methods. It is noteworthy that the precisions of all the methods are extremely low in large networks. For example, the precision median of RLowPC in the GNW3000 networks is around 0.006, which

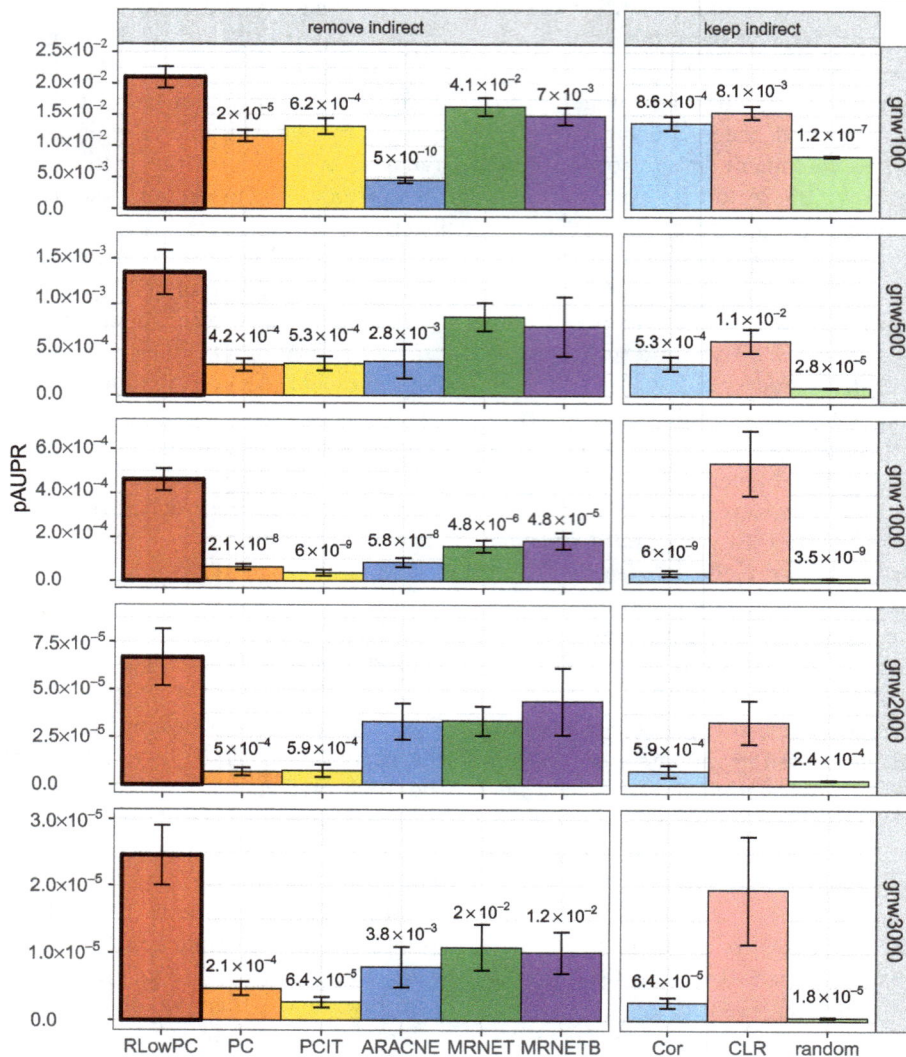

Fig. 2 Comparison of pAUPR values for different methods and different network structures. Each bar in the plots represents mean of pAUPR values from the top 1000 edge predictions. Error bars represent standard error. The differences of pAUPR values between different methods were determined using a Student t-test in pairs between RLowPC and the other eight methods. *P*-values are shown on the top of the bars if it is less than 0.05

indicates that in the top 100 predictions, only 0.6 (0.6%) edges are true predictions.

Clustering before network inference could improve the precision and recall in top predictions

Given that precision and recall is very low among the top predictions for all methods for large networks, we explored whether precision can be improved by dividing the large networks into smaller highly cohesive clusters. Using the time-series data generated for GNW3000 as described above, all genes were clustered into non-overlapping co-expressed modules using the R package Weighted Correlation Network Analysis (WGCNA) with default settings [52, 53]. Then, network inference and evaluation were carried out separately and individually

in each module. Essentially, WGCNA was used to break a big network into smaller non-overlapping subnetworks, at which point we carried out the network inference and evaluations within these smaller networks with the same time-series data. The pAUPR values were averaged across all the modules and it did not include genes that do not fit in any module (grey module). Similar to the simulation settings above, the clustering and evaluation procedures were repeated for five network structures, where five different time-series data were simulated for each structure. The average results were obtained. The average pAUPR values and precision distribution of the top 1000 predictions are presented in Fig. 4. Compared with the results of GNW3000 in Figs. 2 and 3, all methods evaluated have improved when the

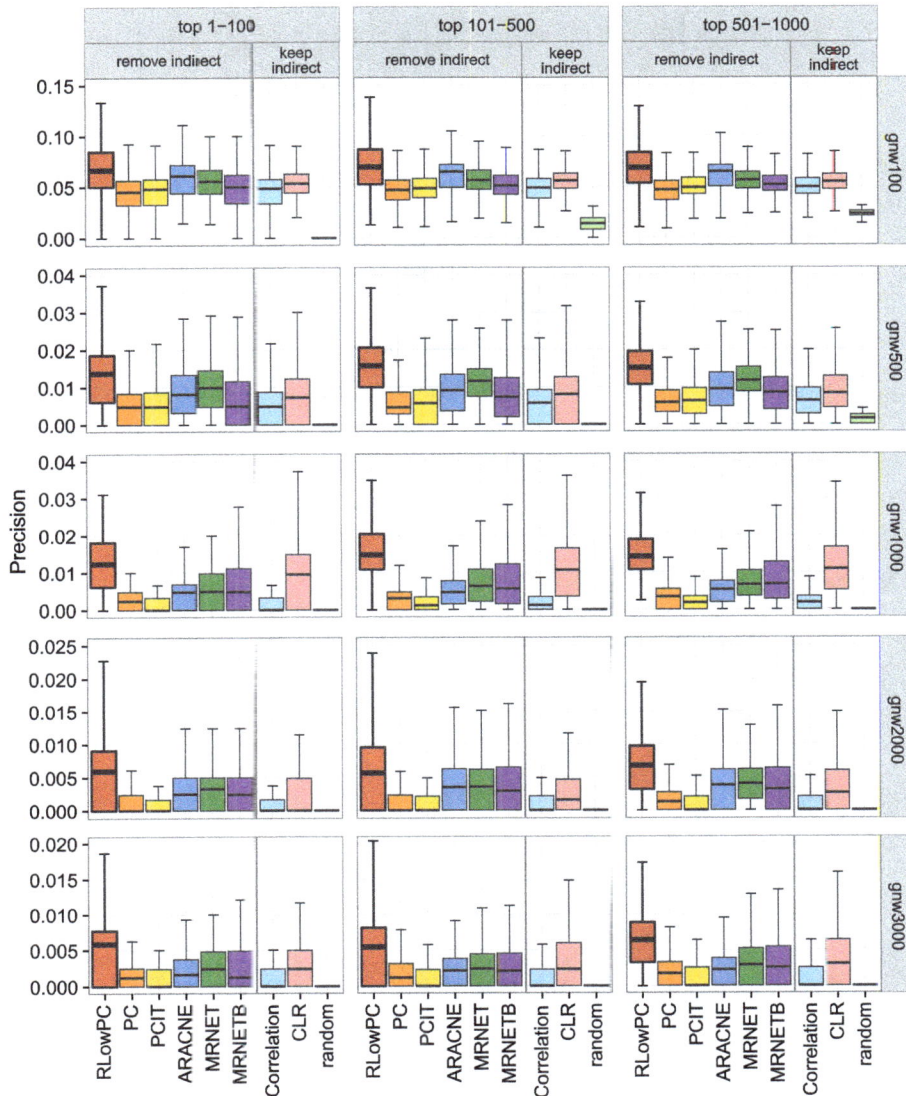

Fig. 3 Precisions within different groups of the top 1000 predicted edges. The top 1000 predicted edges are divided into three groups, top 1–100, 101–500 and 501–1000. Each bin depicts the precision distribution of the method matched to the group and the network structures

WGCNA method was used. This can be seen with the scale of average pAUPR values which increased from 1.0×10^{-5} to 1.0×10^{-3} (Fig. 4a), while the average precision of the top 1000 predictions has changed from 3.1×10^{-3} to 5.7×10^{-3} when the WGCNA method is used (Fig. 4b). The pAUPR value of RLowPC method is again significantly better than PC, PCIT, ARACNE, correlation and random networks. In the groups of top 1–100 and 101–500, the precision of RLowPC is better than the other eight methods and in top 501–1000 it is only better than PC, PCIT, correlation and random networks. The superior performances of RLowPC when the WGCNA method is used are also observed on the pAUROC plots (Additional file 1: Figure S2).

Discussion

The performance of different network inference methods varies according to network structures, data quantity and quality, and methodologies. The insufficiency of sampling and the high complexity of regulation kinetics prevent precise predictions of large gene regulatory networks. As a large regulatory network is often underdetermined using a small number of samples, there exists multiple plausible solutions, which cannot be distinguished by the information presented in the sample. This uncertainty in the inference of gene regulatory networks has been termed in some studies as "inferability" [54, 55]. Although our study mainly focuses on the network inference methods, special attention

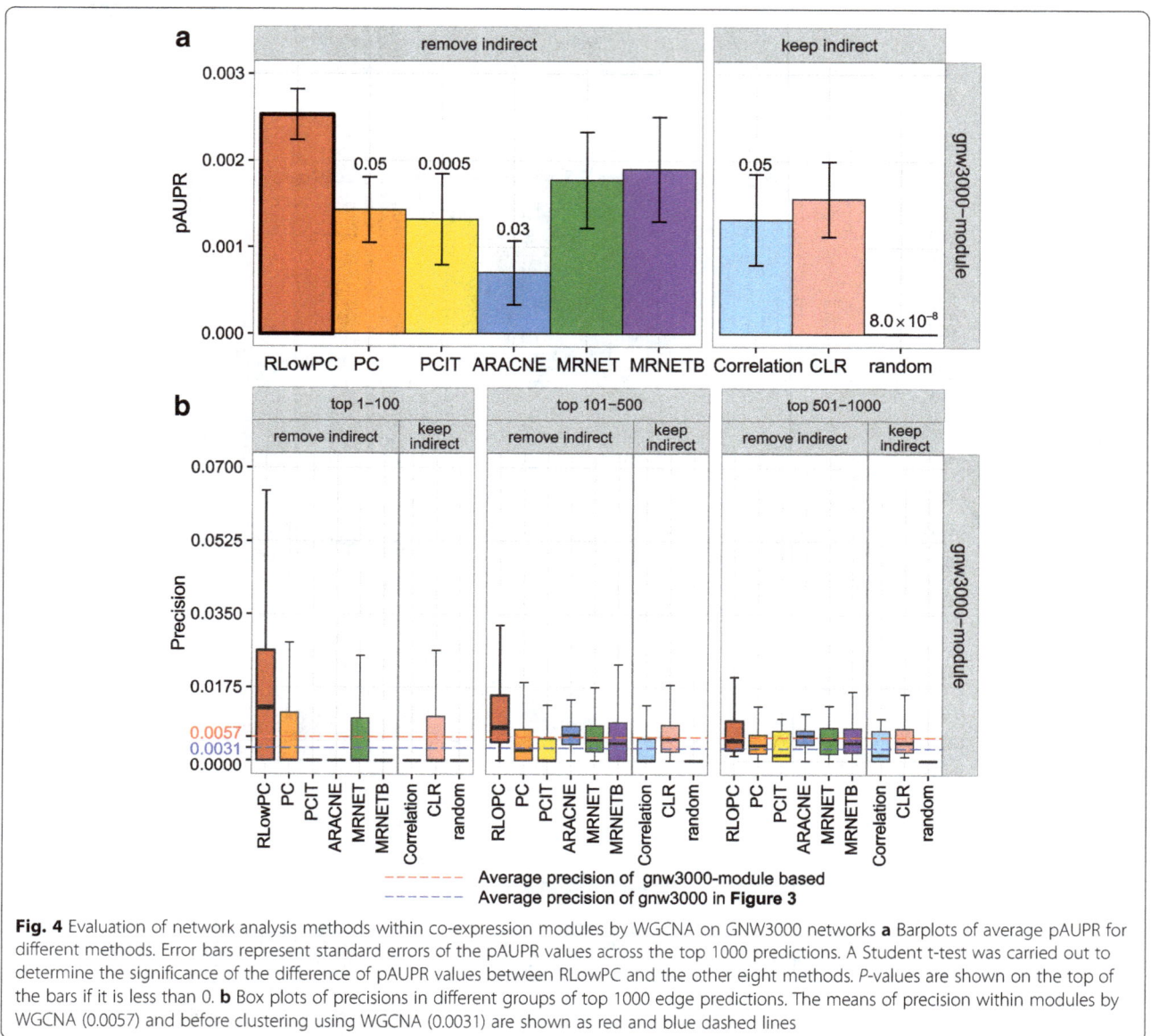

Fig. 4 Evaluation of network analysis methods within co-expression modules by WGCNA on GNW3000 networks **a** Barplots of average pAUPR for different methods. Error bars represent standard errors of the pAUPR values across the top 1000 predictions. A Student t-test was carried out to determine the significance of the difference of pAUPR values between RLowPC and the other eight methods. *P*-values are shown on the top of the bars if it is less than 0. **b** Box plots of precisions in different groups of top 1000 edge predictions. The means of precision within modules by WGCNA (0.0057) and before clustering using WGCNA (0.0031) are shown as red and blue dashed lines

should be paid to generate the most informative data when trying to construct the accurate and comprehensive underlying GRNs.

The co-expression based methods capture the relationships between genes which are perturbed directly or indirectly. Therefore, the multifactorial intervention on the regulators, as discussed in [30], or hub genes rather than on target genes will generate expression data that is more informative for regulatory inference. Results presented here are based on the time-series data corresponding to one perturbation simulation to reflect more typical experimental conditions. When there are more experiments available with different sets of genes being perturbed, the inference accuracy tends to increase with the increased number of gene expression

profiles available [35, 56]. Our data also show that the precision median increases as the experiment size increase (Fig. 5a). Using RLowPC, a precision of 0.014 is achieved in one experiment, while using PC on 10 experiments only leads to a precision of 0.012. Thus refining the top inferred edges using RLowPC is more effective in improving precision than generating data for nine more experiments.

With the number of possible edges growing factorially with increasing number of genes, the sparsity issue in large networks also becomes more prevalent. We observed that precision of the network inference methods increases with the increase of the network density (thus the decrease of network sparsity) as shown in Fig. 5b. Several types of methods have been explored to alleviate

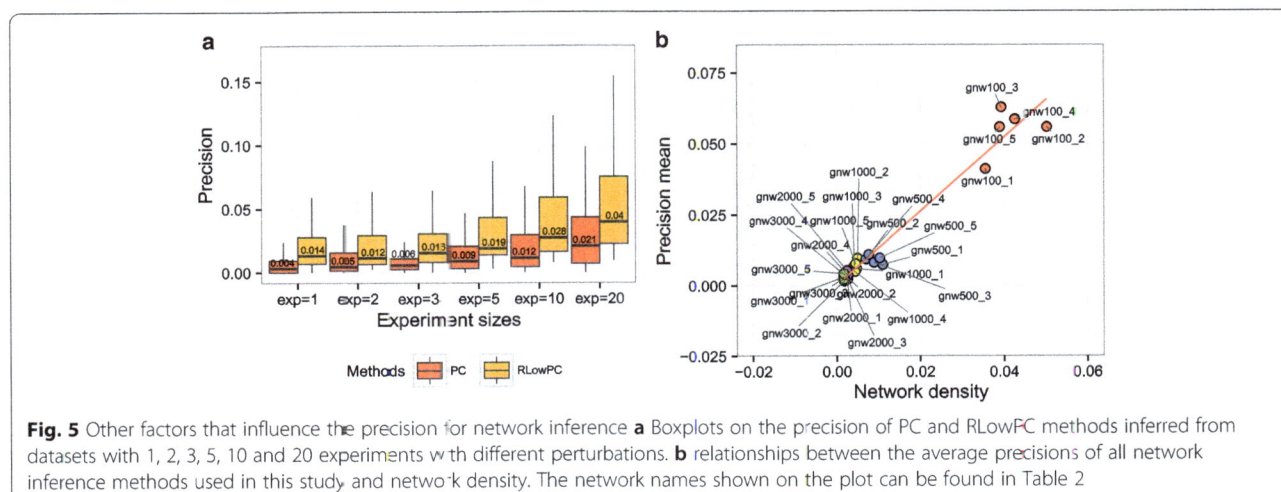

Fig. 5 Other factors that influence the precision for network inference **a** Boxplots on the precision of PC and RLowPC methods inferred from datasets with 1, 2, 3, 5, 10 and 20 experiments with different perturbations. **b** relationships between the average precisions of all network inference methods used in this study and network density. The network names shown on the plot can be found in Table 2

this problem including using network inference methods that allow imposing sparsity constraints [31, 57, 58] or leveraging on multiple datasets on other species that are evolutionary connected [59], or incorporating prior information, such as genetic maps [60], pathways, transcription factor binding, protein-protein interactions, gene ontology, epigenetics, literature, as well as functional association databases to increase the efficiency and reduce the search space by focusing on the top weighted edges [61]. RLowPC method also uses a two-step approach to select on the high-confidence edges first. Thus there is enrichment of true regulatory relationships for the second step of the inference, which explains the improvement of gene regulatory inference performances. Similarly clustering using WGCNA also groups highly correlated and connected genes together, which we see an increase of proportion in the true regulatory relationships. This has a similar effect on the network inference performances.

AUROC and AUPR curves have been popular matrices in the evaluation of network performances [21, 30, 33, 34]. AUROC measures the area under the curve between true positive rate/recall, which is calculated as $(TP/(TP + FN))$ and false positive rate, which is calculated as $(FP/(FP + TN)) = FP/N)$. As in big sparse networks, the negatives (N) greatly exceed the positives (P), thus false positive rate is less discriminative when the network inference methods have very different abilities to largely reduce the false positive predictions.

In the meantime, AUPR measures the area under the curve between precision and recall. Precision, which is calculated as $(TP/(TP + FP)) = 1-FP/(TP + FP))$, captures the impacts of TP or FP in the evaluation of big networks. Studies have shown that AUPR is more informative than AUROC in evaluation on datasets where the TP and TN is imbalanced. Large sparse networks are typical cases [62, 63]. As the purpose of this study is to focus on the utility of co-expression network inferences methods to prioritize the novel regulatory genes pairs for experimental validation from the top ranked edges, we mainly focused on partial AUPR curve to evaluate the accuracies and power of the network inference methods on the top weighted edges, which is more relevant than using the entire area under the curve [64, 65].

One parameter required by the RLowPC method is a number to define the search space for indirect edge reduction. For large networks, a reduction space larger than the size of the top weighted edges under investigation should be applied but has to take into account the computational search space and time required. Table 3 lists the average computational time for different sizes of search space. A useful prior may be to enrich the reduction space with true gene connections. For example, cluster analysis and functional annotation using other experimental data or regulatory databases could be carried out before network inference to investigate the functions and modules of interest.

Table 3 Average computational time of different sizes of reduction space using RLowPC

Top weighted edges	1500	2000	3000	5000	8000	10,000	50,000	100,000
Time	4.71	6.69	11.42	22.62	42.00	54.39	12.97	53.09
Units	secs	secs	secs	secs	secs	secs	mins	mins

The computational time is calculated based on Dell, Windows 7, 64-bit Operating system with 16.0GB RAM and Intel(R) Core (TM) i7–4790 CPU @ 3.60GHz 3.60 GHz processor

Conclusions

In this paper, we present analysis of the evaluation of different regulatory network inference methods with special emphasis on large scale gene regulatory networks with limited sample size. We developed a new method, RLowPC, which improves the precision and recall in the top weighted PC network structures. We evaluated all methods on time-series datasets with only one perturbation for various sizes of networks using a small number of samples, which reflect better the high throughput gene expression data usually generated in laboratory experiments. We also demonstrated that clustering large co-expression networks into functional and informative co-expressed modules, improved the precision and recall of the regulatory inference.

Additional files

Additional file 1: File contains additional Figures and Tables. **Figure S1.** Bar plots of pAUROC values for top 1000 edge predictions. **Figure S2.** Bar plots of pAUROC values of top 1000 predictions for GNW3000 module-based. **Figure S3.** GNW settings for data simulation. **Figure S4.** Examples of evaluation results. **Table S1.** Summaries of evaluation of gene network inference methods. **Table S2.** R packages used to construct and evaluate GRNs.

Additional file 2: Configuration file for GeneNetWeaver (GNW). The settings in the file were load in GNW to generate synthetic data.

Abbreviations

ARACNE: Algorithm for the reconstruction of accurate cellular networks; AUROC: Area under the receiver-operating characteristic curve; AUPR: Area under the precision recall curve; BN: Bayesian network; CLR: Context likelihood or relatedness; DREAM: Dialogue for reverse engineering assessments and methods; FN: False negative; FP: False positive; GCN: Gene co-expression network; GENIE3: Gene network inference with ensemble of trees; GNW: GeneNetWeaver; GRN: Gene regulatory network; MI: Mutual information; MRNET: Minimum redundancy networks; MRNETB: Minimum redundancy networks using backward elimination; pAUROC: Partial area under the receiver-operating characteristic curve; pAUPR: Partial area under the precision-recall curve; PC: Partial correlation; PCIT: Partial correlation coefficient with information theory; RN: Relevance network; RLowPC: Relevance low order partial correlation; SDEs: Stochastic differential eqs.; TN: True negative; TP: True positive; WGCNA: Weighted correlation network analysis

Acknowledgements

We would like to thank Dr. Katherine Denby (University of York) for suggestions on gene network construction and Iain Milne (James Hutton Institute) for technical assistance.

Funding

This project was supported by joint PhD studentship Program from the James Hutton Institute and the University of Dundee [to W.G] and the Scottish Government Rural and Environment Science and Analytical Services division (RESAS) [to J.B., R.W. and R.Z.].

Author's contributions

RZ and WG defined the project and design the simulation experiments. WG carried out the simulations and analyses. WG, RZ, JB and CC wrote the manuscript. All the authors engaged in discussions to improve the project and made contributions to improve the final version of manuscript. All authors read and approved the final manuscript.

Competing interests

The authors declare that they have no competing interests.

Author details

[1]Information and Computational Sciences, The James Hutton Institute, Invergowrie, Dundee, Scotland DD2 5DA, UK. [2]Plant Sciences Division, School of Life Sciences, University of Dundee, Invergowrie, Dundee, Scotland DD2 5DA, UK. [3]Division of Mathematics, University of Dundee, Nethergate, Dundee, Scotland DD1 4HN, UK. [4]Cell and Molecular Sciences, The James Hutton Institute, Invergowrie, Dundee, Scotland DD2 5DA, UK.

References

1. Friedman N, Linial M, Nachman I, Pe'er D. Using Bayesian networks to analyze expression data. J Comput Biol. 2000;7(3–4):601–20.
2. Markowetz F, Spang R. Inferring cellular networks-a review. BMC Bioinformatics. 2007;8(Suppl 6):S5.
3. Murphy KP. Dynamic Bayesian networks: representation, inference and learning. Berkeley: University of California; 2002.
4. Perrin B-E, Ralaivola L, Mazurie A, Bottani S, Mallet J, d'Alche-Buc F. Gene networks inference using dynamic Bayesian networks. Bioinformatics. 2003;19:Iii138–48.
5. Lahdesmaki H, Hautaniemi S, Shmulevich I, Yli-Harja O. Relationships between probabilistic Boolean networks and dynamic Bayesian networks as models of gene regulatory networks. Signal Process. 2006;86(4):814–34.
6. Hache H, Lehrach H, Herwig R. Reverse engineering of gene regulatory networks: a comparative study. EURASIP J Bioinformatics Syst Biol. 2009; 2009:1–12.
7. Bornholdt S. Boolean network models of cellular regulation: prospects and limitations. J R Soc Interface. 2008;5(Suppl 1):S85–94.
8. Martin S, Zhang Z, Martino A, Faulon JL. Boolean dynamics of genetic regulatory networks inferred from microarray time series data. Bioinformatics. 2007;23(7):866–74.
9. de Jong H. Modeling and simulation of genetic regulatory systems: a literature review. J Comput Biol. 2002;9(1):67–103.
10. Linde J, Schulze S, Henkel SG, Guthke R. Data- and knowledge-based modeling of gene regulatory networks: an update. EXCLI J. 2015;14:346–78.
11. Bansal M, Belcastro V, Ambesi-Impiombato A, di Bernardo D: How to infer gene networks from expression profiles. Mol Syst Biol 2007, 3(1): 78.
12. Werhli AV, Grzegorczyk M, Husmeier D. Comparative evaluation of reverse engineering gene regulatory networks with relevance networks, graphical gaussian models and bayesian networks. Bioinformatics. 2006;22(20):2523–31.
13. Zhang B, Horvath S. A general framework for weighted gene co-expression network analysis. Stat Appl Genet Mol Biol. 2005;4:1128. Article17
14. Friedman N. Inferring cellular networks using probabilistic graphical models. Science. 2004;303(5659):799–805.
15. Song L, Langfelder P, Horvath S. Comparison of co-expression measures: mutual information, correlation, and model based indices. BMC Bioinformatics. 2012;13:328.
16. Roy S, Bhattacharyya DK, Kalita JK. Reconstruction of gene co-expression network from microarray data using local expression patterns. BMC Bioinformatics. 2014;15(7):1–14.
17. Ballouz S, Verleyen W, Gillis J. Guidance for RNA-seq co-expression network construction and analysis: safety in numbers. Bioinformatics. 2015;31(13):2123–30.
18. Kogelman LJ, Cirera S, Zhernakova DV, Fredholm M, Franke L, Kadarmideen HN. Identification of co-expression gene networks, regulatory genes and pathways for obesity based on adipose tissue RNA sequencing in a porcine model. BMC Med Genet. 2014;7:57.
19. DiLeo MV, Strahan GD, den Bakker M, Hoekenga OA. Weighted correlation network analysis (WGCNA) applied to the tomato fruit metabolome. PLoS One. 2011;6(10):e26683.
20. Prill RJ, Marbach D, Saez-Rodriguez J, Sorger PK, Alexopoulos LG, Xue X, et al. Towards a rigorous assessment of systems biology models: the DREAM3 challenges. PLoS One. 2010;5(2):e9202.
21. Marbach D, Prill RJ, Schaffter T, Mattiussi C, Floreano D, Stolovitzky G. Revealing strengths and weaknesses of methods for gene network inference. Proc Natl Acad Sci. 2010;107(14):6286–91.

22. Marbach D, Schaffter T, Mattiussi C, Floreano D. Generating Realistic In Silico Gene Networks for Performance Assessment of Reverse Engineering Methods. J. Comput. Biol. 2009;16:229–39.

23. Yip KY, Alexander RP, Yan KK, Gerstein M. Improved reconstruction of in silico gene regulatory networks by integrating knock-out and perturbation data. PLoS One. 2010;5(1):e8121.

24. Schaffter T, Marbach D, Floreano D. GeneNetWeaver: in silico benchmark generation and performance profiling of network inference methods. Bioinformatics. 2011;27(16):2263–70.

25. Young WC, Raftery AE, Yeung KY. Fast Bayesian inference for gene regulatory networks using ScanBMA. BMC Syst Biol. 2014;8:47.

26. Huynh-Thu V, Irrthum A, Wehenkel L, Geurts P. Inferring regulatory networks from expression data using tree-based methods. PLoS One. 2010;5(9):e12776.

27. Meyer P, Marbach D, Roy S, Kellis M. Information-theoretic inference of gene networks using backward elimination. In: BIOCOMP, International Conference on Bioinformatics and Computational Biology: 2010;700–5

28. Margolin AA, Nemenman I, Basso K, Wiggins C, Stolovitzky G, Dalla Favera R, et al. ARACNE: an algorithm for the reconstruction of gene regulatory networks in a mammalian cellular context. BMC Bioinformatics. 2006; 7(Suppl 1):S7.

29. Meyer PE, Kontos K, Lafitte F, Bontempi G. Information-theoretic inference of large transcriptional regulatory networks. EURASIP J Bioinform Syst Biol. 2007;79879

30. Marbach D, Costello JC, Kuffner R, Vega NM, Prill RJ, Camacho DM, et al. Wisdom of crowds for robust gene network inference. Nat Methods. 2012;9(8):796–804.

31. Rogers S, Girolami M. A Bayesian regression approach to the inference of regulatory networks from gene expression data. Bioinformatics. 2005;21(14):3131–7.

32. Van den Bulcke T, Van Leemput K, Naudts E, van Remortel P, Ma H, Verschoren A, et al. SynTReN: a generator of synthetic gene expression data for design and analysis of structure learning algorithms. BMC Bioinformatics. 2006;7:43.

33. Bellot P, Olsen C, Salembier P, Oliveras-Verges A, Meyer PE. NetBenchmark: a bioconductor package for reproducible benchmarks of gene regulatory network inference. BMC Bioinformatics. 2015;16:312.

34. Allen JD, Xie Y, Chen M, Girard L, Xiao G. Comparing statistical methods for constructing large scale gene networks. PLoS One. 2012;7(1):e29348.

35. Steinke F, Seeger M, Tsuda K. Experimental design for efficient identification of gene regulatory networks using sparse Bayesian models. BMC Syst Biol. 2007;1(1):51.

36. Dehghannasiri R, Yoon BJ, Dougherty ER. Efficient experimental design for uncertainty reduction in gene regulatory networks. BMC Bioinformatics. 2015;16(Suppl 13):S2.

37. Cover TM, Thomas JA: Elements of information theory: Wiley-Interscience; 2006.

38. Reverter A, Chan EK. Combining partial correlation and an information theory approach to the reversed engineering of gene co-expression networks. Bioinformatics. 2008;24(21):2491–7.

39. Peng H, Long F, Ding C. Feature selection based on mutual information: criteria of max-dependency, max-relevance, and min-redundancy. IEEE Trans Pattern Anal Mach Intell. 2005 27(8):1226–33.

40. Barabasi AL, Oltvai ZN. Network biology: understanding the cell's functional organization. Nat Rev Genet. 2004;5(2):101–13.

41. Albert R. Scale-free networks in cell biology. J Cell Sci. 2005;118(Pt 21): 4947–57.

42. de la Fuente A, Bing N, Hoeschele I, Mendes P. Discovery of meaningful associations in genomic data using partial correlation coefficients. Bioinformatics. 2004;20(18):3565–74.

43. Zuo Y, Yu G, Tadesse MG, Ressom HW. Biological network inference using low order partial correlation. Methods (San Diego, Calif.) 2014;69(3):266–73.

44. Kim S. Ppcor: an R package for a fast calculation to semi-partial correlation coefficients. Commun Stat Appl Methods 2015;22(6):665–74.

45. Schäfer J, Strimmer K. A shrinkage approach to large-scale covariance matrix estimation and implications for functional genomics. Stat Appl Genet Mol Biol. 2005;4(1):32.

46. Tu Y, Stolovitzky G, Klein U. Quantitative noise analysis for gene expression microarray experiments. Proc Natl Acad Sci U S A. 2002;99(22):14031–6.

47. Madhamshettiwar PB, Maetschke SR, Davis MJ, Reverter A, Ragan MA. Gene regulatory network inference: evaluation and application to ovarian cancer allows the prioritization of drug targets. Genome Med. 2012;4(5):41.

48. Faith JJ, Hayete B, Thaden JT, Mogno I, Wierzbowski J, Cottarel G, et al. Large-scale mapping and validation of Escherichia coli transcriptional regulation from a compendium of expression profiles. PLoS Biol. 2007;5(1):e8.

49. Olsen C, Meyer PE, Bontempi G. On the impact of entropy estimation on transcriptional regulatory network inference based on mutual information. EURASIP J Bioinform Syst Biol. 2009;1:308959.

50. Meyer PE, Lafitte F, Bontempi G: minet: A R/Bioconductor package for inferring large transcriptional networks using mutual information. BMC Bioinformatics 2008, 9:461.

51. Watson-Haigh NS, Kadarmideen HN, Reverter A. PCIT: an R package for weighted gene co-expression networks based on partial correlation and information theory approaches. Bioinformatics. 2010;26(3):411–3.

52. Langfelder P, Horvath S. WGCNA: an R package for weighted correlation network analysis. BMC Bioinformatics. 2008;9:559.

53. Zhao W, Langfelder P, Fuller T, Dong J, Li A, Hovarth S. Weighted gene coexpression network analysis: state of the art. J Biopharm Stat. 2010;20(2):281–300.

54. Ud-Dean SM, Gunawan R. Ensemble inference and inferability of gene regulatory networks. PLoS One. 2014;9(8):e103812.

55. Ud-Dean SM, Heise S, Klamt S, Gunawan R. TRaCE+: ensemble inference of gene regulatory networks from transcriptional expression profiles of gene knock-out experiments. BMC Bioinformatics. 2016;17:252.

56. Altay G. Empirically determining the sample size for large-scale gene network inference algorithms. IET Syst Biol. 2012;6(2):35–43.

57. Slavov N: Inference of Sparse Networks with Unobserved Variables. Application to Gene Regulatory Networks. In: Proceedings of the Thirteenth International Conference on Artificial Intelligence and Statistics; Proceedings of Machine Learning Research: Edited by Yee Whye TMike T. PMLR 2010: 757–764.

58. Sarder P, Schierding W, Cobb JP, Nehorai A. Estimating sparse Gene regulatory networks using a Bayesian linear regression. IEEE Transactions on NanoBioscience. 2010;9(2):121–31.

59. Omranian N, Eloundou-Mbebi JMO, Mueller-Roeber B, Nikoloski Z. Gene regulatory network inference using fused LASSO on multiple data sets. Sci Rep. 2016;6:20533.

60. Flassig RJ, Heise S, Sundmacher K, Klamt S. An effective framework for reconstructing gene regulatory networks from genetical genomics data. Bioinformatics. 2013;29(2):246–54.

61. Studham ME, Tjärnberg A, Nordling TEM, Nelander S, Sonnhammer ELL. Functional association networks as priors for gene regulatory network inference. Bioinformatics. 2014;30(12):i130–8.

62. Saito T, Rehmsmeier M. The precision-recall plot is more informative than the ROC plot when evaluating binary classifiers on imbalanced datasets. PLoS One. 2015;10(3):e0118432.

63. Davis J, Goadrich M: The Relationship Between Precision-Recall and ROC Curves. In ICML '06: Proceedings of the 23rd international conference on Machine learning 2006:233–240.

64. Ma H, Bandos AI, Rockette HE, Gur D. On use of partial area under the ROC curve for evaluation of diagnostic performance. Stat Med. 2013; 32(20):3449–58.

65. Walter SD. The partial area under the summary ROC curve. Stat Med. 2005; 24(13):2025–40.

6

Parameter inference for stochastic single-cell dynamics from lineage tree data

Irena Kuzmanovska[1], Andreas Milias-Argeitis[1,2], Jan Mikelson[1], Christoph Zechner[1,3] and Mustafa Khammash[1]*

Abstract

Background: With the advance of experimental techniques such as time-lapse fluorescence microscopy, the availability of single-cell trajectory data has vastly increased, and so has the demand for computational methods suitable for parameter inference with this type of data. Most of currently available methods treat single-cell trajectories independently, ignoring the mother-daughter relationships and the information provided by the population structure. However, this information is essential if a process of interest happens at cell division, or if it evolves slowly compared to the duration of the cell cycle.

Results: In this work, we propose a Bayesian framework for parameter inference on single-cell time-lapse data from lineage trees. Our method relies on a combination of Sequential Monte Carlo for approximating the parameter likelihood function and Markov Chain Monte Carlo for parameter exploration. We demonstrate our inference framework on two simple examples in which the lineage tree information is crucial: one in which the cell phenotype can only switch at cell division and another where the cell state fluctuates slowly over timescales that extend well beyond the cell-cycle duration.

Conclusion: There exist several examples of biological processes, such as stem cell fate decisions or epigenetically controlled phase variation in bacteria, where the cell ancestry is expected to contain important information about the underlying system dynamics. Parameter inference methods that discard this information are expected to perform poorly for such type of processes. Our method provides a simple and computationally efficient way to take into account single-cell lineage tree data for the purpose of parameter inference and serves as a starting point for the development of more sophisticated and powerful approaches in the future.

Keywords: Parameter inference, Cell lineages, Single cell, Stochastic systems, Monte Carlo methods

Background

Biochemical processes in isogenic cells exhibit substantial heterogeneity [1, 2]. Understanding the latter demands experimental techniques that can resolve such processes at the single-cell level. In contrast to bulk measurements, these techniques provide not only access to the average behavior of intracellular dynamics, but also its variability across cells and over time. Most single-cell techniques, however, reveal only very few components simultaneously that are often multiple steps away from the actual quantities of interest. The dynamics of a promoter, for instance,

may not be accessible directly, but only indirectly through a fluorescent reporter that is expressed upon activation of this promoter [3]. Statistical inference in combination with mathematical models provide a means to reconstruct inaccessible parameters from available measurements, making them instrumental for studying biochemical processes based on single-cell data.

How such inference can be performed depends strongly on the way the data has been collected: flow cytometry measurements, for instance, reveal fluorescence values across a population but individual cells cannot be tracked over time. Consequently, measurements at two different time instances are considered statistically independent. Time-lapse microscopy techniques permit tracking of single-cell trajectories over the duration of a whole experiment [4], which in turn provides a handle also on the

*Correspondence: mustafa.khammash@bsse.ethz.ch
[1]Department of Biosystems Science and Engineering, ETH Zurich, Mattenstrasse 26, 4058 Basel, Switzerland
Full list of author information is available at the end of the article

temporal correlation of the underlying process. This additional degree of information can dramatically improve the inference of unknown process parameters [5].

Most existing inference approaches consider single-cell trajectories to be statistically independent of each other [3, 5–7]. This way, however, important information stemming from the ancestry of a cell is lost: shortly after cell division, for example, two daughter cells are likely to exhibit substantial correlations, which cannot be captured by a model that assumes independence among cells. This can yield incomplete and biased results especially when the time scale of the process under study is slow compared with the cell cycle duration.

In addition, stochastic processes of interest such as epigenetically regulated phase variation in bacteria are often driven by DNA replication just before cell division. Examples in this category are the regulation of *agn43* [8, 9] and *Pap* [10, 11] systems in *E.coli*, and the glucosyltransferase (*gtr*) gene cluster in *Salmonella* [12]. Due to the non-reversibility of the epigenetic modifications, gene replication (and consequently cell division) is crucial for phase variation to happen. Cell lineage information has to be therefore taken into account in single-cell studies of these systems.

Until recently, there existed little work on statistical inference using tree-based single-cell data. In [13], the authors proposed a method for parameter inference from single-cell trajectories based on Approximate Bayesian Computation (ABC). Their approach is applicable to tree-structured data as well, although it requires all trajectories to have the same length and sampling resolution. In [14] the authors proposed an observer-based method for state and parameter estimation in stochastic chemical reaction networks, which is also able to handle lineage tree data. However, its applicability is limited to small systems since it requires the full probability distributions from the solution of the chemical master equation. Another alternative was proposed in [15], which presented an inference algorithm for Hidden Markov Trees using variational Bayesian Expectation Maximization. This class of models is similar to the one considered here, but cannot incorporate dynamic readouts or dynamically evolving single-cell states.

In more recent work, the authors of [16] presented a method for inferring transition dynamics from cell lineages that is best suited to slowly evolving cell states (such as in the case of stem cell lineages) and makes use of end-point smFISH measurements for each cell. Finally, Feigelman et al. [17] proposed a method for exact Bayesian parameter inference from cell lineage data that uses particle filtering to approximate the full joint state and parameter posterior distribution. The method was successfully applied to a stochastic gene expression system that is critical for stem cell differentiation and clearly demonstrated

the strengths of lineage-based inference. On the downside, the computational burden of the method seems to be substantial, while particle degeneracy may arise when trees longer than just a few generations are used because of the way particle sampling and reweighing are carried out.

In this work, we propose an approximate Bayesian parameter inference framework for lineage tree data. The method relies on a combination of Sequential Monte Carlo for likelihood approximation and pseudo-marginal Markov chain Monte Carlo for parameter sampling. To achieve scalability of our method with the number of generations, we make use of a plausible simplifying assumption in the likelihood decomposition which is shown to work well in practice. In contrast to [17], our method allows efficient likelihood calculation and smaller particle degeneracy with increasing tree lengths, which allows us to extract information out of longer lineages. Furthermore, parameter sampling and likelihood approximation are carried out separately from each other, which permits the use of more powerful samplers (such as Population Monte Carlo [18] or Nested Sampling [19]) for the efficient exploration of high-dimensional parameter spaces.

The rest of the manuscript is structured as follows: in 'Methods' section we give a mathematical description of the inference problem and the class of models we consider and we present a detailed description of our method. In 'Results and discussion' section we demonstrate the application of our method to two different example models and in 'Conclusions' section we give some concluding remarks.

Methods
The model class
To introduce the inference problem and the class of models considered here, we refer to the illustration in Fig. 1. Let us consider an intracellular biochemical process of interest modeled by a continuous-time dynamical system S. The system behavior within each cell can be monitored with the help of a dynamic readout, such as the abundance of a fluorescent reporter protein. Through time-lapse microscopy, we assume that a growing population of single cells and their progeny can be tracked over time and measured at multiple time points (green dots in Fig. 1), giving rise to a hierarchical tree data structure that describes the time evolution of the population.

We assume that each tree starts with a single mother at generation 0 and that the population is followed until the final generation N. Without loss of generality we assume that a single mother always gives rise to two daughter cells after division, leading to 2^n cells at generation n. The system S describes the evolution of a set of internal states x (schematically represented by blue curve in Fig. 1). These states can be accessed indirectly at discrete

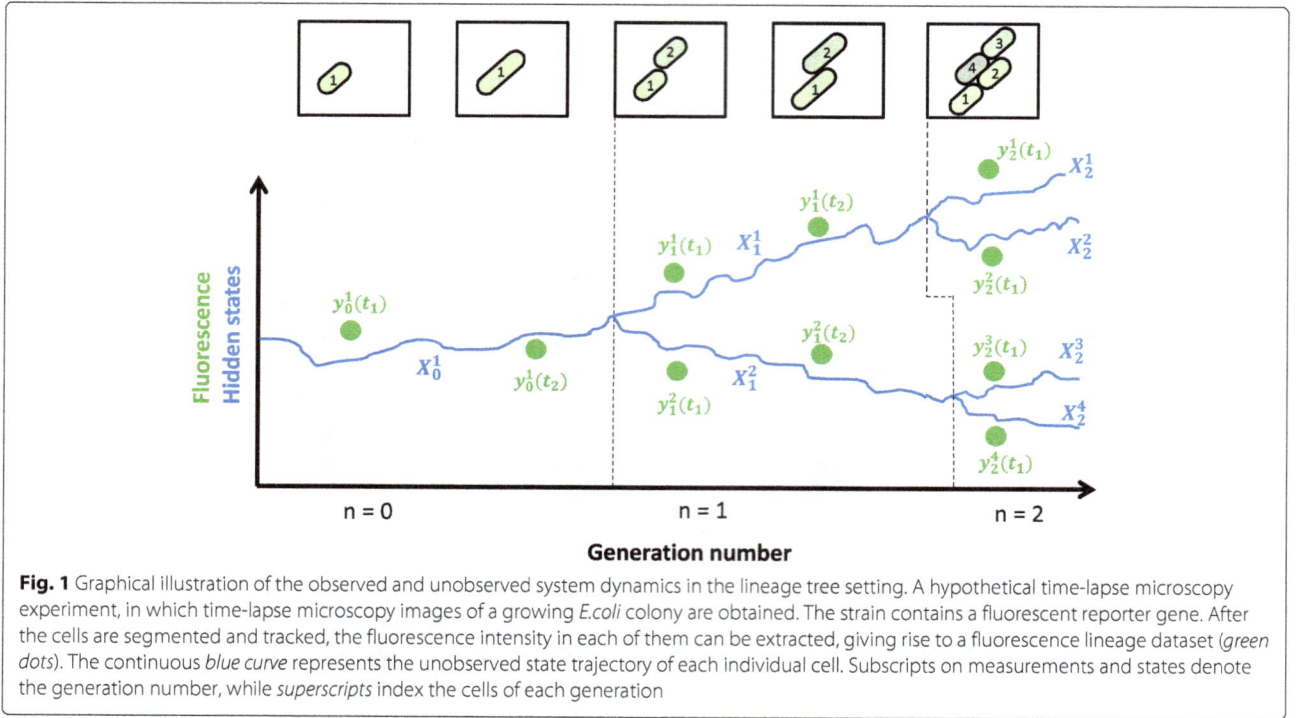

Fig. 1 Graphical illustration of the observed and unobserved system dynamics in the lineage tree setting. A hypothetical time-lapse microscopy experiment, in which time-lapse microscopy images of a growing *E.coli* colony are obtained. The strain contains a fluorescent reporter gene. After the cells are segmented and tracked, the fluorescence intensity in each of them can be extracted, giving rise to a fluorescence lineage dataset (*green dots*). The continuous *blue curve* represents the unobserved state trajectory of each individual cell. Subscripts on measurements and states denote the generation number, while *superscripts* index the cells of each generation

time points through experimental techniques yielding a corresponding readout y. Each cell is assigned a separate time index and a separate time of division, T, which can either be assumed known from the single-cell tracking data, or be inferred based on these data. We denote by X the whole trajectory $\{x(t), t \in [0, T]\}$ from the time of birth of a cell (at $t = 0$) until its division (at $t = T$). The dynamics of S may evolve on a continuous, discrete or hybrid space, and similarly be stochastic, deterministic, or involve components of both types. In any case, we assume that S depends on a set of parameters Θ, which are either assumed to be the same across the population or allowed to vary within the population according to a population-wide distribution.

From this point on, we will distinguish each cell by its generation number, n, and an index i, that ranges from 1 to 2^n (refer to Fig. 1). The ith cell of the nth generation gives rise to two daughters, indexed by $2i - 1$ and $2i$, in generation $n + 1$. Henceforth, all quantities related to a certain cell in a given lineage will be indexed by these two numbers.

Following this notation, we denote by X_n^i the state trajectory of the ith cell in the nth generation; that is,

$$X_n^i := \{x_n^i(t), t \in [0, T_n^i]\}.$$

The state trajectories of the daughters originating from a mother cell with state trajectory X_n^i will therefore be denoted by X_{n+1}^{2i-1} and X_{n+1}^{2i} respectively. The

corresponding discrete set of measurements associated with X_n^i is denoted by Y_n^i. More specifically,

$$Y_n^i := \{y_n^i(t_{n,k}^i), k = 1, \dots, K_n^i\}.$$

This notation reflects the fact that the ith cell of the nth generation is observed at a total number of K_n^i time points (each denoted by $t_{n,k}^i$) during its lifetime, and that the number and location of observation time points will in general be different for every cell.

We will further denote by $P(x_{n+1}^{2i-1}(0), x_{n+1}^{2i}(0) | x_n^i(T_n^i), \Theta)$ the distribution of the daughter initial conditions given the state of the mother just before division, and call this the *transition probability* from one generation to the next. It is reasonable to assume that, once their respective initial conditions are determined based on their mother cell, the two daughters evolve independently of each other. As defined above, the transition probability mechanism may itself contain unknown parameters that need to be estimated from the data.

The inference problem

Our goal is to infer the posterior distribution of Θ given: 1. the set of measured cellular readouts over the whole lineage, 2. our prior knowledge about Θ encoded in a prior distribution $\pi(\Theta)$ and 3. a measurement noise model that describes the likelihood of observing $y_n^i(t)$ given $x_n^i(t)$ (possibly also depending explicitly on unknown parameters contained in Θ). The latter is given by the density $f(y_n^i(t) | x_n^i(t), \Theta)$. With this measurement model, and assuming that measurements at individual time points are independent from each other, the likelihood of the whole measurement set for a single cell can be defined as

$$P(Y_n^i \mid X_n^i, \Theta) = \prod_{k=1}^{K_n^i} f\left(y_n^i\left(t_{n,k}^i\right) \mid x_n^i(t_{n,k}^i), \Theta\right).$$

Setting

$$X_{tree} := \{X_n^i, i = 1, \ldots, 2^n, \, n = 0, \ldots, N\}$$

and

$$Y_{tree} := \{Y_n^i, i = 1, \ldots, 2^n, \, n = 0, \ldots, N\},$$

the joint distribution over states and measurements over a tree starting from a single individual can be written as

$$P(X_{tree}, Y_{tree} \mid \Theta) = P\left(X_0^1 \mid \pi_x\left(X_0^1\right)\right) P\left(Y_0^1 \mid X_0^1, \Theta\right)$$
$$\times \prod_{n=1}^{N} \left[\left(\prod_{i=1}^{2^n} P\left(X_n^{2i-1}, X_n^{2i} \mid X_{n-1}^i, \Theta\right)\right) \right.$$
$$\left. \times \left(\prod_{i=1}^{2^n} P\left(Y_n^i \mid X_n^i, \Theta\right)\right) \right],$$

$$(1)$$

where $\pi_x(X_0^1)$ is the initial distribution of $x_0^1(0)$. The likelihood of the measured outputs given Θ can therefore by obtained by marginalization of (1) over all possible unobserved states:

$$P(Y_{tree} \mid \Theta) = \int P(X_{tree}, Y_{tree} \mid \Theta) dX_{tree}. \quad (2)$$

As can be seen from the above equations, an additional difficulty of our inference problem in comparison to inference based on independent cell trajectories, is the fact that the likelihood $P(X_{tree}, Y_{tree} \mid \Theta)$ does not factorize over the readouts of individual cells, since the tree structure of the population introduces dependencies among the observations coming from different generations. The dependencies are generated through the unobserved state dynamics, which must therefore be taken into account.

Moreover, due to the dependencies introduced by the tree structure of the population, the integral in (2) is analytically intractable already for very simple state dynamics and its numerical evaluation scales exponentially with the number of generations in the tree. To address these difficulties, we employ a sequential Monte Carlo (SMC) scheme as described below to approximate the marginal likelihood (2).

Recursive likelihood and state posterior propagation

The joint likelihood over states and observations given by (1) can be recursively computed, for example by first iterating over generations and then over the individuals of each generation. However, the same cannot be immediately said for (2), where the marginalization complicates

the calculation. Here we propose an iterative calculation of this likelihood that again proceeds sequentially through the tree generations and the daughter pairs of each generation. The dependencies between different daughter pairs of the same generation add to the complexity of the numerical approximation of the likelihood, but, as we will see at the end of the section, this computation can be sped up considerably by making a reasonable simplifying approximation.

Before we derive the exact formulas, we need some additional notation. Let

$$Y_n^{1,\ldots,2^n} := \left\{Y_n^1, \ldots, Y_n^{2^n}\right\}$$

denote the whole dataset of generation n and

$$\mathbb{Y}_{0:n} := \left\{Y_m^{1,\ldots,2^m}, m = 0, \ldots, n\right\}$$

the dataset of all generations up to generation n. Similarly,

$$X_n^{1,\ldots,2^n} := \left\{X_n^1, \ldots, X_n^{2^n}\right\}$$

and

$$\mathbb{X}_{0:n} := \left\{X_m^{1,\ldots,2^m}, m = 0, \ldots, n\right\}.$$

To arrive at the exact formula for the likelihood, we first break up the total likelihood over the generations as follows (the dependence on Θ is suppressed to simplify the notation):

$$P(Y_{tree}) = P(Y_0^1) \prod_{n=1}^{N} P\left(Y_n^{1,\ldots,2^n} \mid \mathbb{Y}_{0:n-1}\right).$$

Assume now that $P(\mathbb{Y}_{0:n})$ (i.e. the likelihood of the subtree consisting of the first n generations) is available, and so is $P(\mathbb{X}_{0:n}\mid\mathbb{Y}_{0:n})$ (the state posterior over the same subtree). Consider the first two individuals of generation $n+1$, with state trajectories X_{n+1}^1 and X_{n+1}^2, descending from the mother cell with state trajectory X_n^1.

Adding the information of this daughter pair to the posterior of the previous generations, we get

$$
\begin{aligned}
&P\left(X_{n+1}^1, X_{n+1}^2, \mathbb{X}_{0:n} | Y_{n+1}^1, Y_{n+1}^2, \mathbb{Y}_{0:n}\right) \\
&= \frac{P\left(Y_{n+1}^1, Y_{n+1}^2, \mathbb{Y}_{0:n} | X_{n+1}^1, X_{n+1}^2, \mathbb{X}_{0:n}\right) P\left(X_{n+1}^1, X_{n+1}^2, \mathbb{X}_{0:n}\right)}{P\left(Y_{n+1}^1, Y_{n+1}^2, \mathbb{Y}_{0:n}\right)} \\
&= \frac{P\left(Y_{n+1}^1 | X_{n+1}^1\right) P\left(Y_{n+1}^2 | X_{n+1}^2\right) P\left(X_{n+1}^1, X_{n+1}^2 | \mathbb{X}_{0:n}\right)}{P\left(Y_{n+1}^1, Y_{n+1}^2 | \mathbb{Y}_{0:n}\right)} \\
&\quad \times \frac{P\left(\mathbb{Y}_{0:n} | \mathbb{X}_{0:n}\right) P\left(\mathbb{X}_{0:n}\right)}{P\left(\mathbb{Y}_{0:n}\right)} \\
&= \frac{P\left(Y_{n+1}^1 | X_{n+1}^1\right) P\left(Y_{n+1}^2 | X_{n+1}^2\right) P\left(X_{n+1}^1, X_{n+1}^2 | X_n^1\right)}{P\left(Y_{n+1}^1, Y_{n+1}^2 | \mathbb{Y}_{0:n}\right)} P(\mathbb{X}_{0:n} | \mathbb{Y}_{0:n}).
\end{aligned}
$$

$$(3)$$

The denominator of Eq. (3) extends $P(\mathbb{Y}_{0:n})$ with the daughter pair of the next generation:

$$
\begin{aligned}
&P(Y_{n+1}^1, Y_{n+1}^2 | \mathbb{Y}_{0:n}) = \\
&\iiint \left(P\left(Y_{n+1}^1, Y_{n+1}^2 | X_{n+1}^1, X_{n+1}^2\right) \right. \\
&\quad \left. \times P\left(X_{n+1}^1, X_{n+1}^2 | X_n^1\right) dX_{n+1}^1 dX_{n+1}^2 \right) \\
&\quad \times P(X_n^1 | \mathbb{Y}_{0:n}) dX_n^1
\end{aligned}
$$

$$(4)$$

Equations. (3) and (4) allow us to update the starting posterior and likelihood with the first daughter pair from generation $n+1$. However, to add the second daughter pair (cells 3 and 4 of generation $n+1$, descending from cell 2 of generation n), we need to take into account the information provided by the first pair. To see this, we proceed as above (3) to arrive at:

$$
\begin{aligned}
&P\left(X_{n+1}^{3,4}, X_{n+1}^{1,2}, \mathbb{X}_{0:n} | Y_{n+1}^{3,4}, Y_{n+1}^{1,2}, \mathbb{Y}_{0:n}\right) \\
&= P\left(X_{n+1}^{1,2}, \mathbb{X}_{0:n} | Y_{n+1}^{1,2}, \mathbb{Y}_{0:n}\right) \\
&\quad \times \frac{P\left(Y_{n+1}^{3,4} | X_{n+1}^{3,4}\right) P\left(X_{n+1}^{3,4} | X_{n+1}^{1,2}, \mathbb{X}_{0:n}\right)}{P(Y_{n+1}^{3,4} | Y_{n+1}^{1,2}, \mathbb{Y}_{0:n})}.
\end{aligned}
$$

$$(5)$$

The above expression can be simplified by noting that $P(X_{n+1}^{3,4} | X_{n+1}^{1,2}, \mathbb{X}_{0:n}) = P(X_{n+1}^{3,4} | \mathbb{X}_{0:n})$, i.e. daughter pairs of the same generation are conditionally independent given the parent states. However, the term $P\left(X_{n+1}^{1,2}, \mathbb{X}_{0:n} | Y_{n+1}^{1,2}, \mathbb{Y}_{0:n}\right)$ implies that, by taking into account the measurements of the first daughter pair, our posterior belief about the n-th generation states also needs to be updated before proceeding to the next pair. This leads to the creation of dependencies between the tree branches and means that they cannot be treated independently of each other, a feature than can create computational difficulties when one attempts to approximate the joint posterior by simulation. We thus make the *simplifying assumption* that

$$
P(X_n^i | Y_{n+1}^{2i-1,2i}, \mathbb{Y}_{0:n}) \approx P(X_n^i | \mathbb{Y}_{0:n})
$$

$$(6)$$

In words, we assume that *the additional state information transferred from the measurement of a daughter pair at generation $n + 1$ to their corresponding mother at generation n is negligible in comparison to the information provided by the previous generations to the mother*. This is especially the case when frequent observations of a cell population are available. In such a setting, the state of the mother can already be well constrained by measurements of itself and its ancestors, making the additional information provided by the daughters less significant. In case of very sparse measurements which are only available for the daughter cells right after cell division it is not certain to what extent the assumption will hold, since these measurements will also carry information about the state of the mother cell. However, frequent observations of the cells in the lineage can be easily achieved with currently used time-lapse microscopy methods.

As we will show below, the aforementioned assumption allows us to treat each mother-daughters triplet within a generation independently from the rest. Continuing the analysis of the first two daughter pairs from above, we have that

$$
\begin{aligned}
&P\left(X_{n+1}^{1,2}, \mathbb{X}_{0:n} | Y_{n+1}^{1,2}, \mathbb{Y}_{0:n}\right) \\
&= P\left(X_{n+1}^{1,2} | \mathbb{X}_{0:n}, Y_{n+1}^{1,2}, \mathbb{Y}_{0:n}\right) P\left(\mathbb{X}_{0:n} | Y_{n+1}^{1,2}, \mathbb{Y}_{0:n}\right) \\
&\approx P\left(X_{n+1}^{1,2} | \mathbb{X}_{0:n}, Y_{n+1}^{1,2}, \mathbb{Y}_{0:n}\right) P\left(\mathbb{X}_{0:n} | \mathbb{Y}_{0:n}\right) \\
&= P\left(X_{n+1}^{1,2} | \mathbb{X}_{0:n}, Y_{n+1}^{1,2}\right) P\left(\mathbb{X}_{0:n} | \mathbb{Y}_{0:n}\right).
\end{aligned}
$$

$$(7)$$

This fact therefore leads to a simplification of the conditional likelihood, $P(Y_{n+1}^{3,4} | Y_{n+1}^{1,2}, \mathbb{Y}_{0:n})$:

$$
\begin{aligned}
&P(Y_{n+1}^{3,4} | Y_{n+1}^{1,2}, \mathbb{Y}_{0:n}) \\
&= \iiint \left(P(Y_{n+1}^{3,4} | X_{n+1}^{3,4}) P(X_{n+1}^{3,4} | X_n^2) dX_{n+1}^{3,4} \right) \\
&\quad \times P(X_{n+1}^{1,2} | X_n^2, Y_{n+1}^{1,2}) P\left(X_n^2 | \mathbb{Y}_{0:n}\right) dX_{n+1}^{1,2} dX_n^2 \\
&= \iint \left(P(Y_{n+1}^{3,4} | X_{n+1}^{3,4}) P(X_{n+1}^{3,4} | X_n^2) dX_{n+1}^{3,4} \right) P\left(X_n^2 | \mathbb{Y}_{0:n}\right) \\
&\quad \times dX_n^2 = P\left(Y_{n+1}^{3,4} | \mathbb{Y}_{0:n}\right),
\end{aligned}
$$

$$(8)$$

and the total likelihood of generation $n + 1$ (conditioned on $\mathbb{Y}_{0:n}$) can be decomposed as a product of likelihoods over the individual daughter pairs.

Finally, the joint posterior (5) can be also decomposed as:

$$
\begin{aligned}
&P\left(X_{n+1}^{3,4}, X_{n+1}^{1,2}, \mathbb{X}_{0:n} | Y_{n+1}^{3,4}, Y_{n+1}^{1,2}, \mathbb{Y}_{0:n}\right) \\
&= P\left(X_{n+1}^{1,2} | Y_{n+1}^{1,2}, \mathbb{X}_{0:n}\right) P\left(X_{n+1}^{3,4} | Y_{n+1}^{3,4}, \mathbb{X}_{0:n}\right) P\left(\mathbb{X}_{0:n} | \mathbb{Y}_{0:n}\right).
\end{aligned}
$$

$$(9)$$

These facts will be put in use in the next section, where a sequential Monte Carlo algorithm for the approximation of the tree likelihood will be presented.

Recursive likelihood approximation

Our SMC scheme is used to approximate $P(Y_{tree} \mid \Theta)$, i.e. the likelihood of a set of measurements over a tree starting from a single individual, given a set of parameters Θ, under the simplifying assumption presented above. Our algorithm uses this assumption to exploit the conditional independence structure of the tree dynamics it generates in order to break down the likelihood computation. More concretely, the idea is to start at the root of the tree (i.e., a single cell) and recursively propagate the data likelihood from one generation to the next, treating the mother-daughter triplets of each generation independently from each other. This can be understood as a generalization of recursive filtering for tree-structured data. To illustrate the idea better, we present the treatment of a single mother-daughter triplet in detail.

Given data up to generation n, assume that L samples (particles) from the already estimated posterior $P(x_n^i(T_n^i) \mid \mathbb{Y}_{0:n})$ of the i-th mother in the n-th generation are available. First, a pair of daughter cells is generated according to the transition probabilities $P\left(x_{n+1}^{2i-1}(0), x_{n+1}^{2i}(0) | x_n^i(T_n^i), \Theta\right)$ for each particle. Given the daughters' initial conditions, we next simulate each daughter until its own division time and calculate the likelihoods $P\left(Y_{n+1}^{2i-1} \mid X_{n+1}^{2i-1,l}, \Theta\right)$ and $P\left(Y_{n+1}^{2i} \mid X_{n+1}^{2i,l}, \Theta\right)$ for $l = 1, \dots, L$.

By assigning to the l-th particle a weight

$$w_{n+1}^{i,l} = P\left(Y_{n+1}^{2i-1} \mid X_{n+1}^{2i-1,l}, \Theta\right) P\left(Y_{n+1}^{2i} \mid X_{n+1}^{2i,l}, \Theta\right),$$

we next compute the marginal likelihood of the ith mother-daughter triplet by averaging the weights for all the particles:

$$P\left(Y_{n+1}^{2i-1}, Y_{n+1}^{2i} | \Theta\right) = \frac{1}{L} \sum_{l=1}^{L} w_{n+1}^{i,l}.$$

After normalizing the particle weights to sum up to one, we have obtained weighted samples from the posteriors $P(X_{n+1}^{2i-1} \mid Y_{n+1}^{2i-1}, \Theta)$ and $P(X_{n+1}^{2i} \mid Y_{n+1}^{2i}, \Theta)$. The samples are subsequently unweighted by resampling L particles from each posterior according to the normalized weights. These samples will serve as starting points for the daughters of the next generation. The same process is repeated for the rest of the nth generation mothers, before moving on to generation $n+1$. This very general procedure is summarized in Algorithm 1, in which for simplicity we assume X_0^1 to be known.

Algorithm 1: The SMC algorithm for tree likelihood calculation

---

Result: Estimate of $P(Y_{tree}|\Theta)$

1 Create L replicates of X_0^1, $\{X_0^{1,l}\}_{l=1}^{L}$;
2 Set $P(Y_{tree}|\Theta) = 1$;
3 **for** $n = 0$ to $N-1$ **do**
4   **for** $i=1$ to 2^n **do**
5   **for** $l = 1$ to L **do**
6   Simulate a pair of daughter cells, $X_{n+1}^{2i-1,l}$ and $X_{n+1}^{2i,l}$ with initial conditions drawn from $P(x_{n+1}^{2i-1}(0), x_{n+1}^{2i}(0)|x_n^i(T_n^i), \Theta)$;
7   Compute the weights $w_{n+1}^{i,l} = P(Y_{n+1}^{2i-1,l} \mid X_{n+1}^{2i-1,l}, \Theta)P(Y_{n+1}^{2i,l} \mid X_{n+1}^{2i,l}, \Theta)$;
8   **end**
9   Compute marginal likelihood of the triplet: $P(Y_{n+1}^{2i-1}, Y_{n+1}^{2i}|\Theta) = L^{-1} \sum_{l=1}^{L} w_{n+1}^{i,l}$;
10   Update tree likelihood estimate: $P(Y_{tree}|\Theta) = P(Y_{tree}|\Theta)P(Y_{n+1}^{2i-1}, Y_{n+1}^{2i}|\Theta)$;
11   Compute normalized weights: $\widetilde{w}_{n+1}^{i,l} := w_{n+1}^{i,l} / \sum_{l=1}^{L} w_{n+1}^{i,l}$;
12   Resample $\{X_{n+1}^{2i-1,l}\}_{l=1}^{L}$ and $\{X_{n+1}^{2i,l}\}_{l=1}^{L}$ according to $\{\widetilde{w}_{n+1}^{i,l}\}_{l=1}^{L}$;
13   **end**
14 **end**

Notes

1. To simplify the presentation of the algorithm, the state trajectory of the initial mother cell was assumed known. In practice it is not and the initial conditions for the particles $\{x_0^{1,l}(0)\}_{l=1}^{L}$ would be drawn from a prior distribution. Then, a classical filtering step [7, 20] can be employed to obtain the final particle conditions for the first mother.

2. The most computationally intensive step of the algorithm lies between lines 5-9, where the marginal likelihood of each mother-daughter triplet needs to be computed. Depending on the type of the unobserved state dynamics, accurate marginalization may require the use of very large particle numbers and greatly increase the computational cost of the algorithm. Typically, the situation is worse when the hidden state contains components driven by stochastic dynamics. This challenge has already been recognized and addressed in the literature, since it also appears in the parameter inference problem from independent single-cell trajectories [7, 21]. One can thus employ one of the several available alternatives at this step, such as sequential

computation of the likelihood [7], or the use of approximating dynamics [7, 22, 23].

3. Since we assume that the measurements from individual trees are independent from each other, the joint likelihood of a dataset consisting of several trees is simply a product of the likelihood of the individual trees. The likelihoods of individual trees can be thus estimated in parallel. Moreover, looking at the algorithm structure for a single tree, the likelihood calculation can be parallelized at two levels: 1) the mother cells of a given generation can be treated independently of each other 2) individual particle calculations for a given mother-daughter triplet can be done in parallel.

A pseudo-marginal MCMC sampler for parameter inference

The goal of Bayesian inference is to compute or approximate via sampling the posterior distribution of the parameters of the system Θ, $P(\Theta|Y_{tree}) \propto P(Y_{tree}|\Theta)\pi(\Theta)$. To this end, we follow the "pseudo-marginal" MCMC approach [24], according to which a Markov chain Monte Carlo (MCMC) sampler makes use of the noisy marginal likelihood estimates provided by the SMC algorithm of the previous section to generate samples from the posterior of Θ (Algorithm 2).

Algorithm 2: The pseudo-marginal MCMC sampler for parameter inference

---

Result: $\{\Theta_m\}_{m=1}^M \sim P(\Theta|Y_{tree})$

1 Draw an initial point Θ_1 from the prior $\pi(\Theta)$;
2 Estimate the likelihood $P(Y_{tree}|\Theta_1)$ using Algorithm 1;
3 **for** $m = 1$ *to* $M - 1$ **do**
4   Propose a parameter vector Θ^* according to a proposal distribution $q(\cdot|\Theta_m)$;
5   Calculate the likelihood $P(Y_{tree} | \Theta^*)$ using Algorithm 1 ;
6   Sample $u \sim \mathcal{U}([\,0, 1\,])$;
7   If

$$u < \min\{1, \frac{P(Y_{tree} | \Theta^*)\pi(\Theta^*)q(\Theta_m|\Theta^*)}{P(Y_{tree} | \Theta_m)\pi(\Theta_m)q(\Theta^*|\Theta_m)}\},$$

8   accept the proposed parameters and set $\Theta_{m+1} = \Theta^*$ and $P(Y_{tree} | \Theta_{m+1}) = P(Y_{tree} | \Theta^*)$; else, set $\Theta_{m+1} = \Theta_m$ and $P(Y_{tree} | \Theta_{m+1}) = P(Y_{tree} | \Theta_m)$.
9 **end**

It should be noted that the use of very noisy SMC estimates may considerably slow down the mixing of the sampler, since the chain may get trapped at a point with artificially large likelihood value. However, as the variance

of the estimator decreases (e.g. through the use of larger particle sample sizes), it is expected that the mixing speed of our sampler will converge to that of a sampler with perfect (i.e. noiseless) marginal likelihood information.

Results and discussion

In the following sections we will consider two possible examples for the dynamical system S and demonstrate the application of our inference method on these cases. We use the first example primarily to verify and characterize the performance of our algorithm and the second example to convincingly demonstrate its application on a more complex problem. In both examples, we assume that a cell is characterized by a discrete state, $x_d(t)$. Over time and across generations, cells stochastically adopt a certain *type* (which for example corresponds to cell phenotype) determined by $x_d(t)$. The cell type in turn determines the evolution of a continuous state vector, which may for example correspond to the immature and mature molecule types of a fluorescent reporter. In abstract terms, $x_d(t)$ may be thought of as the state of a gene, whose activity affects the cell phenotype.

In the first example, the discrete state dynamics is described by a generalized two-type branching process. According to this scenario, the type of a cell is fixed throughout its lifetime and may change only at cell division, since the types of the daughters depend probabilistically on the type of the mother. In the second example, the cell type may stochastically vary throughout the cell lifetime according to a two-state continuous-time Markov chain (CTMC), while the two daughters are assumed to inherit the type of the mother. To test the performance of our inference framework, we generated simulated datasets for the two example systems (Additional file 1: Figure S1) and used them to infer parameters of interest in each case. Details about some of the parameters used in the data generation process are provided in Additional file 1: Table S1. The results for each example system are summarized below.

Example 1: a two-type branching process with dynamic readouts

In this example, cells can adopt one of two possible types (ON or OFF) and maintain their type throughout their lifetime, which, for simplicity, we assume to be the same and equal to T for every cell. At cell division, the daughter cell types are determined based on the type of the mother cell, according to a set of transition probabilities, as illustrated in Fig. 2a. In turn, the type of each cell is assumed to determine the production rate of a fluorescent reporter protein (such as GFP) which can then be observed using fluorescence time-lapse microscopy.

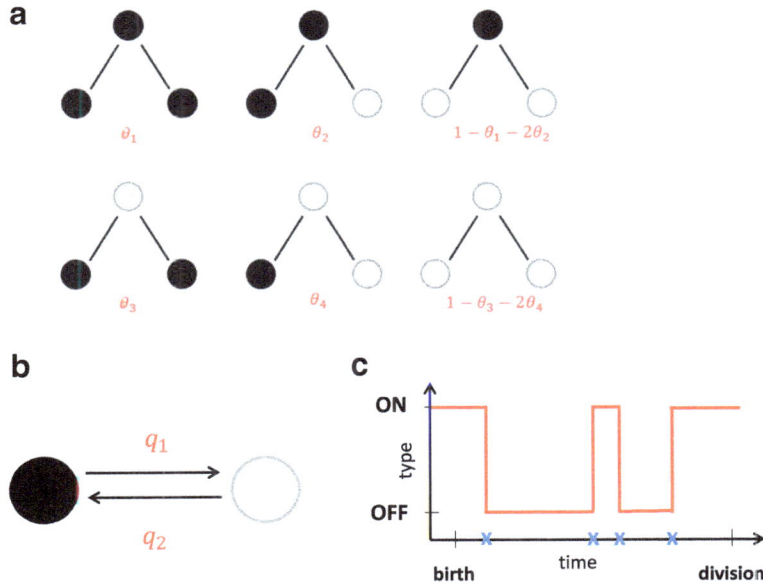

Fig. 2 Graphical illustration of the two systems considered. **a** Model 1: All possible daughter pairs from a single mother and the corresponding probabilities of obtaining those pairs. The cells can be found in two possible states OFF (*black*) and ON (*white*). Note that the probability of the first daughter cell being OFF and the second ON is equal as the probability of the first daughter being ON and the second OFF. **b** Model 2: Cell types switch during the cell lifetime according to a two state continuous-time Markov chain with rates q_1 and q_2, **c** Model 2: An example of the evolution of a cell type during its lifetime. The *blue crosses* on the time axis indicate switching point from ON-OFF or vice versa

The state vector of each cell is thus defined as $x = [x_d(t) \ D(t) \ F(t)]$, where $x_d(t)$ contains the cell type, while $D(t)$ and $F(t)$ correspond to the concentrations of the immature (dark) and mature (fluorescent) forms of the fluorescent reporter, respectively. Out of these, we assume that we can only obtain noisy measurements of $F(t)$ at discrete points in time. Contrary to the cell type, the concentrations of the two reporter species are carried over from the mother to the daughters unchanged. That is, $D_{n+1}^{2i-1}(0) = D_n^i(T)$, $F_{n+1}^{2i-1}(0) = F_n^i(T)$ and similarly for the second daughter. This is a reasonable modeling assumption, given that a daughter cell has half the volume of the mother and receives roughly half of its protein content as well.

The fluorescent reporter dynamics of every cell evolves according to the following set of linear ODEs:

$$\dot{D}(t) = \alpha(x_d(t)) - \delta \cdot D(t) - m \cdot D(t) \quad (10)$$

$$\dot{F}(t) = -\delta \cdot F(t) + m \cdot D(t), \quad (11)$$

where $\alpha(x_d(t))$, δ and m are reporter protein production, dilution and maturation rates respectively. The production rate is determined by the cell type: for an OFF-type cell, $\alpha(OFF) = \alpha_{OFF}$, while a cell of the ON type has $\alpha(ON) = \alpha_{ON} > \alpha_{OFF}$.

As described above, we assume that noise-corrupted measurements proportional to the $F(t)$ species are available at M points, $t_1, ..., t_M$, during the life of every cell. The

readout of a single cell at a given time is therefore assumed to be a scaled and noisy version of the $F(t)$ concentration:

$$y(t_m) \sim \mathcal{N}(c \cdot F(t_m), \sigma^2 \cdot F(t_m)),$$

where the scaling constant c and the measurement variance σ^2 are known. The intuition behind this noise model is that a single concentration unit of mature GFP emits fluorescence which is normally distributed with mean c and variance σ^2. Therefore, for $F(t)$ concentration units of mature GFP, the overall fluorescence emitted will be normally distributed with mean $c \cdot F(t)$ and variance $\sigma^2 \cdot F(t)$. The issue of mapping from protein concentrations to fluorescence intensities is still not well-established in the literature, but several possible approaches have been proposed. The method presented in [25] exploits the deviations in daughter cell fluorescence levels from the average at each cell division. An alternative approach, suggested in [5] is based on recording a calibration curve with several proteins of known abundance fused to the same fluorescence tag.

Given that individual measurements for each cell are independent from each other, the expression for the likelihood $P(Y_n^i \mid X_n^i, \Theta)$, where $Y_n^i = \{y_n^i(t_m), m = 1, \ldots, M\}$, is given by

$$P\left(Y_n^i \mid X_n^i\right) = \prod_{m=1}^{M} P\left(y_n^i(t_m) \mid F_r^i(t_m)\right). \quad (12)$$

Using this type of reporter measurements for every cell belonging to a fully observed tree, our goal is to infer the transition probabilities that govern cell type switching ($\theta_1, \ldots, \theta_4$ in Fig. 2a).

If the type of each cell was readily measurable, the use of simple maximum likelihood estimators for branching processes would suffice to obtain all the necessary discrete state statistics from a fully observed tree, making the use of the reporter model unnecessary. However, the intervening reporter maturation step, the slow dilution dynamics and the sparse, noisy sampling, make inference much more challenging and require the use of the sophisticated computational machinery presented in this work.

To test the performance of our algorithm on this system we simulated a synthetic dataset comprising of a single tree seven generations long (Additional file 1: Figure S1). The lifetime of each cell in the dataset was fixed at 30 min and the measurement interval was 5 min. We generated the daughter types according to the transition probabilities shown in Table 1. The rest of the parameters used for the data generation are summarized in Additional file 1: Table S1.

Note that due to symmetry, the second and third entries of each row are equal. Moreover, the values of the first and second entries in each row determine the rest of the entries, since every row sums to one. We therefore considered θ_1, θ_2, θ_3 and θ_4 as unknown.

We ran the pseudo-marginal MCMC sampler (Algorithm 2) to generate samples from the posterior distribution of $\Theta = [\theta_1\ \theta_2\ \theta_3\ \theta_4]$. The transition probabilities θ_1 and θ_2 were sampled with the help of a Dirichlet distribution (more details are given in Additional file 1) and similarly for θ_3 and θ_4. For all of the parameters we considered flat priors supported on the interval [0,1]. The initial values of $\theta_1 - \theta_4$ were chosen to be all equal to 0.25, with the initial assumption that there are equal probabilities for all of the four possible transitions from a single mother. The number of SMC particles used for the inference procedure was 1000.

The estimated posterior distributions $P(\Theta|Y)$ based on 16604 MCMC samples are given in Fig. 3, where it can be clearly seen that the inferred posterior means (black dashed lines) are located close to the true parameter

values (red lines). On Additional file 1: Figure S2 we can see that the sampler takes very few iterations to find the high-log-likelihood region. The movement of the chain is shown in Additional file 1: Figure S3, the autocorrelation of the samples is given in Additional file 1: Figure S4, while their pairwise scatter plots are given in Additional file 1: Figure S5. Sufficient number of independent samples (low autocorrelation) need to be obtained in order to be confident that they are representative of the true posterior distribution. To check this, we also thinned the chain by using every 10th sample, after discarding the first 1000 burn-in samples. While the thinned chain exhibits much lower sample autocorrelation than the original chain (Additional file 1: Figure S6), the obtained posteriors in Additional file 1: Figure S7 are visually identical to those in Fig. 3. The raw data files generated by the sampler, which were used to generate some of the aforementioned figures, are given in Additional file 2.

To assess the variability of the likelihood SMC estimator as a function of the number of particles we estimated the log likelihood $P(Y_{tree}|\Theta_{true})$ of the same dataset with different numbers of particles, given the true parameter values Θ_{true}. As it can be seen on Fig. 4 (top), the average of 100 log-likelihood estimates converges as the particle number increases. With the increase of the number of particles the coefficient of variation of the log-likelihood calculations also drops quickly (Fig. 4 (bottom)). To avoid negative values we computed the coefficient of variation by dividing the standard deviation by the absolute value of the mean of the log-likelihood. We used the convergence properties of the estimator to get an initial sense of the order of magnitude for the particle number in the SMC algorithm. The particle number choice was refined empirically, based on the mixing and convergence behavior of the pseudo-marginal MCMC sampler. It can be seen in Fig. 4 that the estimator starts converging when around 750 particles are used. For the above-mentioned inference run we used 1000 particles.

To verify computationally that the simplifying assumption we employ in Eq. 6 does not lead to considerable bias in the estimated posteriors, we performed an inference run in which the assumption was not employed and the full likelihood was calculated using a particle filter based on the exact formulas presented above. The obtained posterior distributions (Additional file 1: Figure S7) are visually identical to the ones in Fig. 3, indicating that the simplifying assumption does not create significant bias of the inference results for the example considered here. The mean, variance and the median of the two sets of marginal posterior distributions of each inferred parameter are compared side-by-side in Additional file 1: Table S3. Their similarity indicates that by employing the approximation the basic features of the distributions are maintained.

Table 1 Transition probabilities for the cell types considered in Example 1 and depicted on Fig. 2

		Daughter Types			
		(OFF,OFF)	(OFF,ON)	(ON,OFF)	(ON,ON)
Mother type	OFF	$\theta_1 = 0.6$	$\theta_2 = 0.1$	θ_2	$1 - \theta_1 - 2\theta_2$
	ON	$\theta_3 = 0.1$	$\theta_4 = 0.05$	θ_4	$1 - \theta_3 - 2\theta_4$

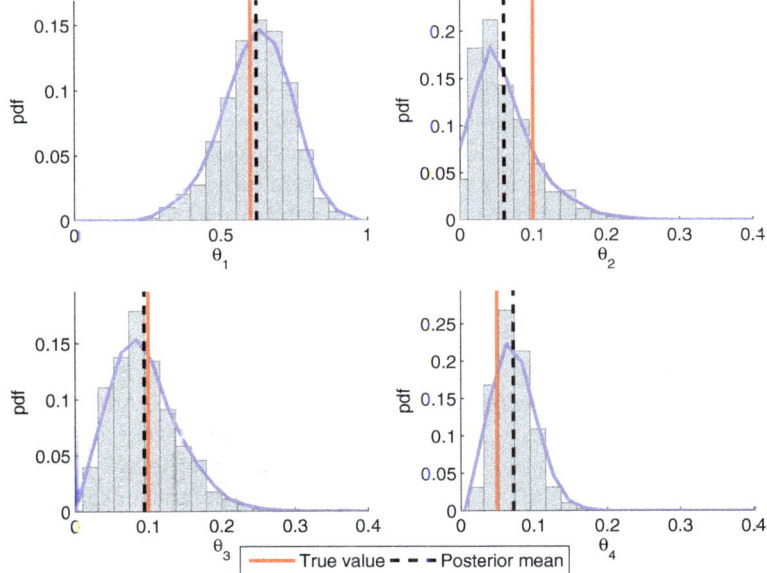

Fig. 3 The proposed algorithm successfully infers parameters involved in the cell division process. Posterior distributions of the inferred parameters described in Example 1. The posteriors were obtained using 16,604 samples, after 2000 burn-in samples had been discarded. The *red vertical bar* is positioned at the true parameter value (the one used for data generation), while the *black dashed line* is positioned at the estimated posterior mean. The *blue curves* are obtained by smoothing of the normalized histograms of the samples

Example 2: stochastic cell type switching

In the second example, we assume that the cell type evolves according to a two-state CTMC with rates q_1 and q_2 for the OFF-to-ON and ON-to-OFF transitions respectively. At division, each daughter inherits the type of its mother (together with the reporter concentrations, as before), but subsequently evolves independently from other cells according to the CTMC dynamics, as shown on Fig. 2b and c. In this case, during the cell lifetime the reporter production rate

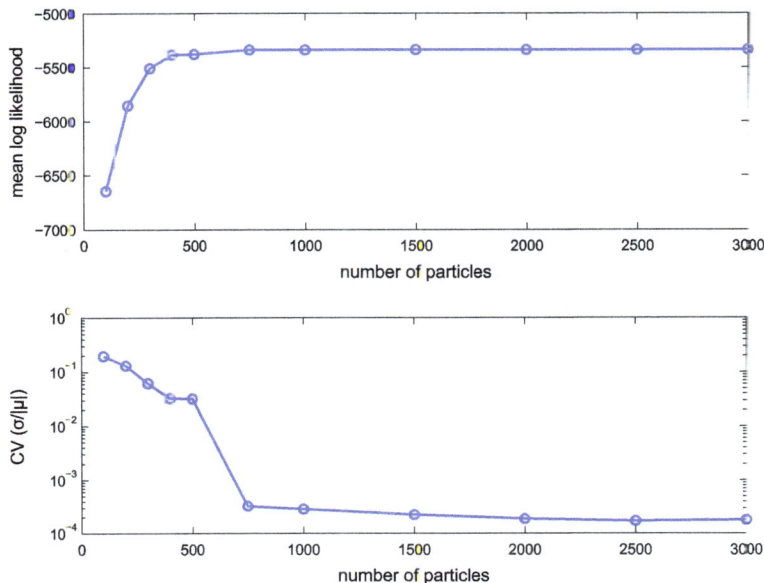

Fig. 4 Convergence of the SMC likelihood estimator with increasing numbers of particles. Mean (*top*) and coefficient of variation (*bottom*) of the log-likelihood vs. number of particles used in the our likelihood estimation algorithm. To avoid negative values, the coefficient of variation was calculated by dividing the standard deviation by the absolute value of the mean, and plotted in $\log_{10}$ scale for better visualization. The results were based on 100 repetitions of the likelihood estimation for each particle count

alternates between α_{ON} and α_{OFF} in accordance with the cell type.

Furthermore, to simulate a more biologically realistic scenario, we also incorporated extrinsic variability in our model by considering different GFP production rates for different cells. When a single cell was born, the GFP production rates α_{OFF} and α_{ON} were drawn in a correlated fashion from lognormal distribution with log-mean at $[\mu_{\alpha_{OFF}} \ \mu_{\alpha_{ON}}]$. The value of the log-standard deviation of each marginal distribution σ_{ext} (henceforth referred to as 'extrinsic noise') was chosen to be 0.3. More concretely, the production rates were drawn by first drawing z from $\log\mathcal{N}(1, \sigma_{ext}^2)$ and then defining α_{OFF} and α_{ON} as described below. These production rates were subsequently used throughout the cell lifetime.

$$z \sim \log\mathcal{N}(1, \sigma_{ext}^2) \tag{13}$$
$$\alpha_{OFF} = \mu_{\alpha_{OFF}} \cdot z$$
$$\alpha_{ON} = \mu_{\alpha_{ON}} \cdot z$$

For the inference, the production rates were drawn at the beginning of the cell lifetime in a similar fashion for each particle.

Using the same type of reporter model dynamics and readouts as in the previous example, our goal in this case was to infer the CTMC transition rates q_1 and q_2 (Fig. 2b), the production rate log-mean of the ON cells $\mu_{\alpha_{ON}}$, the

dilution rate δ, the extrinsic noise σ_{ext} and the measurement variance σ^2, assuming the rest of the parameters to be known (i.e. $\Theta = \begin{bmatrix} q_1 & q_2 & \mu_{\alpha_{ON}} & \delta & \sigma_{ext} & \sigma^2 \end{bmatrix}$).

While in the first example system the tree-structure information was essential for inference as the cell type can only change at cell division, it was not equally obvious that our method would outperform traditional inference on independent single-cell trajectories in this example system, where each daughter inherits the state of its mother and then evolves independently. To verify this, we additionally performed parameter inference (using the same MCMC sampler) by breaking up the tree into individual cell trajectories and considering each cell independently from the others (see Additional file 1), as is usually done in conventional inference based on single-cell data.

To ensure that the MCMC chains converge to the same region of the parameter space, we performed several independent inference runs for each type of data by using the same sampler settings. After obtaining sufficiently long MCMC chains (Additional file 1: Figures S9 and S10) we thinned the chains as described in Additional file 1. The posterior distributions obtained from the thinned chains from a single tree-based and trajectory-based inference run are plotted and compared in Fig. 5. The posterior distribution sets obtained from the multiple independent MCMC runs are overlayed in Additional file 1: Figure S12. It can be easily seen that posterior distributions based on individual cell trajectories are in several cases biased,

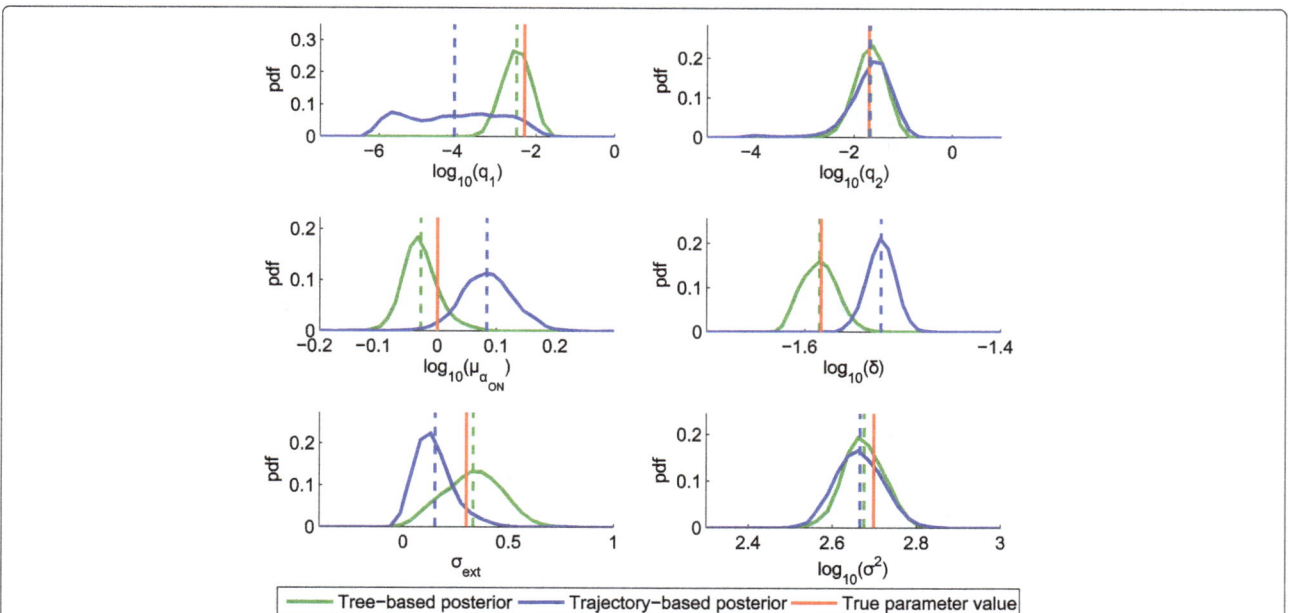

Fig. 5 The proposed tree-based inference method outperforms the traditionally used inference on independent cell trajectories. Posterior distributions of the six unknown parameters presented in Example 2, obtained both with tree-based (*green*) and trajectory-based inference (*blue*). The *red bars* are positioned at the true parameter values (i.e. the ones used for data generation), while the *dashed lines* indicate the estimated posterior means. The curves are obtained by smoothing of the normalized histograms of the MCMC samples. One can clearly observe the bias in the parameter estimates when trajectory-based inference was used

which indicates a potential disadvantage of the traditionally used trajectory-based inference. More details about the inference runs, including the prior distribution for the parameters and the proposal kernels used in the MCMC are given in Additional file 1, along with the running log-likelihoods, autocorrelation plots and the pairwise scatter plots of some of the samples. The raw data files produced by some of the MCMC runs (including all the accepted parameters and their corresponding log-likelihood values) are given in Additional file 2.

The advantage of using lineage data instead of individual trajectories for inference is that the uncertainty regarding the initial conditions of the system states $(x_d(t), D(t), F(t))$ of each cell is greatly reduced, since the prior of each daughter state is based on the posterior of its mother. On the contrary, when the cell trajectories are assumed to be independent, the state of every cell has to be independently initialized according to an assumed prior (described in detail in Additional file 1). Moreover, when the switching rates correspond to mean holding times that exceed the lifetime of a single cell, inference based on a relatively small number of single-cell trajectories will tend to produce biased estimates, as our results show.

Conclusions

In this work we proposed a parameter inference method for stochastic single-cell dynamics from tree-structured data. More specifically, we considered a class of systems with one or more unobserved states and fluorescent reporter readouts, observed through time-lapse microscopy, which allows tracking individual cells and their progeny over time. Our goal was to estimate the posterior distribution of the unknown system parameters given such readouts. To calculate the likelihood of the data for a given parameter set, the hidden state trajectories had to be integrated out. This marginalization was accomplished with the help of a sequential Monte Carlo method, which recursively computes a sampling-based estimate of this analytically intractable quantity. To sample the system parameter space, we employed an MCMC scheme, which was able to target the correct parameter posterior despite the noisy likelihood estimates. The application of our method to two simple examples showed that it can correctly infer the parameters of interest and approximate their posterior distribution. Our algorithm is currently implemented in C++ and an inference run for the second example presented, consisting of 111,000 MCMC iterations, took about 24 h.

Our inference framework extends to more complex applications in a straightforward manner (for instance larger stochastic chemical reaction networks), although its computational complexity increases with the state and parameter dimensionality. More complex networks will require heavier stochastic simulations, which take a crucial part of the computations carried out in our method. They will also require a larger number of particles to achieve a reasonable accuracy of the SMC-based likelihood estimator. If the latter is too noisy, one may observe slow mixing of the MCMC sampler and in turn poor posterior estimates.

The choice of the MCMC sampler is also a critical issue and depends on the features of the problem at hand, such as the complexity of the target distribution. The Metropolis-Hastings sampler was sufficiently well-suited for the examples presented in this study, but may not be the best choice for every problem. To apply our inference framework on more complex problems, different samplers might need to be employed, with better and more effective convergence and parameter exploration properties. In some cases the model structure might also not be completely known a priori and model selection should be performed to discriminate among several candidate model structures. Bayesian model selection requires the computation of the evidence (marginal likelihood) for each candidate model, which will demand much more powerful and sophisticated samplers than the MCMC sampler presented here. Model selection is, however, beyond the scope of this work.

The inference framework presented here could be very useful in the case of systems where accurate tracking of single-cell dynamics across cell lineages plays an important role. These are, for example, systems involved in stem cell fate decisions, or stochastic phenotype switching in bacteria [8, 10]. In many such cases, stochastic fluctuations of key factors over long timescales and/or stochastic events taking place at cell division create strong mother-daughter and daughter-daughter correlations that play a crucial role in determining the overall behavior of a colony. In such cases, treatment of the measured single-cell trajectories independently from each other will result a large loss of information and biased parameter estimates. We believe that proper incorporation of the population lineage information into the parameter inference problem will thus provide the right framework for treating this type of systems and may reveal important insights into their function.

Abbreviations
ABC: Approximate Bayesian Computation; CTMC: Continuous-time Markov chain; MCMC: Markov chain Monte Carlo; SMC: Sequential Monte Carlo

Acknowledgments
Not applicable.

Funding
Not applicable.

Authors' contributions

IK and MK conceived the study. IK, AM and CZ designed the method and examples presented in the manuscript. JM assisted in the design of the examples, implemented the method and carried out the inference runs for the first example presented. IK and JM carried out the inference runs for the second example presented. IK, AM and CZ wrote the manuscript. MK supervised the study in all of its stages. All authors read and approved the final version of the manuscript.

Competing interests

The authors declare that they have no competing interests.

Author details

[1] Department of Biosystems Science and Engineering, ETH Zurich, Mattenstrasse 26, 4058 Basel, Switzerland. [2] Groningen Biomolecular Sciences and Biotechnology, University of Groningen, Nijenborgh 4, 9747 AG Groningen, Netherlands. [3] Max Planck Institute of Molecular Cell Biology and Genetics and Center for Systems Biology, Pfotenhauerstrasse 108, 01307 Dresden, Germany.

References

1. McAdams HH, Arkin A. Stochastic mechanisms in gene expression. Proc Natl Acad Sci USA. 1997;94(3):814–9.
2. Elowitz MB, Levine AJ, Siggia ED, Swain PS. Stochastic gene expression in a single cell. Science. 2002;297(5584):1183–6.
3. Hansen AS, O'Shea EK. Promoter decoding of transcription factor dynamics involves a trade-off between noise and control of gene expression. Mol Syst Biol. 2013;9(1):704.
4. Young JW, Locke JC, Altinok A, Rosenfeld N, Bacarian T, Swain PS, Mjolsness E, Elowitz MB. Measuring single-cell gene expression dynamics in bacteria using fluorescence time-lapse microscopy. Nat Protoc. 2012;7(1):80–8.
5. Zechner C, Unger M, Pelet S, Peter M, Koeppl H. Scalable inference of heterogeneous reaction kinetics from pooled single-cell recordings. Nat Methods. 2014;11(2):197–202.
6. Llamosi A, Gonzalez-Vargas AM, Versari C, Cinquemani E, Ferrari-Trecate G, Hersen P, Batt G. What population reveals about individual cell identity: single-cell parameter estimation of models of gene expression in yeast. PLoS Comput Biol. 2016;12(2):e1004706.
7. Andrew Golightly DJW. Bayesian parameter inference for stochastic biochemical network models using particle Markov chain Monte Carlo. Interface Focus. 2011;1(6):807–20.
8. van der Woude MW, Henderson IR. Regulation and function of Ag43 (Flu). Annu Rev Microbiol. 2008;62:153–69.
9. Lim HN, van Oudenaarden A. A multistep epigenetic switch enables the stable inheritance of DNA methylation states. Nat Genet. 2007;39(2):269–75.
10. Munsky B, Hernday A, Low D, Khammash M. Stochastic modeling of the pap-pili epigenetic switch. Proc FOSBE. 2005:145–8.
11. O'Hanley P, Low D, Romero I, Lark D, Vosti K, Falkow S, Schoolnik G. Gal-gal binding and hemolysin phenotypes and genotypes associated with uropathogenic Escherichia coli. N Engl J Med. 1985;313(7):414–20.
12. Broadbent SE, Davies MR, van der Woude MW. Phase variation controls expression of Salmonella lipopolysaccharide modification genes by a DNA methylation-dependent mechanism. Mol Microbiol. 2010;77(2):337–53. doi:10.1111/j.1365-2958.2010.07203.x.
13. Loos C, Marr C, Theis FJ, Hasenauer J. Approximate Bayesian Computation for Stochastic Single-Cell Time-Lapse Data Using Multivariate Test Statistics. In: Proceedings of the 13th International Conference on Computational Methods in Systems Biology. Springer International Publishing; 2015. p. 52–63.
14. Thorsley D, Klavins E. Estimation and discrimination of stochastic biochemical circuits from time-lapse microscopy data. PLoS ONE. 2012;7(11):47151.
15. Olariu V, Coca D, Billings SA, Tonge P, Gokhale P, Andrews PW, Kadirkamanathan V. Modified variational bayes EM estimation of hidden Markov tree model of cell lineages. Bioinformatics. 2009;25(21):2824–30.
16. Hormoz S, Singer ZS, Linton JM, Antebi YE, Shraiman BI, Elowitz MB. Inferring cell-state transition dynamics from lineage trees and endpoint single-cell measurements. Cell Syst. 2016;3(5):419–33.
17. Feigelman J, Ganscha S, Hastreiter S, Schwarzfischer M, Filipczyk A, Schroeder T, Theis FJ, Marr C, Claassen M. Analysis of cell lineage trees by exact Bayesian inference identifies negative autoregulation of Nanog in mouse embryonic stem cells. Cell Syst. 2016;3(5):480–90.
18. Cappe O, Guillin A, Marin JM, Robert CP. Population Monte Carlo. J Comput Graph Stat. 2004;13:907–29.
19. Skilling J. Nested sampling for general Bayesian computation. Bayesian Anal. 2006;1(4):833–59.
20. Doucet A, Johansen AM. A tutorial on particle filtering and smoothing: fifteen years later. Handb Nonlinear Filtering. 2009;12:656–704.
21. Amrein M, Künsch HR. Rate estimation in partially observed Markov jump processes with measurement errors. Stat Comput. 2012;22(2):513–26.
22. Stathopoulos V, Girolami MA. Markov chain Monte Carlo inference for Markov jump processes via the linear noise approximation. Philos Trans R Soc Lond A Math Phys Eng Sci. 2012;371(1984):20110541.
23. Golightly A, Wilkinson DJ. Bayesian inference for stochastic kinetic models using a diffusion approximation. Biometrics. 2005;61(3):781–8.
24. Andrieu C, Roberts GO. The Pseudo-Marginal Approach for Efficient Monte Carlo Computations. Ann Stat. 2009;37(2):697–725.
25. Rosenfeld N, Perkins TJ, Alon U, Elowitz MB, Swain PS. A fluctuation method to quantify in vivo fluorescence data. Biophys J. 2006;91(2):759–66.

7

Bayesian network model for identification of pathways by integrating protein interaction with genetic interaction data

Changhe Fu[1,2]*, Su Deng[2], Guangxu Jin[3], Xinxin Wang[2] and Zu-Guo Yu[1]*

Abstract

Background: Molecular interaction data at proteomic and genetic levels provide physical and functional insights into a molecular biosystem and are helpful for the construction of pathway structures complementarily. Despite advances in inferring biological pathways using genetic interaction data, there still exists weakness in developed models, such as, activity pathway networks (APN), when integrating the data from proteomic and genetic levels. It is necessary to develop new methods to infer pathway structure by both of interaction data.

Results: We utilized probabilistic graphical model to develop a new method that integrates genetic interaction and protein interaction data and infers exquisitely detailed pathway structure. We modeled the pathway network as Bayesian network and applied this model to infer pathways for the coherent subsets of the global genetic interaction profiles, and the available data set of endoplasmic reticulum genes. The protein interaction data were derived from the BioGRID database. Our method can accurately reconstruct known cellular pathway structures, including SWR complex, ER-Associated Degradation (ERAD) pathway, N-Glycan biosynthesis pathway, Elongator complex, Retromer complex, and Urmylation pathway. By comparing N-Glycan biosynthesis pathway and Urmylation pathway identified from our approach with that from APN, we found that our method is able to overcome its weakness (certain edges are inexplicable). According to underlying protein interaction network, we defined a simple scoring function that only adopts genetic interaction information to avoid the balance difficulty in the APN. Using the effective stochastic simulation algorithm, the performance of our proposed method is significantly high.

Conclusion: We developed a new method based on Bayesian network to infer detailed pathway structures from interaction data at proteomic and genetic levels. The results indicate that the developed method performs better in predicting signaling pathways than previously described models.

Keywords: Protein interaction, Genetic interaction, Biological pathway, Bayesian model

Background

A cellular biological system is controlled by the molecules at different levels, such as protein phosphorylation or genetic variations, and their interactions. Protein interaction (i.e., protein-protein interaction) refers to physical interconnection between two or more proteins that occur in a cell, by which protein components can carry out most of cellular molecular processes [1]. Genetic interaction refers to functional relationship between two genes, which can be measured by the difference between the phenotype levels of double gene mutations and the expected neutral level evaluated by the corresponding single mutant phenotype level [2, 3]. The publicly available data sets, such as Biological General Repository for Interaction Datasets (BioGRID, https://thebiogrid.org/), Saccharomyces Genome Database (SGD, http://www.yeastgenome.org/), Human Protein Reference Database (HPRD, http://www.hprd.org/), Search Tool for the Retrieval of

* Correspondence: fuch@synu.edu.cn; yuzuguo@aliyun.com
[1]School of Mathematics and Computational Science, Xiangtan University, Xiangtan 411105, China
Full list of author information is available at the end of the article

Interacting Genes/Proteins (STRING: http://www.string-db.org/) and so on, collect thousands of proteins and a few genetic interactions from several of species.

Given a great deal of these interaction data collected, it is of challenges to elucidate biological meaning behind the data, especially to identify biological pathways underlying the data [4]. A few methods and tools have been developed to predict signaling pathways based on protein interaction networks [5–8]. Several different studies utilized various biological data to discover regulatory networks [9–12] and reconstruct metabolic networks [13–16]. There are other methods that uncover pathway networks by integrating protein-protein interaction data and gene expression data [17–19]. In genetic interaction studies, the most important method is cluster analysis, grouping genes by the similar genetic interaction profiles [20–22]. Some other studies focus on aggravating or alleviating relationships between related gene groups [23–25]. In order to automatically identify detailed pathway structures using high-throughput genetic interaction data, the activity pathway network (APN) was developed [26]. However, these available approaches cannot fully take advantages of the complementarity between protein and genetic interaction data to infer the biological pathway structures.

In this paper, we present a Bayesian model that integrates high-throughput protein and genetic interaction data to reconstruct detailed biological pathway structures. The model can organize related genes into the corresponding pathways, arrange the order of genes within each pathway, and decide the orientation of each interconnection. Based on protein interaction network, the model predicts detailed pathway structures by using genetic interaction information to delete redundancy edges and reorient the kept edges in the network. Similar to APN [26], our model represents a biological pathway network as a Bayesian network [27], in which each node presents the activity of a gene product. Different from APN that drew network sample from complete network, our method introducing protein interaction networks as underlying pathway structures. In addition, a scoring function is defined by gene pairwise score, which can avoid the unadjusted balance between gene pairwise score and edge score in the APN. Thus, our model is able to improve computational efficiency of stochastic simulation algorithm and overcome the limitation of APN that some edges in the results are difficult to interpret. In our model, each edge in the network can capture physical docking, and represent functional dependency.

Methods
Bayesian network
We model a pathway network as a Bayesian network that is a directed acyclic graph. The activity of a gene is

assigned to a node in the network [26]. The edge in the network is an interaction in protein interaction network. Additionally, it presents the conditional dependency between the nodes connected as well. The experiments of genetic interaction are not for detection of the influence between pairwise genes but for measurement of impact of mutation of these two genes on phenotype of interest. Thus, it is impossible to evaluate conditional probability distribution between the nodes of the Bayesian network, and the standard Bayesian learning methods lost their efficacy. Here, we only utilize conditional independence assumptions of the Bayesian network theory to construct a network that can represent independence assumptions hidden in the gene interaction data. As in Ref. [26], based on the independence assumptions, it is elucidated that given the activity level of X, the fitness level is independent of the activity level of Y, if gene X is fully epistatic to gene Y. The constructed network can encode a linear pathway substructure between X and Y, in which Y must be the father node of X, that is, the direction of edge between is decided.

Scoring
For a candidate pathway network (Fig. 1b) sampled from protein interaction network, we score it in term of genetic interaction quantitative measurement using method in Ref. [26]. For every pair of genes, there are four topological structures and their local scores shown in Fig. 2. Despite the larger score indicating the more possible local structure for each gene pair, we still need every one of four scores to find the optimal global structure. We computed the four possible scores for each pair of genes before all the steps to improve computation efficiency.

Using the scoring methods in Fig. 2 and dataset D of genetic interaction and protein interaction, we can compute a local score for every pair of genes in a candidate pathway network N, and sum up all of the scores for all pairs to define the global score function $f(N)$, to which the Bayesian network posterior probability distribution $p(N|D)$ is proportional, shown as eq. (1). In Bayesian network theory [27], a network N with the higher posterior probability or global score should be more accord with the data set.

$$f(N) = \exp\left(\sum_{x*y \ in \ N} Score(x, y) \right) \quad (1)$$

Different from study of Ref. [26], we do not include every edge score in $f(N)$, because the edge in our network represents protein interaction that insures its existence. Then, it avoids the dilemma how to adjust the balance between the two scores.

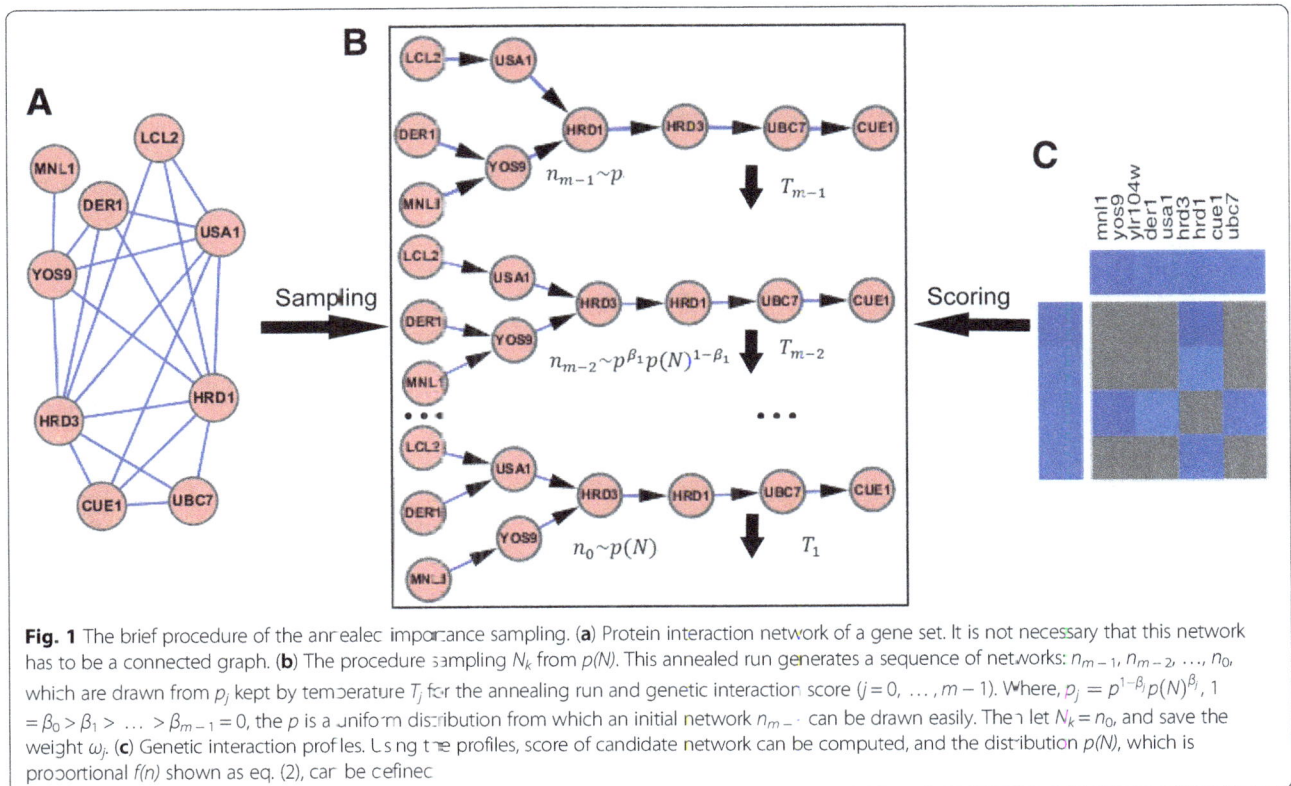

Fig. 1 The brief procedure of the annealed importance sampling. (**a**) Protein interaction network of a gene set. It is not necessary that this network has to be a connected graph. (**b**) The procedure sampling N_k from $p(N)$. This annealed run generates a sequence of networks: $n_{m-1}, n_{m-2}, ..., n_0$, which are drawn from p_j kept by temperature T_j for the annealing run and genetic interaction score ($j = 0, ..., m-1$). Where, $p_j = p^{1-\beta_j} p(N)^{\beta_j}$, $1 = \beta_0 > \beta_1 > ... > \beta_{m-1} = 0$, the p is a uniform distribution from which an initial network n_{m-1} can be drawn easily. Then let $N_k = n_0$, and save the weight ω_j. (**c**) Genetic interaction profiles. Using the profiles, score of candidate network can be computed, and the distribution $p(N)$, which is proportional $f(n)$ shown as eq. (2), can be defined

Sampling

We utilized annealed importance sampling [26, 28] to learn the pathway structure by the above distribution p $(N|D) \propto f(N)$. The annealed importance sampling approach can assign weights to pathway networks sampled by simulated annealing schedules, then to evaluate that converge to the real network structure. The approach is appropriate for sampling N from multi-modal distributions p $(N|D)$ or abbreviated to $p(N)$, since its independent sampling method

Topological structure	Score
	$Score(x,y) = -\dfrac{1}{2\sigma^2}(D(x,y) - S(y))^2 + \log(p)$
	$Score(x,y) = -\dfrac{1}{2\sigma^2}(D(x,y) - S(x))^2 + \log(p)$
	$Score(x,y) = -\dfrac{1}{2\sigma^2}(D(x,y) - E(x,y))^2 + \log(p)$
	$Score(x,y) = -\dfrac{1}{2\sigma^2}\Big(D(x,y) - \dfrac{1}{2}(E(x,y)$ $+ \ max(S(x),S(y)))\Big)^2 + \log(p)$

Fig. 2 Topological structure and score of gene pair. For a pair of genes (X, Y), node P represents the measured quantitative phenotype. Dotted line means that the connection may be not a direct edge in the pathway network. Based on Gaussian distribution, the score for each topological structure is defined by the probability of actual double mutation measured phenotype value (fitness). Where, $D(x, y)$ is (x, y) double mutant fitness, $S(.)$ is single mutant fitness, and $E(x, y)$ is typical genetic interaction value defined as [22]. The variance σ^2 is the empirical variance of repeated experiments, and p is the prior probability for each given topological structure

can overcome some problems of convergence and autocorrelation in general Markov chain Monte Carlo (MCMC) samplers. Figure 1 presents the brief procedure of an annealing run of the annealed importance sampling.

Pooling

After K annealing runs, the sampler generates K pathway networks and their weights. Then we can compute the confidence for any given substructure s, shown as

$$C(s) = \frac{\sum_{k=1}^{K} \omega_k I(s \subset N_k)}{\sum_{k=1}^{K} \omega_k}$$

(2)

Where $I(\cdot)$ is the indicator function, N_k is the sample at the kth annealing run, and ω_k is the important weight. Based on the theory of annealed importance sampling, we can compute confidences of all structure forms of an interesting gene subset, and choose the maximal one as the possible detailed pathway structure of the subset.

Pseudo-code for pathway network reconstruction

Input: Matrix P: protein interaction network

Vector S: signal mutation levels
Matrix D: double mutation levels

Matrix E: typical value for double mutation levels
Vector T: temperatures for the annealing run
Integer K: number of parallel annealing runs
Some optional parameters

Output: Matrix of directed pathway networks and their weights
Procedure:
Compute all scores for every possible gene pair by inputs of genetic interaction data
Compute $p(N)$ by scores of gene pairs in N
m = length(TV)
Design distributions $p_j (j = 0, \dots, m - 1)$ (as Fig. 1) to approach $P(N)$
For $i = 1$ to K:

Sample initial network n_{m-1} from uniform distribution p_{m-1}
For $j = m\text{-}2$ to 0:
   Generate n_j from n_{j-1} by uniform distribution over P
   Accept n_j according to Metropolis–Hastings algorithm by T_j and p_j
   Update importance weight
Save network N_i and its weight ω_i

Return networks N_i ($i = 1, \dots, K$) and their importance weights

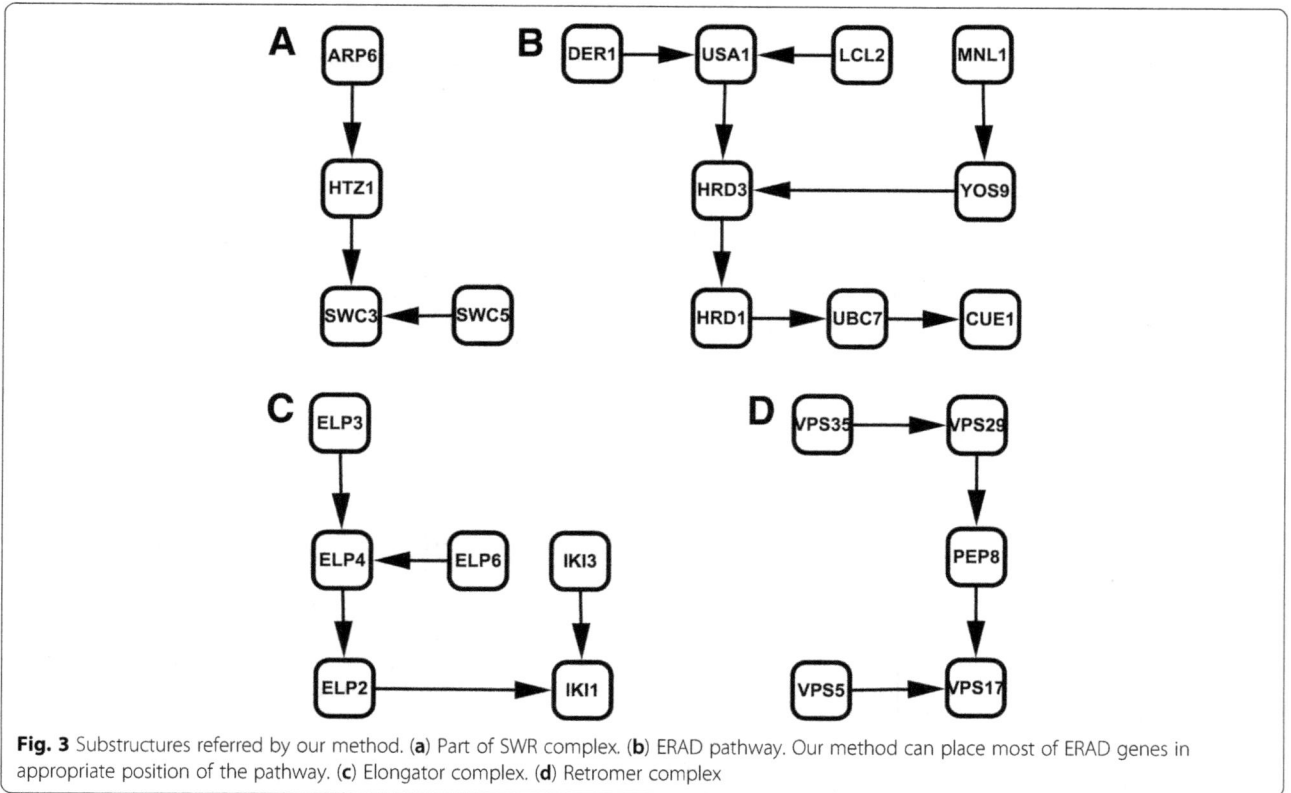

Fig. 3 Substructures referred by our method. (**a**) Part of SWR complex. (**b**) ERAD pathway. Our method can place most of ERAD genes in appropriate position of the pathway. (**c**) Elongator complex. (**d**) Retromer complex

Specify interesting pathways and compute their confidence

The MATLAB codes of our algorithm can be freely downloaded at [29].

Results and discussion

We applied our developed method to the genetic interaction measured by the protein folding in the endoplasmic reticulum [22] and the corresponding protein interaction network. The genetic interaction data set contains 444 queries crossed to the same 444 array strains from the budding yeast, *Saccharomyces cerevisiae*, and keeps available 86,396 raw double mutants from the 444 × 444 genetic interaction pairs [22]. Another genetic interaction data are from the coherent subsets of the global genetic interaction network [30], including 198 single mutants and 30,256 double mutants. We used regression method to predict the missing genetic interaction data from known genetic interaction profiles. The protein interaction data of the above gene set are downloaded from BioGRID till December 2016.

Due to the fact that the raw measurements of genetic interaction data are limited in publicly available databases, we applied our developed method to an available data set from Ref. [22]. Though there are some raw measurement data sets in Refs. [2, 30], either smaller number of samples or the higher sparsity makes it infeasible to apply our method to these data sets. That also explains why few available methods were designed to reconstruct pathways by integrating genetic interaction and protein interaction data. We compared our results with those predicted by the APN to validate the advantage of our method.

In our method, we modeled the pathway network as a Bayesian network. The sampling algorithm of annealed importance is applied to curate networks with the probability distribution defined by genetic interaction data, and simultaneously assign weights to them. And the corresponding protein interaction network of the genes in genetic interactions was used to represent underlying sample population, interpreting existence of potential edges in the sampled networks. Using these sampled networks and their assigned weights, we can estimate the detailed structure of the gene subset with high confidence (see Methods). Two substructures reconstructed by our method are shown in Fig. 3. Though the genetic

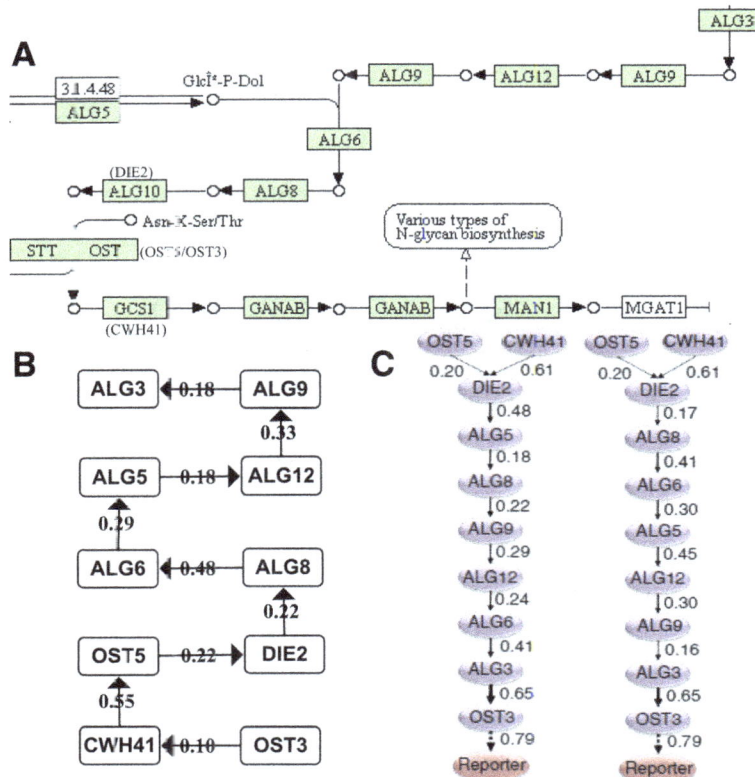

Fig. 4 Comparison of N-Glycan biosynthesis pathways. The mechanisms of genetic interaction dependency can make the reverse ordering in (**b**) and (**c**). (**a**) The true ordering of N-Glycan biosynthesis pathway substructure from KEGG. (**b**) The linear structure of the pathway discovered by our model with high confidence. It can be found that DIE2 is not connected with CWH41 directly, because there is not the protein interaction in current database. And almost of all genes are on the correct place. (**c**) The structure from APN

interaction data for SWR complex are not complete, our approach still pools the existing genes together (Fig. 3a). It precisely reconstructs the known functional dependencies of ERAD pathway (Fig. 3b).

We compared N-Glycan biosynthesis pathway substructure reconstructed by our model with the result of APN (Fig. 4). The detailed structures of the pathway from our model (Fig. 4b) and APN (Fig. 4c) [26] are very similar. Both of them are similar to the true one in Kyoto Encyclopedia of Genes and Genomes (KEGG, http://www.kegg.jp). One obvious difference is the place of OST3 that is incorrectly placed in APN (Fig. 4c). It may be due to the scoring function of APN based on edge score that strengthens the confidence of edge *(ALG3, OST3)*. The edge from ALG3 to OST3 has a high confidence, 0.65, indicating that APN really cannot interpret some edges in its result. Moreover, the orders of genes from our model and APN may be reverse to the true one [31] because the mechanisms of genetic interaction dependency are represented by phenotype (the unfolded protein response or fitness). Intriguingly, the OST3 position is correctly predicted in our method. It indicates the power of our developed method by integration of protein interaction data. However, we still found the limitation of the protein interaction data. The edge *(CWH41, DIE2)* is not presented in our result, because the corresponding protein interaction is not found in currently used protein interaction databases. In future, we are planning to include more predicted protein

interaction data from STRING, and design parallel computing in high-performance computers to improve the performance.

We also applied our model to infer pathways from another available data set of a global genetic interaction profiles [30]. From about 5.4 million gene pairs, we only selected coherent subsets in which the gene pairs have the high Pearson correlation coefficients, for our method based on annealed importance sampling is not suitable for so large data set. Using our model, we reconstructed three substructures, that is Urmylation pathway (Fig. 5c), Elongator complex (Fig. 3c), and Retromer complex (Fig. 3d). In Fig. 5, we compared our developed method with APN. The edge (NFS1, NCS2) presented in results of APN, as shown in Fig. 5b is difficult to interpret. However, our result in Fig. 5c is consistent with protein information from BioGrid as shown in Fig. 5a. The interactions of UBA4, NFS4, and NCS2 were predicted by our method. The edge (UBA4, AHP1) in Fig. 5d is not inferred by these two methods. For our model, the reason may be the incompleteness of protein interaction network that is the main weakness of our model.

Conclusions

In this paper, we propose a Bayesian network model to identify pathway structures by integrating protein interaction with genetic interaction data. Our approach makes use of the complementarity between protein (physical) and genetic (functional) interaction data to

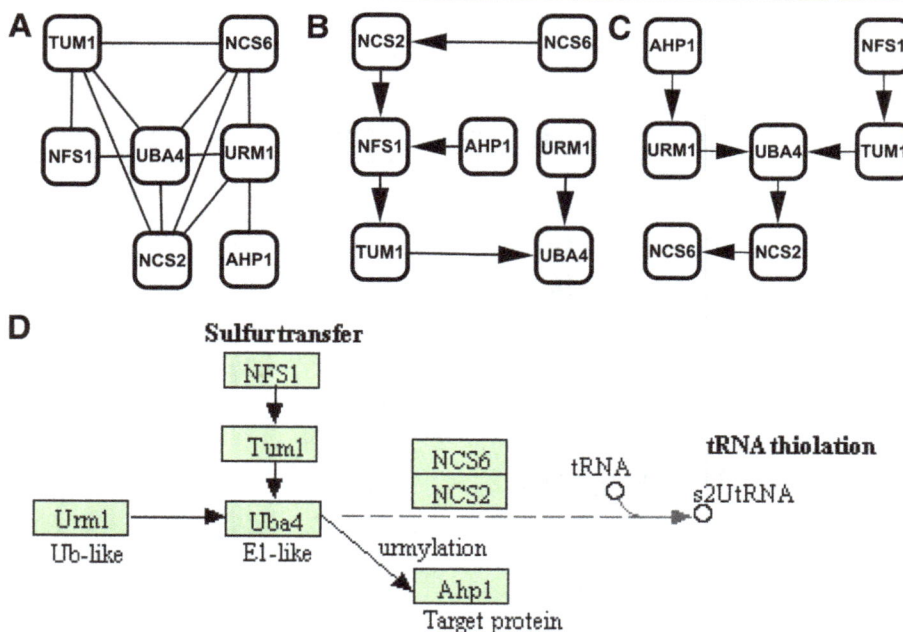

Fig. 5 Comparison of Urmylation pathways. (**a**) The protein interaction network of the pathway from BioGRID. (**b**) The substructure of the pathway from APN. (**c**) The substructure of the pathway reconstructed by our model with high confidence. (**d**) The true ordering of the pathway substructure from KEGG

refer the biological pathway structures. We define a scoring function by which the sampling algorithm of annealed importance can draw some pathway networks and their weights that are used to evaluate the candidate pathway structures. The results show that our model can predict the pathway structures more accurately.

Abbreviations
APN: Activity pathway networks; BioGRID: Biological general repository for interaction datasets; ERAD: ER-Associated degradation; HPRD: Human protein reference database; KEGG: Kyoto encyclopedia of genes and genomes.; MCMC: Markov chain Monte Carlo; SGD: Saccharomyces genome database; STRING: Search tool for the retrieval of interacting genes/proteins

Acknowledgements
Not applicable.

Funding
Support for the authors was provided by the National Natural Science Foundation of China (#11371016), the Chinese Program for Changjiang Scholars and Innovative Research Team in University (PCSIRT) (#IRT_15R58). The publication costs were funded by the National Natural Science Foundation of China (#11371016).

About this supplement
This article has been published as part of BMC Systems Biology Volume 11 Supplement 4, 2017: Selected papers from the 10th International Conference on Systems Biology (ISB 2016). The full contents of the supplement are available online at <https://bmcsystbiol.biomedcentral.com/articles/supplements/volume-11-supplement-4>.

Authors' contributions
CHF came up with the idea of the study built the model of the study, designed the algorithm, and wrote the manuscript; SD debugged the program, and performed functional and statistical analyses; GXJ assisted in functional, statistical and data analyses, and revised the manuscript; XXW gathered the data, and performed data analyses; ZGY supervised the model building, and statistical computational approaches, and revised the manuscript. All authors read and approved the final version of the manuscript.

Competing interests
The authors declare that they have no competing interests.

Author details
[1]School of Mathematics and Computational Science Xiangtan University, Xiangtan 411105, China. [2]School of Mathematics and System Science, Shenyang Normal University, Shenyang 110034, China. [3]Center of Systems Biology and Bioinformatics, Wake Forest School of Medicine, Winston-Salem, NC 27157, USA.

References
1. De Las RJ, Fontanillo C. Protein-protein interactions essentials: key concepts to building and analyzing interactome networks. PLoS Comput Biol. 2010;6: e1000807.
2. Mani R, St Onge RP, Hartman JL, Giaever G Roth FP. Defining genetic interaction. Proc Natl Acad Sci U S A. 2008;105:3461–6.
3. Beltrao P, Cagney G, Krogan NJ. Quantitative genetic interactions reveal biological modularity. Cell. 2010;141:739–45.
4. Wang Y, Zhang XS, Chen L. Modelling biological systems from molecules to dynamical networks. BMC Syst Biol. 2012;6(Suppl 1):S1.
5. Gitter A, Klein-Seetharaman J, Gupta A, Bar-Joseph Z. Discovering pathways by orienting edges in protein interaction networks Nucleic Acids Res. 2011; 39:e22.
6. Bebek G, Yang J. PathFinder: mining signal transduction pathway segments from protein-protein interaction networks. BMC Bioinformatics. 2007;8:335.
7. Scott J, Ideker T, Karp RM, Sharan R. Efficient algorithms for detecting signaling pathways in protein interaction networks. J Comput Biol. 2006;13: 133–44.
8. Shlomi T, Segal D, Ruppin E, Sharan R. QPath: a method for querying pathways in a protein-protein interaction network. BMC Bioinformatics. 2006;7:199.
9. Segal E, Shapira M, Regev A, Pe'er D, Botstein D, Koller D, Friedman N. Module networks: identifying regulatory modules and their condition-specific regulators from gene expression data. Nat Genet. 2003;34:166–76.
10. Margolin AA, Nemenman I, Basso K, Wiggins C, Stolovitzky G, Dalla Favera R, Califano A. ARACNE: an algorithm for the reconstruction of gene regulatory networks in a mammalian cellular context. BMC Bioinformatics. 2006;7 Suppl 1:S7.
11. Grzegorczyk M, Husmeier D. Improvements in the reconstruction of time-varying gene regulatory networks: dynamic programming and regularization by information sharing among genes. Bioinformatics. 2011;27: 693–9.
12. Ravcheev DA, Best AA, Sernova NV, Kazanov MD, Novichkov PS, Rodionov DA. Genomic reconstruction of transcriptional regulatory networks in lactic acid bacteria. BMC Genomics. 2013;14:94.
13. Barba M, Dutoit R, Legrain C, Labedan B. Identifying reaction modules in metabolic pathways: bioinformatic deduction and experimental validation of a new putative route in purine catabolism. BMC Syst Biol. 2013;7:99.
14. Guillen-Gosalbez G, Sorribas A. Identifying quantitative operation principles in metabolic pathways: a systematic method for searching feasible enzyme activity patterns leading to cellular adaptive responses. BMC Bioinformatics. 2009;10:386.
15. Shirshin E, Cherkasova O, Tikhonova T, Berlovskaya E, Priezzhev A, Fadeev V. Native fluorescence spectroscopy of blood plasma of rats with experimental diabetes: identifying fingerprints of glucose-related metabolic pathways. J Biomed Opt. 2015;20:051033.
16. Wang Y, Wu QF, Chen C, Wu LY, Yan XZ, Yu SG, Zhang XS, Liang FR. Revealing metabolite biomarkers for acupuncture treatment by linear programming based feature selection. BMC Syst Biol. 2012;6(Suppl 1):S15.
17. Liu Y, Zhao H. A computational approach for ordering signal transduction pathway components from genomics and proteomics data. BMC Bioinformatics. 2004;5:158.
18. Zhao XM, Wang RS, Chen L, Aihara K. Uncovering signal transduction networks from high-throughput data by integer linear programming. Nucleic Acids Res. 2008;36:e48.
19. Steffen M, Petti A, Aach J, D'Haeseleer P Church G. Automated modelling of signal transduction networks. BMC Bioinformatics. 2002;3:34.
20. Tong AH, Lesage G, Bader GD, Ding H, Xu H, Xin X, Young J, Berriz GF, Brost RL, Chang M, et al. Global mapping of the yeast genetic interaction network. Science. 2004;303:808–13.
21. Schuldiner M, Collins SR, Thompson NJ, Denic V, Bhamidipati A, Punna T, Ihmels J, Andrews B, Boone C, Greenblatt JF, et al. Exploration of the function and organization of the yeast early secretory pathway through an epistatic miniarray profile. Cell. 2005;123:507–19.
22. Jonikas MC, Collins SR, Denic V, Oh E, Quan EM, Schmid V, Weibezahn J, Schwappach B, Walter P, Weissman JS, et al. Comprehensive characterization of genes required for protein folding in the endoplasmic reticulum. Science. 2009;323:1693–7.
23. Segre D, Deluna A, Church GM, Kishony R. Modular epistasis in yeast metabolism. Nat Genet. 2005;37:77–83.
24. Kelley R, Ideker T. Systematic interpretation of genetic interactions using protein networks. Nat Biotechnol. 2005;23:561–6.
25. Qi Y, Suhail Y, Lin YY, Boeke JD, Bader JS. Finding friends and enemies in an enemies-only network: a graph diffusion kernel for predicting novel genetic interactions and co-complex membership from yeast genetic interactions. Genome Res. 2008;18:1991–2004.
26. Battle A, Jonikas MC, Walter P, Weissman JS, Koller D. Automated identification of pathways from quantitative genetic interaction data. Mol Syst Biol. 2010;6:379.
27. Pearl J. Probabilistic reasoning in intelligent systems: networks of plausible inference. Artif Intell. 1991;48:117–24.
28. Neal R. Annealed importance sampling. Stat Comput. 1998;11:125–39.
29. MATLAB codes [http://www.fupage.org/downloads/bmipi.zip] May 15th 2016.

8

Single-cell study links metabolism with nutrient signaling and reveals sources of variability

Niek Welkenhuysen[1], Johannes Borgqvist[2], Mattias Backman[1], Loubna Bendrioua[1], Mattias Goksör[3], Caroline B Adiels[3], Marija Cvijovic[2*] and Stefan Hohmann[1,4*]

Abstract

Background: The yeast AMPK/SNF1 pathway is best known for its role in glucose de/repression. When glucose becomes limited, the Snf1 kinase is activated and phosphorylates the transcriptional repressor Mig1, which is then exported from the nucleus. The exact mechanism how the Snf1-Mig1 pathway is regulated is not entirely elucidated.

Results: Glucose uptake through the low affinity transporter Hxt1 results in nuclear accumulation of Mig1 in response to all glucose concentrations upshift, however with increasing glucose concentration the nuclear localization of Mig1 is more intense. Strains expressing Hxt7 display a constant response to all glucose concentration upshifts. We show that differences in amount of hexose transporter molecules in the cell could cause cell-to-cell variability in the Mig1-Snf1 system. We further apply mathematical modelling to our data, both general deterministic and a nonlinear mixed effect model. Our model suggests a presently unrecognized regulatory step of the Snf1-Mig1 pathway at the level of Mig1 dephosphorylation. Model predictions point to parameters involved in the transport of Mig1 in and out of the nucleus as a major source of cell to cell variability.

Conclusions: With this modelling approach we have been able to suggest steps that contribute to the cell-to-cell variability. Our data indicate a close link between the glucose uptake rate, which determines the glycolytic rate, and the activity of the Snf1/Mig1 system. This study hence establishes a close relation between metabolism and signalling.

Keywords: Microfluidics systems, Glucose uptake, Non-linear mixed effect modelling, Dynamical modelling

Background

Cells have developed an extensive network of signalling pathways in order to mediate appropriate responses to varying nutrient availability and concentrations in the surrounding environment. Glucose, a rapidly fermentable carbon source, is the preferred carbon source of *Saccharomyces cerevisiae* cells and regulates numerous nutrient signalling pathways [1, 2]. Glucose is taken up by the yeast cell through multiple hexose transporters, which have a broad range of different affinities and transport capacities. This enables the yeast cell to respond to a wide range of glucose concentrations [3, 4]. Glucose sensing pathways

that employ membrane-localized receptors, such as in the Snf3-Rgt2 pathway, are relatively well understood. However, the sensing mechanism of intracellular glucose or metabolites of glycolysis is poorly explained [5, 6]. To study how metabolism is connected to these signalling pathways has proven to be a major challenge since it is difficult to uncouple signalling from metabolism.

The AMPK/SNF1 system controls energy homeostasis and is best known for its function in glucose signalling. The SNF1 protein kinase complex, which consists of three subunits, is activated by glucose depletion through phosphorylation [7, 8]. It is not well known how metabolism is connected with the activity of the SNF1 complex. When rapidly-fermentable sugars are available Snf1 becomes dephosphorylated. For the establishment of glucose repression only uptake and phosphorylation of glucose is required, but no further glucose metabolism

* Correspondence: marija.cvijovic@chalmers.se; stefan.hohmann@chalmers.se
[2]Department of Mathematical Sciences, Chalmers University of Technology and the University of Gothenburg, SE-412 96 Gothenburg, Sweden
[1]Department of Chemistry and Molecular Biology, University of Gothenburg, SE-412 96 Gothenburg, Sweden
Full list of author information is available at the end of the article

[9]. Glucose repression is regulated, at least, at two different steps, i.e. control of Snf1 activation and its function on downstream targets such as Mig1 [10]. It has been suggested that the Snf1-Mig1 pathway works in a continuous on-off manner [11]. However, evidence has emerged that Mig1 shuttles in and out of the nucleus and shows transient behaviour at a single cell level [12, 13]. This indicates that the dynamics of the Snf1-Mig1 at single cell level is less simple than previously assumed.

To study the influence of glucose metabolism on the complex single cell dynamics of the Snf1 pathway we decided to control the uptake of glucose into the cell while employing a microfluidic system to control the extracellular glucose concentration. We show that glucose repression is regulated by glucose flux rather than the absolute glucose concentrations and that the Snf1-Mig1 system is closely regulated by glycolytic flux. In our experiments we observed cell-to-cell variability. To explain this variability, we developed a dynamical and nonlinear mixed effect model. Dynamical models of signalling pathways in yeast have previously been employed to describe the behaviour of populations of cells [14–19]. Nonlinear mixed effects (NLME) modelling is a theoretical approach that provides a framework to account for cell-to-cell variability [20]. NMLE modelling is traditionally used in pharmacokinetic and pharmacodynamics studies since it allows for the analysis of sparse and unbalanced datasets [21, 22]. NMLE has been proposed and used to model dynamic single cell data [20, 23]. Recently, a simple phenomenological model describing the Snf1-Mig1 pathway using NLME approach has been constructed, capturing dynamics of Mig1 localization, without taking into account parameter variabilities [24]. Our integrative approach reveals that the main source of variability is linked to transport of Mig1 in and out of the nucleus. Our experimental data indicate that rapid degradation and cell size cause no or little contribution to the cell-to-cell variability, while variation in expression and translation of the hexose transporters is a possible source of cell-to-cell variability.

Results

Single cell time-scale fluorescence microscopy enables dynamic studies of the Snf1-Mig1 pathway

A high control of the cell environment is needed in order to study nutrient responses on single cells. A microfluidics device allows for a fast and precise switch between different media and enables the nutrients composition in the media to be kept constant. Here we used a three inlet-channel microfluidics setup [25], to achieve a high control of the cell environment and to study the influence of glucose concentration on *Saccharomyces cerevisiae*. Upon deactivation of Snf1, Mig1 is dephosphorylated and subsequently moves into the nucleus and

is therefore a suitable marker for real-time Snf1 activity [26, 27]. The nucleo-cytosolic shuttling of the transcription factor Mig1 fused to a Green Fluorescent Protein (GFP) served as single cell readout (Additional file 1: Figure S1).

In the wild type (WT) Mig1 localized to the nucleus after being exposed to a glucose concentration of 2.75 mM (Fig. 1a and Additional file 2: Figure S2a). It has been shown that Mig1 is phosphorylated in this range of glucose concentration [28]. Following shifts to higher glucose concentrations more cells respond to the upshift and the Mig1 fluorescence intensity in the nucleus is stronger indicating that higher glucose concentrations result in a higher proportion of Mig1 molecules in the nucleus (Fig. 1a). The Snf1-Mig1 system responds to exposure to glucose and the degree of response is sensitive for the absolute glucose level which the cell is exposed to. These results are consistent with those of a similar study in *Saccharomyces cerevisiae* with a different genetic background [12]. The response to increased glucose concentrations occurs rapidly after upshift, pointing to the fast adaption of cells to nutrients in the environment.

Glucose uptake through only low affinity transporters results in a strong response in the Snf1-Mig1 pathway

The data obtained from the wild-type strain raised the question whether glycolytic flux correlates directly with Snf1-Mig1 pathway activity. This would imply that the Snf1-Mig1 pathway is controlled by glucose metabolism by a quantitative sensor system. To address this question we chose to control the flux through glycolysis via the glucose uptake into the cell. A large set of isogenic strains expressing only a single hexose transporter is available [29]. We employed strains that express either a low affinity or high affinity glucose transporter, respectively, under the control of the promotor of a high affinity transporter.

Hxt1 is a low affinity transporter that is normally highly expressed under high glucose conditions; yeast cells expressing only *HXT1* displays a high glucose transport capacity and a higher V_{max} then the wild type strain [28, 29]. The HXT1 cells already display Mig1 nuclear accumulation when they were exposed to an upshift from 0 to 2.75 mM glucose as in the WT however the fraction of the whole population displaying nuclear localization never exceeds 50%. while for an upshift to 11 mM glucose a higher fraction of cells display nuclear accumulation but this nuclear accumulation never exceeds 80% (Additional file 3: Figure S3a). While in the WT for both the upshift to 2.75 mM and 11 mM more than 80% of the population displays nuclear accumulation. For the other upshift to higher glucose concentration nearly all cells display nuclear

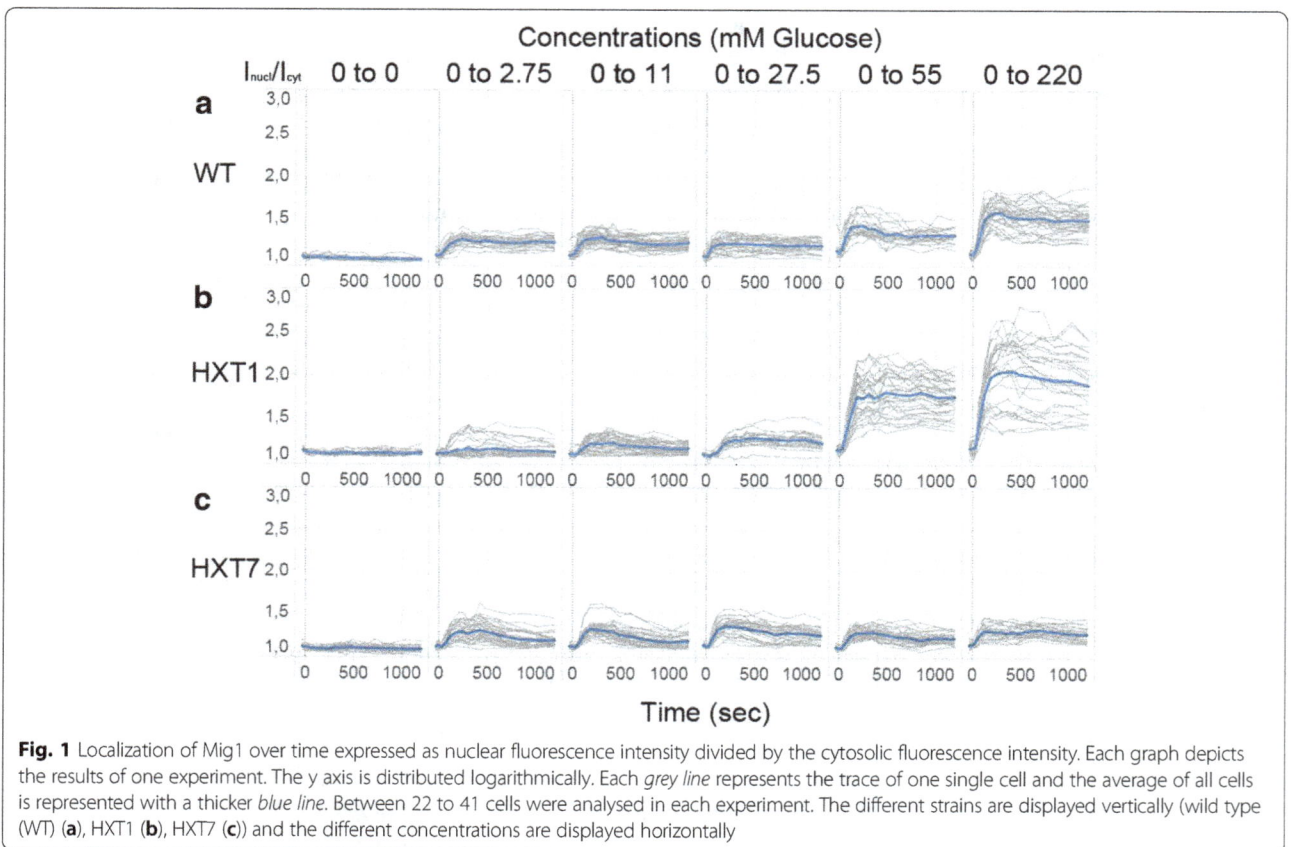

Fig. 1 Localization of Mig1 over time expressed as nuclear fluorescence intensity divided by the cytosolic fluorescence intensity. Each graph depicts the results of one experiment. The y axis is distributed logarithmically. Each *grey line* represents the trace of one single cell and the average of all cells is represented with a thicker *blue line*. Between 22 to 41 cells were analysed in each experiment. The different strains are displayed vertically (wild type (WT) (**a**), HXT1 (**b**), HXT7 (**c**)) and the different concentrations are displayed horizontally

localization. The HXT1 strain displayed a large cell-to-cell variability of cells either responding or not responding to an upshifts of 2.75 mM, 11 mM, and 27.5 mM in glucose concentration in comparison with the WT cells (Fig. 1b). Only after an upshift to 27.5 mM glucose almost all the cells of the population show Mig1 localization in the nucleus. At higher glucose concentrations, 55 mM or 220 mM glucose, a higher proportion of Mig1 is localized in the nucleus than at lower concentrations. The upshifts to 55 mM and 220 mM result in a higher Mig1 nuclear accumulation compared to the WT and displays a higher cell variability compared to the other strains (Fig. 1b and Additional file 3: Figure S3b). The response times of Mig1 nuclear accumulation appear to be remarkably similar for all responding cells under all glucose concentrations, after 1 min the max is reached for the fraction of cells displaying Mig1 nuclear localization of all strains and all upshift conditions (Additional file 3: Figure S3a). Overall, the data shows that the response characteristics of the Snf1-Mig1 system correlates well with the kinetic characteristics of the Hxt1 transporter as the Snf1-Mig1 strain displays low response at an upshift to low glucose concentration but a strong response to upshift to high glucose concentrations.

The high affinity transporter causes a weak Mig1 response to all glucose concentration upshifts
The Hxt7 high affinity transporter is highly expressed at very low glucose concentrations. The HXT7 strain displays a lower glucose uptake capacity than the HXT1 strain. Therefore the glucose uptake capacity is saturated at low glucose concentrations [28]. The majority of the population shows Mig1 nuclear localization after the cells are exposed to growth media containing glucose, in contrast with the low fraction of responders in the HXT1 strain (Fig. 1c and Additional file 2: Figure S2c). However, unlike in the WT and the HXT1 strain the response is very similar for all glucose concentrations, and the intensity of Mig1 in the nucleus is the same for all upshifts. Already at an upshift from 0 to 2.75 mM glucose Mig1 has reached a maximum Mig1 nuclear localization for the HXT7 strain. Hence, even in the HXT7 strain the Snf1-Mig1 response characteristic corresponds to the properties of the transport system with high affinity but low capacity.

Neither regulated degradation nor cell size is a major contributor to cell heterogeneity
Cell-to-cell variability in Mig1 localization upon changing glucose concentration has been reported [12, 13, 30].

These studies however do not examine the source(s) of the observed cell-to-cell variability, therefore we set out to explore the source(s) of the variation. It is known that the high affinity transporters Hxt7 and Hxt6 are internalized and degraded when cells are exposed to high concentration of glucose [31]. Degradation of Hxt7 requires inactivation of TORC1 [32, 33]. Also Hxt1 is actively internalized and degraded if glucose is depleted, an effect possibly mediated by downregulation of PKA [34]. Internalization of the hexose transporters for catabolic degradation could lead to a decrease of glucose import. We observed in the upshift experiments to higher glucose concentrations a slight decrease of the median after the nuclear localization reached its maximum value for all strains (Additional file 2: Figure S2). Therefore, we reasoned that rapid activity adjustments of the hexose transporters could impact the Mig1-Snf1 pathway. Since this drop in Mig1 nuclear localization differed between cells this mechanism could be a contributing factor to the observed cell-to-cell variability. We exposed yeast cells expressing Hxt7-GFP under the native promotor grown on 3% ethanol to 220 mM glucose and followed the localization of Hxt7-GFP for 15 min (Fig. 2a, Additional file 4: Figure S4). The data was quantified by measuring the fluorescence intensity along an intersection through the cell. We observed no significant change in the localization of Hxt7-GFP during the experiment (Fig. 2b).

We next asked whether cell size could influence the cell-to-cell variability. Fluctuations in cellular states, such as cell size, can cause extrinsic noise which could lead to the observed cell-to-cell variability [35]. We therefore decided to test the influence of cell size by plotting the response of the Snf1-Mig1 pathway over the cell size. As measurement for the Snf1-Mig1 pathway we used the Mig1 fluorescence intensity ratio 15 min after the upshift. The final ratio for the HXT1-strain did not show any correlation between the cell size and the Snf1-Mig1 pathway activity (Fig. 2c). Instead, the final ratio showed an even distribution around the average cell size with the values for the upshifts towards higher glucose concentration position higher along the y-axes. This

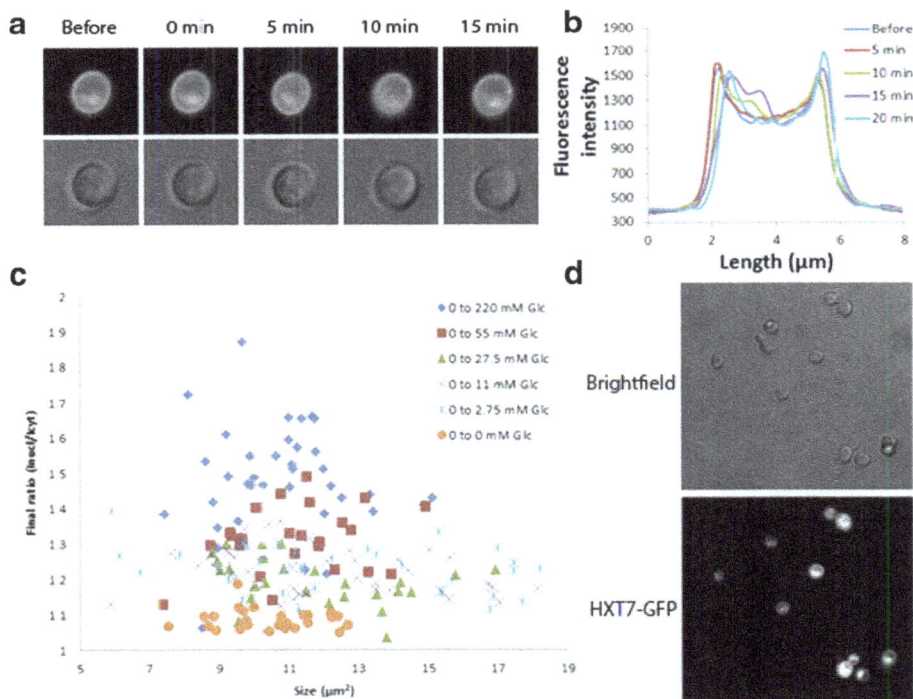

Fig. 2 Study of the cell-to-cell variability observed in the Snf1/Mig1 system. (**a**)(**b**) Hxt7-GFP before and following a switch from ethanol media to media containing 220 mM glucose. (**a**) Time lapse microscopic images, *upper* images show HXT7-GFP, the *lower* images show brightfield. (**b**) Fluorescence intensity along an intersection through the yeast cells. The fluorescence intensity is higher at the points the *intersection line* crosses the cell membrane and does not change over time. The result of only one cell is displayed but multiple cells were analyzed and none of the cells showed a decrease in membrane localization of the Hxt7 transporter after 15 min following the shift to glucose media. (**c**) The ratio 15 min after glucose upshift plotted over the cell size for the HXT1 strain. The cell size is plotted on the x-axes. As a measurement for the Mig1-Snf1 pathway response we chose the Mig1 nuclear/cytosolic ratio. Upshifts to higher glucose concentration; 0 to 220 mM (*blue diamonds*), 0 to 55 mM (*red squares*) and 0 to 27.5 mM glucose (*green triangle*) result in higher final ratio while upshifts to lower glucose concentration 0 to 11 mM glucose (*purple crosses*), 0 to 2.75 mM (*blue stars*) and 0 to 0 mM glucose (*orange dots*) display a lower final ratio. (**d**) Hxt7-GFP pregrown overnight in 3% ethanol media. *Upper* image shows the bright field image, *lower* image shows the cellular distribution of Hxt7-GFP

result excludes cell size as a major determinant for the cell-to-cell variability on the relatively short time frame of the experiments.

Stochastic expression is a plausible cause of cell-to-cell variability in the Mig1-Snf1 system

Cell-to-cell variability in dynamic adaptation responses might be caused by, among others, stochastic transcription activity [36]. Therefore, we reasoned that the expression pattern of the hexose transporters could lead to the cell-to-cell variability. We therefore grew a strain expressing Hxt7-GFP on 3% ethanol and followed the population distribution of Hxt7-GFP. The fluorescence intensity of Hxt7-GFP differed significantly between the cells (Fig. 2d). The lowest observed Hxt7-GFP fluorescence intensity was only 10% of the maximal observed fluorescence intensity of Hxt7-GFP. The amount of Hxt7 transporter molecules within each single cell varies and can therefore be a major contributor to the observed cell-to-cell variability. These results show that, under our experimental conditions it is likely that expression and translation of hexose transporters is a major contributor to the observed cell-to-cell variability.

A mixed effect model suggests Mig1 dephosphorylation as a new regulatory step

To better understand the effect of glucose upshift on the Snf1-Mig1 pathway we developed a mathematical model of glucose flux which controls Snf1 phosphorylation and consequently Mig1 localization (Fig. 3a). The aim was to investigate if glucose uptake was able to regulate Mig1 localization by controlling only one step in the Snf1-

Mig1 regulatory system. We assumed this step to be dephosphorylation of Snf1, since several publications identified this step to be controlled by the ADP/ATP ratio [37–40]. The ADP/ATP ratio is indirectly determined by glucose uptake and glycolysis, therefore the binding of ADP to the SNF1 complex could be the connection between glycolysis and the Snf1-Mig1 system. NLME modelling was implemented in order to simulate the dynamics of Mig1 localization for different yeast strains in various experimental conditions (Additional file 5: Figure S5). The model captures the characteristics of our experimental data (Fig. 1). By simulating parameters for multiple cells we could produce a distribution of the parameters and we compared this distribution between the Wild type, HXT1 and HXT7 strains (Additional file 6: Figure S6). The model predicts that the Vmsi, the parameter for Snf1 dephosphorylation, increases with upshifts to increasing glucose concentrations (Additional file 6: Figure S6a). However, there is no significant difference between the different strains. The model suggests Snf1 dephosphorylation to be active immediately after glucose is imported into the cell, but this process is influenced neither by glucose concentration nor by the strain it was simulated in. This suggests that Snf1 dephosphorylation is regulated more in an on/off fashion rather than in a dynamic fashion. However, the parameter for Mig1 dephosphorylation, Vmd, did display the characteristics of the different strain (Additional file 6: Figure S6b). At low glucose upshift the simulated Vmd parameters of the HXT7 strain were higher than the HXT1 parameters. Only at the higher upshift concentrations the Vmd parameters simulated for HXT1 strain

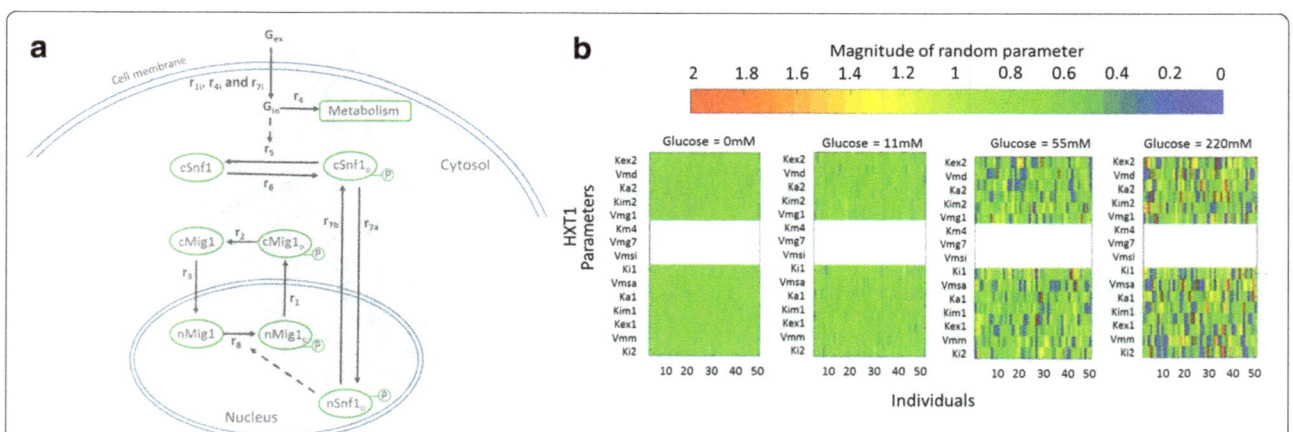

Fig. 3 Dynamic and NLME modelling of the Snf1/Mig1 pathway. **a** Schematic representation of the model. The model consists of three main parts, namely the activity of glucose, the activity of Snf1 and the activity of Mig1. **b** Simulation of the distribution of the random parameters for the HXT1 strain. The columns indicate the extracellular glucose concentration, ranging from 0, 11, 55 to 220 mM which are illustrated from the *left* to the *right* in the figure. Each heat map is generated by drawing 50 mixed effect random terms, that is $\eta \sim N(0, \sigma)$, corresponding to the parameter vectors from the generated parameter distributions for the various strains and glucose concentrations. The heat map displays various parameters on the y-axis, the individuals on the x-axis and the magnitude of the random terms are indicated by the colour scale shown above the figure. The colour scale ranges from 0 to 2 where a *red* colour corresponds to a high random term and a *blue* colour correspond to a low value of the random term. The white fields correspond to the parameters connected to the hexose transporters that are not active in HXT1 strains

where much higher than the Vmd parameters simulated for HXT7 strain. From this, the model suggested Mig1 dephosphorylation as a regulatory step which is controlled by glucose flux.

A mixed effect model identifies hypothetical sources of variability in the Snf1-Mig1 regulatory module

Since our model takes cell-to-cell variability into account we could use it to identify which parameters display highest cell-to-cell variability and under which conditions. The wild type strain displayed increasing variance with upshift to the higher glucose concentrations 55 mM and 220 mM (Additional file 7: Figure S7). The highest variability was observed in the HXT1 strain with the upshift to the higher glucose concentrations (Fig. 3 and Additional file 7: Figure S7). The HXT1 transporter strain displayed a large cell-to-cell variability following shift to high glucose but a small cell-to-cell variability after upshift to low glucose. The simulated variance of the HXT7 strain was lower in comparison with the wild type and HXT1 strains and did not increase with upshift to higher glucose levels (Additional file 7: Figure S7). The simulated variance is in correlation with the observed variance seen in the experimental data (Additional file 3: Figure S3b). The model also allows us to predict the most important parameters that are the major contributor to the cell-to-cell variability. We compared the magnitude of perturbation of each parameter for the simulation of the wild type strain in the upshift from 0 to 220 mM glucose (Additional file 8: Table S1). A parameter which displays a higher perturbation error has a higher variability in that parameter. Parameters involved in dephosphorylation events have been ranked as low significance (Additional file 8: Table S1), suggesting that the dephosphorylation of Mig1 and Snf1 after glucose upshift is a minor contributor to the observed cell-to-cell variability. The parameters which display the highest variance are Kex2 and Kim2, which account for transport of Mig1 in and out of the nucleus. The respective perturbation errors for these two parameters are in the order of 10^{-3} while the other parameters have perturbation errors in the order of 10^{-7} and smaller. Indeed, it has been shown that movement of Mig1 in and out of the nucleus shows considerable variability between cells [12]. Those data suggest that variability in the nucleocytoplasmic transport of Mig1 would be the major contributor to cell-to-cell variability and not the dephosphorylation events following glucose upshifts.

Discussion

It is well established that the Snf1-Mig1 system and hence the nuclear accumulation of Mig1 are controlled by the level of glucose in the growth medium. We have previously reported that glucose derepression senses glucose concentration in a highly dynamic fashion [12]. But

it remained unclear whether the observed dynamics were correlated with adaptations of sugar transport and glycolytic metabolism. Previous modelling approaches have suggested the importance of the kinetics of the glycolytic flux in signalling of glucose [41]. To elucidate the influence of sugar transport on the Snf1-Mig1 pathway we studied the response of three different *S. cerevisiae* strains with different glucose uptake capacity when exposed to an upshift in glucose concentration. Our data show that yeast cells response rapidly to changes in glucose concentration and that there is little to no cell-to-cell variability in response time. Even between the strains and between glucose concentrations we observed no significant difference in response time. This indicates that yeast cells are programmed and determined to rapidly respond to a change in glucose concentration. This behaviour allows yeast cells to rapidly adapt to new environmental conditions and thereby to potentially outcompete other species. Although single yeast cells induced a response at almost the same time, there was difference between the cells in the magnitude of their response. Our data showed that the Mig1 localization pattern after glucose upshift correlates well with the glucose uptake characteristics of the respective yeast strains. It has already been shown that establishment of glucose repression is driven by sensing of an intracellular metabolite rather than extracellular glucose [42, 43]. Our single cell data confirms that glucose repression is sensed through an intracellular metabolite rather than extracellular glucose. It has been shown that in the strains we tested the reduced glucose uptake capacity results in reduced glycolytic rate [41]. The data indicate a closer link between the glucose uptake rate, which determines the glycolytic rate, and the activity of the Snf1/Mig1 system than previously anticipated. This indicates that the signal which originates from the glycolytic flux is very dynamic in response to the changing glycolytic rate. A source for this signal might be the ADP/ATP ratio [37, 38, 40]. Since, it has been reported that ADP binds to the regulatory subunit Snf4 and this binding leads to protection of the catalytic subunit Snf1 from dephosphorylation, which leads to increased Snf1 activity [38]. The turn-over of ADP to ATP or vice versa could be a sensor for the glycolytic rate. Also Hexokinase 2, an enzyme part of the glycolytic pathway, has been suggested to serve as a sensor for internal glucose by serving as a threshold for the interaction between Mig1 and Snf1 [44, 45].

Our data showed a considerable cell-to-cell variability in glucose sensing. This variability could have considerable impact on the Snf1/Mig1 pathway. We investigated the causes of the cell-to-cell variability that was observed in our initial experiments. Our results indicate that this behaviour is not caused by the size of the cell or rapid activity adjustments of the hexose transporters. A large variation is observed between the concentrations of Hxt7 in the cells and hence the high cell-to-cell variability can be caused by the expression and/or translation of hexose transporter Hxt7. In a yeast single cell study of the shift between sulphur sources it was observed that the transcriptional adaptation displayed a large cell-to-cell variability [36]. The variability in expression and translation of hexose transporters may cause a different uptake capacity within a population and consequently variability in further glucose metabolism. Such variability in glycolysis might lead to cells responding differently to nutritional changes and different subpopulations. Systems to restore unbalanced dynamics in glycolysis have already been reported [46].

Data obtained by single cell techniques coupled with mathematical modelling offer an opportunity to understand the variability within a population of cells. This work employs NLME Modelling in larger dynamical models providing a framework to deeper investigate and identify source of cell-to-cell variability in Snf1-Mig1 signalling pathway.

The variability in our model corresponded to that observed in our experimental data. Therefore, we can conclude that the implemented model can in fact account for the cell-to-cell variability of the nuclear/cytoplasmic Mig1 ratio. Our approach classifies certain parameters as low significant, having very small perturbation errors indicating that they could potentially be neglected without losing the predictive capability of the model (Additional file 8: Table S1). This partly explains that the simplified model proposed by Almqvist and colleagues [24] can capture the dynamics of the Mig1 localization, but fails to provide more information about the relationship between the parameters. To our knowledge, this is the first Mixed Effect Model which is complex enough to allow identification of sources of cell-to-cell variability in signalling pathways. Our model suggested a new regulatory step at the level of Mig1 dephosphorylation and this step would be controlled by glycolytic flux. It is known that Mig1 is dephosphorylated upon glucose addition leading to glucose derepression [28]. However, the glucose activated process which dephosphorylates Mig1 has not yet been clearly identified. For instance, it is unclear if Mig1 dephosphorylation is regulated by glucose or by a constitutive phosphatase counteracting Snf1 activity. Our computational approach suggests that Mig1 dephosphorylation is regulated and therefore probably an active step. The model suggest that dephosphorylation of Snf1 is regulated by absolute glucose concentrations while Mig1 dephosphorylation is regulated in a more dynamic way related to glucose flux. We were able to identify import and export of Mig1 in and out of the nucleus as a

possible source for cell-to-cell variability. Both Mig1 import and export are regulated by glucose through phosphorylation and dephosphorylation of Mig1 [27]. Mig1 shuttles in and out of the nucleus regardless of the glucose concentration and FLIP/FRAP experiments have shown that there is a considerable cell-to-cell variability in Mig1 nuclear translocation [12]. Therefore, it is not entirely unexpected that this step is predicted to encompass a high cell-to-cell variability.

Conclusions

This work links the glucose flux to Snf1-Mig1 signalling. Although the control of the Snf1-Mig1 regulatory module is complicated by crosstalk with other glucose sensing signalling systems, we suggest that glycolytic metabolic reactions are playing a major role in the regulation of Mig1 localization. We show that the initial response time of Snf1-Mig1 pathway displays no cell-to-cell variability. We further developed and presented a modelling approach which can model the cell variability observed in the data. Most importantly, we demonstrate the close correlation between glycolytic metabolism and glucose signalling metabolism.

Methods

Yeast strains

The strains employed were transformed with GFP-KanMx and mCherry hphNT1 using standard methods for yeast genetics and transformation: Yeast strains were grown to mid-log phase at 30 °C in YNB synthetic complete medium containing 1.7 g/l yeast nitrogen base, 5 g/l ammonium sulphate, 670 mg/l complete supplement mix and supplied with 3% ethanol. For the glucose upshift serial dilutions of a stock of 220 mM glucose YNB complete medium were made in order to ensure final glucose concentration ranging between 220 mM and 2.75 mM. A complete overview of the used strain can be found in Additional file 9: Table S2.

Microfluidics

We employed Nrd1-RFP, a nuclear RNA-binding protein, as a nuclear marker. The ratio of Mig1-GFP within the nucleus to the Mig1-GFP throughout cells was used a quantifiable measure of Mig1 localization [47]. Strains were first pregrown in 3% ethanol, loaded on the microfluidic device and exposed to an upshift in glucose concentration, from 0 mM to 0 mM, 2.75 mM, 11 mM, 27.5 mM, 55 mM or 220 mM glucose which corresponds to respectively 0 to 0%, 0.05, 0.2, 0.5, 1 or 4%. To control the spatial and temporal changes of extracellular glucose concentration in the environment of the yeast cell we applied a three-channel microfluidic system that merge into one single channel [25]. We attached single cells to the surface of the single channel. By maintaining a constant flow of media through one of the three inlet channel we could expose the yeast cells to a certain concentration of glucose while acquiring time-lapse images. By analysing individual cells in these images with the Cellstat and CellStress software [48, 49], we could track the nuclear localization dynamics of Mig1 over time. For more details in on the microfluidics, imaging and data analysis see *Bendrioua* et al., *2014* [12] and Additional file 10.

Data analysis

The boxplots, fraction and experimental coefficient of variation plots were generated in Matlab (MathWorks, MA).

Model description

The presented model (Fig. 3a) consists of three modules: (1) activity of glucose, (2) the activity of Mig1 and (3) the activity of Snf1. The activity of Glucose includes the import of extracellular Glucose (Gex) into the cell and the degradation of intracellular Glucose (Gin) through the events of metabolism. The activity of Mig1 consists of the import and export of Mig1 into and out of the nucleus and the phosphorylation and dephosphorylation of Mig1. This activity is an irreversible cycle in which the Mig1 alternates between four different forms namely cytosolic Mig1 (cMig1), phosphorylated located in the cytosol Mig1 (cMig1p), Mig1 located in the nucleus (nMig1) and phosphorylated Mig1 located in the nucleus (nMig1p). The activity of Snf1 is divided into two subevents, firstly the phosphorylation and desphosphorylation of cytosolic Snf1 (cSnf1) resulting in phosphorylated cytosolic Snf1 (cSnf1p) and secondly the import and export of phosphorylated Snf1 located in the nucleus (nSnf1p). The set of Ordinary Differential Equations (ODEs) describing the dynamics of the system is listed below. Note that in Eq. (1), the three scalars HXT1a, HXT4a and HXT7a are introduced in order to account for the three data sets: HXT1, the HXT7 and the WT. They are represented as binary variables, where the value of 1 indicates their presence in the given reaction and 0 otherwise (for details see Additional file 10).

$$\frac{dG_{in}}{dt} = (HXT1a \cdot r_{1i}) + (HXT4a \cdot r_{4i}) + (HXT7a \cdot r_{7i}) - r_4 \tag{1}$$

$$\frac{dcSnf1}{dt} = r_5 - r_6 \tag{2}$$

$$\frac{dcSnf1p}{dt} = r_6 - r_5 - (r_{7a} - r_{7b}) \tag{3}$$

$$\frac{dnSnf1p}{dt} = r_{7a} - r_{7b} \tag{4}$$

$$\frac{dcMig1p}{dt} = r_1 - r_2 \tag{5}$$

$$\frac{dcMig1}{dt} = r_2 - r_3 \tag{6}$$

$$\frac{dnMig1}{dt} = r_3 - r_8 \tag{7}$$

$$\frac{dnMig1p}{dt} = r_8 - r_1 \tag{8}$$

The collection of all the parameters, their connection to their respective reaction and the meaning of each reaction is listed in Additional file 11 Table S3.

Non-linear mixed effect modelling

The dynamical model consists of 8 species, 12 reactions and 18 parameters. The vector consisting of these 18 parameters represents the overall dynamics of the entire population of cells and is denoted as fixed effect vector $\bar{\theta} \in \mathbb{R}_+^{18}$. However, in order to account for the individuality we constructed a lognormal distribution from which each parameter vector $\theta \in \mathbb{R}_+^{18}$ representing the dynamics of an individual cell is drawn from. Thus by introducing a multivariate normally distributed variable, denoted as mixed *effect* vector $\eta \in \mathbb{R}^{18}$ with zero mean and the corresponding covariance matrix $\sigma \in \mathbb{R}^{18 \times 18}$ ($\eta \sim N(0, \sigma)$), it is possible to construct the lognormal distribution which is summarized in Eq. 9:

$$\theta = \bar{\theta} \cdot \exp(\eta), \quad \eta \sim N(0, \sigma) \tag{9}$$

Parameter estimation

For parameter estimation a continuous optimization method was implemented and has been conducted using the AMIGO toolbox [50] a software package within MATLAB that utilizes the built in function fmincon in combination with a multiple shooting technique [50]. For more information about the parameter estimation see Additional file 10.

Parameter perturbation

Each parameter in the fixed effect parameter vector $\bar{\theta}$ has been perturbed individually by multiplying the parameter of interest with a scalar of the value $\exp(s^2)$, denoted as $\bar{\theta}_{per_i}$ where $i = 1, \ldots, 18$ is the index of the parameter that is being perturbed. Given the above notation, a measure of the change in the output in response to the perturbation in the model is given by

$$e_i = \frac{||\hat{y}(\bar{\theta}) - \hat{y}(\bar{\theta}_{per_i})||}{l}, \quad i \in \{1, \ldots, 18\} \tag{10}$$

where e_i is the mean perturbation error of parameter i and l is the number of time points for which the output has been measured. Thus, a larger value of e_i would correspond to a greater significance of parameter i in the vector θ in terms of explaining the spread of the measured output.

Additional files

> **Additional file 1: Figure S1.** Sequential images of typical experiment.
>
> **Additional file 2: Figure S2.** Boxplot overview of the upshift experiments on all the strains.
>
> **Additional file 3: Figure S3.** Data analysis on upshift data.
>
> **Additional file 4: Figure S4.** Study of the cell-to-cell variability observed in the Snf1/Mig1 system.
>
> **Additional file 5: Figure S5.** The dynamics of the Mig1-quotient.
>
> **Additional file 6 Figure S6.** Simulation of the spread in the two parameters Vmsi and Vmd.
>
> **Additional file 7: Figure S7.** Simulation of the distribution.
>
> **Additional file 8: Table S1.** The perturbation error.
>
> **Additional file 9: Table S2.** *S. cerevisiae* strains used in this study.
>
> **Additional file 10:** Supplementary methods on microfluidics device, microscopy, cell imaging, data analysis, model description, parameter estibation, and covariance matrix estimation and construction
>
> **Additional file 11: Table S3.** A collection of all parameters in the model.

Abbreviations
ADP: Adenosine Diphosphate; AMPK: AMP-activated protein kinase; ATP: Adenosine Triphosphate; GFP: Green fluorescent protein; NLME: Non-linear mixed effect; PKA: Protein kinase A; RFP: Red fluorescent protein; WT: Wild-type

Acknowledgements
We thank George van der Merwe for generous gifts of yeast strains and Eva Balsa Canto for the support in parameter estimation. We would like to thank the PDMS Microfluidic Fabrication Facility, supported by the science faculty (University of Gothenburg), for assistance and infrastructure.

Funding
The work was supported by the European Commission via the Marie Curie Initial training network ISOLATE (Grant agreement nr: 289995) to NW and SH. MC and JB are supported by the Swedish foundation for Strategic Research (SFF) (Grant agreement nr: 253140901).

Authors' contributions
NW constructed strains, designed the study, collected the data, performed the analysis, and wrote the manuscript. JB and MB have constructed the deterministic model, JB has constructed the NMLE model and performed the model analysis. LB constructed strains and helped collected the data. MG and CBA provided and supported the microscopy experiments and microfluidics systems. MC and SH have designed the study, supervised the work and edited the manuscript. All authors have read and approved the final version of the manuscript.

Competing interests
The authors declare that they have no competing interests.

Author details
[1]Department of Chemistry and Molecular Biology, University of Gothenburg, SE-412 96 Gothenburg, Sweden. [2]Department of Mathematical Sciences, Chalmers University of Technology and the University of Gothenburg, SE-412 96 Gothenburg, Sweden. [3]Department of Physics, University of Gothenburg, SE-412 96 Gothenburg, Sweden. [4]Department of Biology and Biological Engineering, Chalmers University of Technology, SE-412 96 Gothenburg, Sweden.

References
1. Gancedo JM. The early steps of glucose signalling in yeast. FEMS Microbiol rev. 2008;32(4):673–704.
2. Gancedo JM. Yeast carbon catabolite repression. Microbiol Mol Biol rev. 1998;62(2):334–61.
3. Özcan S, Johnston M. Function and regulation of yeast hexose transporters. Microbiol Mol Biol rev. 1999;63(3):554–69.
4. Reifenberger E, Boles E, Ciriacy M. Kinetic characterization of individual hexose transporters of *Saccharomyces cerevisiae* and their relation to the triggering mechanisms of glucose repression. Eur J Biochem/FEBS. 1997;245(2):324–33.
5. Conrad M, Schothorst J, Kankipati HN, Van Zeebroeck G, Rubio-Texeira M, Thevelein JM. Nutrient sensing and signaling in the yeast *Saccharomyces cerevisiae*. FEMS Microbiol rev. 2014;38(2):254–99.
6. Broach JR. Nutritional control of growth and development in yeast. Genetics. 2012;192(1):73–105.
7. Schmidt MC. McCartney RR: beta-subunits of Snf1 kinase are required for kinase function and substrate definition. Embo j. 2000;19(18):4936–43.
8. Jiang R, Carlson M. The Snf1 protein kinase and its activating subunit, Snf4, interact with distinct domains of the Sip1/Sip2/Gal83 component in the kinase complex. Mol Cell Biol. 1997;17(4):2099–106.
9. Rose M, Albig W, Entian KD. Glucose repression in *Saccharomyces cerevisiae* is directly associated with hexose phosphorylation by hexokinases PI and PII. Eur J Biochem/FEBS. 1991;199(3):511–8.
10. Garcia-Salcedo R, Lubitz T, Beltran G, Elbing K, Tian Y, Frey S, et al. Glucose de-repression by yeast AMP-activated protein kinase SNF1 is controlled via at least two independent steps. FEBS J. 2014;281(7):1901–17.
11. Zhang Y, McCartney RR, Chandrashekarappa DG, Mangat S, Schmidt MC. Reg1 protein regulates phosphorylation of all three Snf1 isoforms but preferentially associates with the Gal83 isoform. Eukaryot Cell. 2011;10(12):1628–36.
12. Bendrioua L, Smedh M, Almquist J, Cvijovic M, Jirstrand M, Goksor M, et al. Yeast AMP-activated protein kinase monitors glucose concentration changes and absolute glucose levels. J Biol Chem. 2014;289(18):12863–75.
13. Dalal CK, Cai L, Lin Y, Rahbar K, Elowitz MB. Pulsatile dynamics in the yeast proteome. Curr Biol. 2014;24(18):2189–94.
14. Klipp E. Modelling dynamic processes in yeast. Yeast (Chichester, England). 2007;24(11):943–59.
15. Neves SR, Iyengar R. Modeling of signaling networks. Bioessays. 2002; 24(12):1110–7.
16. Klipp E, Nordlander B, Kruger R, Gennemark P, Hohmann S. Integrative model of the response of yeast to osmotic shock. Nat Biotechnol. 2005; 23(8):975–82.
17. Kofahl B, Klipp E. Modelling the dynamics of the yeast pheromone pathway. Yeast (Chichester, England). 2004;21(10):831–50.
18. Klipp E, Schaber J: Modelling of signal transduction in yeast – sensitivity and model analysis. In: Understanding and exploiting systemy biology in bioprocesses and biomedicine. Edited by In M. Cánovas JLI, & A. Manjón: Murcia: Fundación Cajamurcia.; 2006: 15-30.
19. Klipp E, Liebermeister W. Mathematical modeling of intracellular signaling pathways. BMC Neurosci. 2006;7(Suppl 1):S10.
20. Karlsson M, Janzen DL, Durrieu L, Colman-Lerner A, Kjellsson MC, Cedersund G. Nonlinear mixed-effects modelling for single cell estimation: when, why, and how to use it. BMC Syst Biol. 2015;9:52.
21. Niepel M, Spencer SL, Sorger PK. Non-genetic cell-to-cell variability and the consequences for pharmacology. Curr Opin Chem Biol. 2009;13(5-6):556–61.
22. Ribba B, Holford NH, Magni P, Troconiz I, Gueorguieva I, Girard P, et al. A review of mixed-effects models of tumor growth and effects of anticancer drug treatment used in population analysis. CPT Pharmacometrics Syst Pharmacol. 2014;3:e113.
23. Zechner C, Unger M, Pelet S, Peter M, Koeppl H. Scalable inference of heterogeneous reaction kinetics from pooled single-cell recordings. Nat Methods. 2014;11(2):197–202.
24. Almquist J, Bendrioua L, Adiels CB, Goksor M, Hohmann S, Jirstrand M. A nonlinear mixed effects approach for modeling the cell-to-cell variability of Mig1 dynamics in yeast. PLoS One. 2015;10(4):e0124050.
25. Eriksson E, Sott K, Lundqvist F, Sveningsson M, Scrimgeour J, Hanstorp D, et al. A microfluidic device for reversible environmental changes around single cells using optical tweezers for cell selection and positioning. Lab Chip. 2010;10(5):617–25.
26. Treitel MA, Kuchin S, Carlson M. Snf1 protein kinase regulates phosphorylation of the Mig1 repressor in *Saccharomyces cerevisiae*. Mol Cell Biol. 1998;18(11):6273–80.

27. DeVit MJ, Johnston M. The nuclear exportin Msn5 is required for nuclear export of the Mig1 glucose repressor of *Saccharomyces cerevisiae*. Curr Biol. 1999;9(21):1231–41.

28. Elbing K, Stahlberg A, Hohmann S, Gustafsson L. Transcriptional responses to glucose at different glycolytic rates in *Saccharomyces cerevisiae*. Eur J Biochem/FEBS. 2004;271(23-24):4855–64.

29. Elbing K, Larsson C, Bill RM, Albers E, Snoep JL, Boles E, et al. Role of hexose transport in control of glycolytic flux in *Saccharomyces cerevisiae*. Appl Environ Microbiol. 2004;70(9):5323–30.

30. Lin Y, Sohn CH, Dalal CK, Cai L, Elowitz MB. Combinatorial gene regulation by modulation of relative pulse timing. Nature. 2015;527(7576):54–8.

31. Krampe S, Stamm O, Hollenberg CP, Boles E. Catabolite inactivation of the high-affinity hexose transporters Hxt6 and Hxt7 of *Saccharomyces cerevisiae* occurs in the vacuole after internalization by endocytosis. FEBS Lett. 1998; 441(3):343–7.

32. Snowdon C, Hlynialuk C, van der Merwe G. Components of the Vid30c are needed for the rapamycin-induced degradation of the high-affinity hexose transporter Hxt7p in *Saccharomyces cerevisiae*. FEMS Yeast res. 2008;8(2):204–16.

33. Snowdon C, van der Merwe G. Regulation of Hxt3 and Hxt7 turnover converges on the Vid30 complex and requires inactivation of the Ras/cAMP/PKA pathway in *Saccharomyces cerevisiae*. PLoS One. 2012;7(12):e50458.

34. Roy A, Kim YB, Cho KH, Kim JH. Glucose starvation-induced turnover of the yeast glucose transporter Hxt1. Biochim Biophys Acta. 2014;1840(9):2878–85.

35. Raser JM, O'Shea EK. Control of stochasticity in eukaryotic gene expression. Science. 2004;304(5678):1811–4.

36. Schwabe A, Bruggeman FJ. Single yeast cells vary in transcription activity not in delay time after a metabolic shift. Nat Commun. 2014;5:4798.

37. Xiao B, Sanders MJ, Underwood E, Heath R, Mayer FV, Carmena D, et al. Structure of mammalian AMPK and its regulation by ADP. Nature. 2011; 472(7342):230–3.

38. Mayer FV, Heath R, Underwood E, Sanders MJ, Carmena D, McCartney RR, et al. ADP regulates SNF1, the *Saccharomyces cerevisiae* homolog of AMP-activated protein kinase. Cell Metab. 2011;14(5):707–14.

39. Chandrashekarappa DG, McCartney FR, Schmidt MC. Subunit and domain requirements for adenylate-mediated protection of Snf1 kinase activation loop from dephosphorylation. J Biol Chem 2011;286(52):44532–41.

40. Chandrashekarappa DG, McCartney FR, Schmidt MC. Ligand binding to the AMP-activated protein kinase active site mediates protection of the activation loop from dephosphorylation. J Biol Chem. 2013;288(1):89–98.

41. Bosch D, Johansson M, Ferndahl C, Franzen CJ, Larsson C, Gustafsson L. Characterization of glucose transport mutants of *Saccharomyces cerevisiae* during a nutritional upshift reveals a correlation between metabolite levels and glycolytic flux. FEMS Yeast res. 2008;8(1):10–25.

42. Ye L, Kruckeberg AL, Berden JA, van Dam K. Growth and glucose repression are controlled by glucose transport in *Saccharomyces cerevisiae* cells containing only one glucose transporter. J Bacteriol. 1999;181(15):4673–5.

43. Meijer MM, Boonstra J, Verkleij AJ, Verrips CT. Glucose repression in *Saccharomyces cerevisiae* is related to the glucose concentration rather than the glucose flux. J Biol Chem. 1998;273(37):24102–7.

44. Vega M, Riera A, Fernandez-Cid A, Herrero P, Moreno F. Hexokinase 2 is an intracellular glucose sensor of yeast cells that maintains the structure and activity of Mig1 repressor complex. J Biol Chem. 2016;291(14):7267–85.

45. Moreno F, Vega M, Herrero P. The nuclear Hexokinase 2 acts as a glucose sensor in *Saccharomyces cerevisiae*. J Biol Chem. 2016;291(32):16478.

46. van Heerden JH, Wortel MT, Bruggeman FJ, Heijnen JJ, Bollen YJ, Planque R, et al. Lost in transition: start-up of glycolysis yields subpopulations of nongrowing cells. Science. 2014;343(6174):1245114.

47. Conrad NK, Wilson SM, Steinmetz EJ, Patturajan M, Brow DA, Swanson MS, et al. A yeast heterogeneous nuclear ribonucleoprotein complex associated with RNA polymerase II. Genetics. 2000;154(2):557–71.

48. Smedh M, Beck C, Sott K, Goksör M: CellStress - open source image analysis program for single-cell analysis. In: 2010. 77622N-77622N-77611.

49. Kvarnstrom M, Logg K, Diez A, Bodvard K, Kall M. Image analysis algorithms for cell contour recognition in budding yeast. Opt Express. 2008;16(17):12943–57.

50. Balsa-Canto E, Banga JR. AMIGO, a toolbox for advanced model identification in systems biology using global optimization. Bioinformatics. 2011;27(16):2311–3.

9

Modeling and analysis of the Delta-Notch dependent boundary formation in the *Drosophila* large intestine

Fei Liu[1,2*†], Deshun Sun[1†], Ryutaro Murakami[3] and Hiroshi Matsuno[3]

Abstract

Background: The boundary formation in the *Drosophila* large intestine is widely studied as an important biological problem. It has been shown that the Delta-Notch signaling pathway plays an essential role in the formation of boundary cells.

Results: In this paper, we propose a mathematical model for the Delta-Notch dependent boundary formation in the *Drosophila* large intestine in order to better interpret related experimental findings of this biological phenomenon. To achieve this, we not only perform stability analysis on the model from a theoretical point of view, but also perform numerical simulations to analyze the model with and without noises, the phenotype change with the change of Delta or Notch expression, and the perturbation influences of binding and inhibition parameters on the boundary formation.

Conclusions: By doing all these work, we can assure that our model can better interpret the biological findings related to the boundary formation in the *Drosophila* large intestine.

Keywords: Boundary formation, *Drosophila* large intestine, Delta-Notch signaling pathway, Local stability analysis, Simulation validation, Perturbation analysis

Background

The large intestine of *Drosophila* embryo is a middle and large region of the *Drosophila* hindgut and is subdivided into dorsal and ventral domains, between which a one-cell-wide strand of boundary cells forms bilaterally for wild-type embryos [1–4]. The large intestine is a multicellular system that involves a number of cells composed of three cell types, dorsal, ventral and boundary cells, organized in a single-layered epithelial tube [5]. For such developmental patterning problems, different kinds of computational strategies [6] have been proposed, e.g., signaling gradients [7] and activator-inhibitor systems [8],

and many computational techniques are adopted, e.g., ordinary/partial differential equation [9] and colored Petri nets [10]. See [11] for a review. However, for the boundary formation in the *Drosophila* large intestine, the mechanism has been widely explored in vivo, e.g., [1–4, 12], but rarely from the computational point of view, e.g., [13].

It has been shown that the Delta-Notch signaling pathway plays an essential role in the boundary formation of the *Drosophila* large intestine [1, 12]. In fact, the Delta-Notch pathway is considered as one of the six major signaling pathways in cells, which is active in developing embryos at different phases [14]. Both Notch and Delta proteins are transmembrane proteins, where Notch proteins act as receptors and Delta proteins as ligands. When Delta ligands in a cell bind to Notch receptors in neighboring cells, all the cells in a system may evolve and finally form different types of patterns [15–18].

*Correspondence: liufei@hit.edu.cn

†Equal contributors

[1]Control and Simulation Center, Harbin Institute of Technology, West Dazhi Street 92, 150001 Harbin, People's Republic of China

[2]School of Software Engineering, South China University of Technology, Building B7, 510006 Guangzhou, People's Republic of China

Full list of author information is available at the end of the article

In order to understand the mechanism of the Delta-Notch pathway, there have been some mathematical models proposed. For example, Collier et al. proposed a simple ordinary differential equation (ODE) model for the Delta-Notch signaling pathway, and discussed numerical simulation of multiple-cell systems [19]. Boareto et al. devised a theoretical framework that includes a couple of ODEs to explore the effects of Jagged in cell-fate determination [20]. Sprinzak et al. gave a model of a set of ODEs for describing mutual inactivation of Notch and Delta and used this model to illustrate how cis-interactions between Notch and Delta generate mutually exclusive signaling states [21]. Specifically, Matsuno et al. analyzed the mechanism of the Notch-dependent boundary formation in the *Drosophila* large intestine and built a hybrid Petri net model to numerically explore how the Delta-Notch pathway affects the boundary formation in two-dimensional space [13].

In this paper, we propose a mathematical model for the Delta-Notch dependent boundary formation in the *Drosophila* large intestine based on the work of [13], aiming at better interpreting related biological findings and further making predictions. Compared with the existing work in this area, our work has the following main contributions.

(1) We give a mathematical model for the Delta-Notch dependent boundary formation in the *Drosophila* large intestine, which can better interpret relevant biological findings, so far obtained in the lab. Moreover, this model can be numerically simulated efficiently. To make the model both mathematically and biologically sound, we do the following analysis work.

(2) We perform local stability analysis on the two-cell model to make the model mathematically sound. The analysis confirms that the model would reach an equilibrium, which corresponds to that the system would result in a stable pattern in either normal or mutant conditions. The model is also stable when some conditions are satisfied, which means even if there are small disturbances of parameters (e.g., small environmental noises), the system would still converge to a stable state after a period of time. We use the local stability analysis result to determine appropriate parameter values that make the system stable.

(3) We perform numerical simulations with and without noises of Notch expression, which shows that both deterministic and random simulation results are consistent with experimental observations. We further analyze how phenotypes change with the change of Delta or Notch expression.

(4) We analyze the perturbation influences of binding and inhibition parameters on the boundary formation. As the boundary formation is affected by many environmental factors or noises, to make the model more realistic, we need to consider noises into the model by finding those

parameters that have significant influences on the dynamics of the model when they vary. According to the analysis result, we add appropriate random noise items to key parameters of the model.

This paper is structured as follows. In the section of methods, we introduce relevant biological background of the Delta-Notch dependent boundary formation in the *Drosophila* large intestine, and present a mathematical model for this biological phenomenon. In the section of results and discussions, we give stability analysis, simulation analysis, and parameter perturbation analysis of the model. Finally, the conclusion is given.

Methods

In this section, we introduce relevant biological background and give a description of the mathematical model we develop, as well as the simulation and analysis methods which we use.

Boundary formation in the *Drosophila* large intestine

The large intestine of *Drosophila* embryo is a major middle region of the *Drosophila* hindgut and is subdivided into dorsal and ventral domains (see Fig. 1), each of which is characterized by different cell types, dorsal or ventral. Between these two domains, a one-cell-wide strand of boundary cells forms in wild-type embryos.

The Delta-Notch signaling pathway plays an essential role in the formation of boundary cells [1, 12]. The main processes [1] are shown in Fig. 2. When the Delta ligands on the ventral cell surface bind to the Notch receptors of a neighboring Delta-negative dorsal cell, a Delta-Notch signaling cascade is activated. First the site 2 cleavage of the Notch proteins occurs to generate a transmembrane form ($N^{\Delta E}$), followed by the Presenilin-dependent site 3 cleavage, producing an active Notch intracellular fragment (N^{intra}). N^{intra} then activates the target genes, inducing boundary cell differentiation. Moreover, Delta autonomously blocks the Presenilin-mediated site 3 cleavage and thus inhibits Notch signal transduction within Delta-positive ventral cells.

In this paper, we aim to interpret the publicated findings about the boundary cell patterning in the large intestine [1, 4, 13] (see Fig. 3) and hypothesize the following scenarios for our model.

- **Scenario 1.** In a wild-type embryo, a one-cell-wide strand of boundary cells forms at the interface of dorsal and ventral domains. See Fig. 3a for an illustration.
- **Scenario 2.** In an embryo where over-expression of Notch or Delta proteins happens, the number of boundary cells would change. If we fix over-expression of Notch proteins, the number of boundary cells would increase with the decrease of

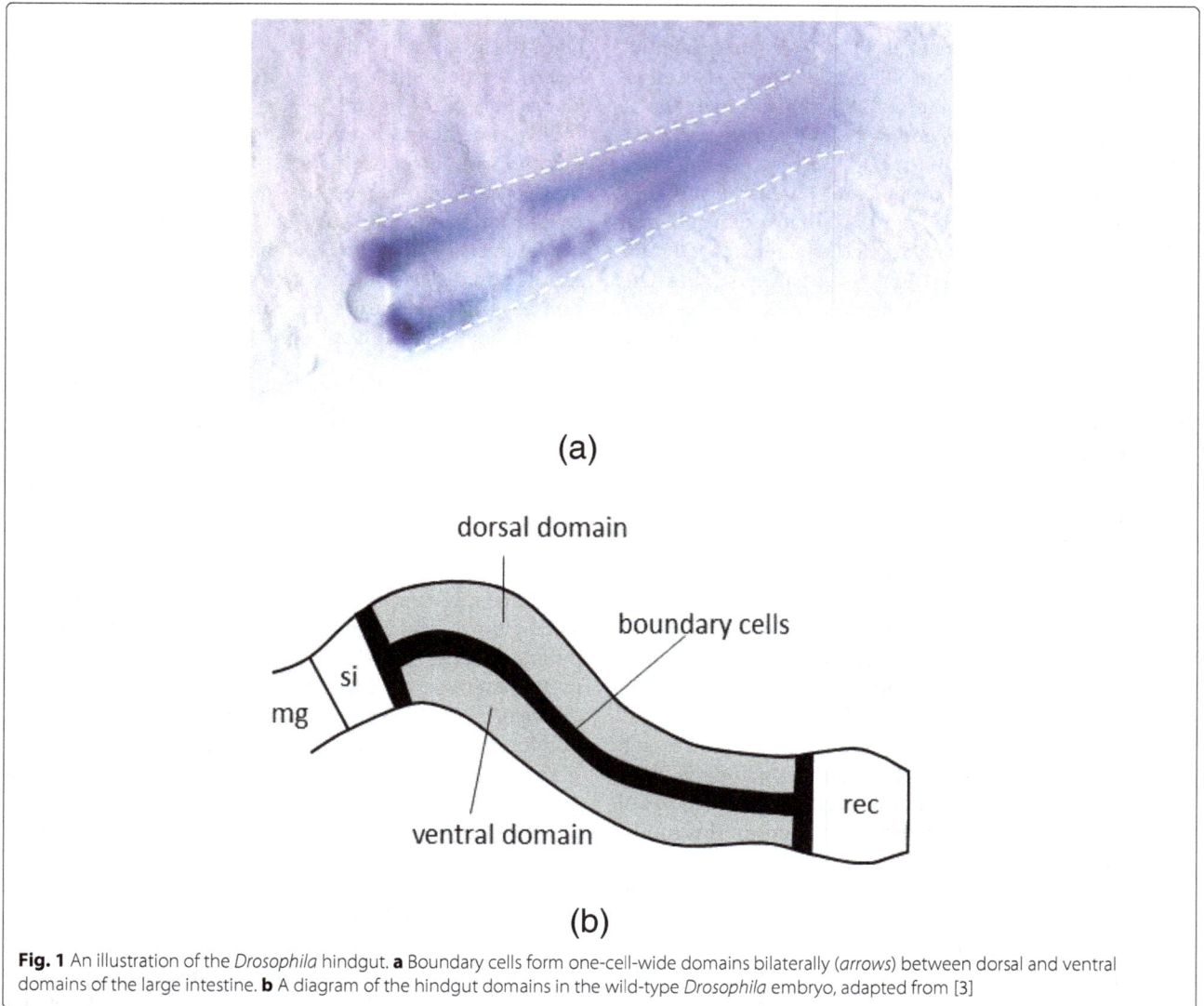

Fig. 1 An illustration of the *Drosophila* hindgut. **a** Boundary cells form one-cell-wide domains bilaterally (*arrows*) between dorsal and ventral domains of the large intestine. **b** A diagram of the hindgut domains in the wild-type *Drosophila* embryo, adapted from [3]

Delta background. See Fig. 3b1 to b2 for an illustration.

- **Scenario 3.** However, if we fix the background of Delta proteins, the number of boundary cells would increase with the increase of over-expression of Notch proteins. See Fig. 3c1 to c2 for an illustration.
- **Scenario 4.** During the boundary formation in the *Drosophila* large intestine, environmental factors such as temperature may heavily affect the boundary formation, thus resulting in different (random) phenotypes. Therefore, the model to be built should incorporate such random noises into some parameters of interest.

In what follows, we will construct and analyze our model step by step.

Mathematical model

Cell-to-cell interaction mediated by the Delta-Notch signaling pathway plays an essential role in the development of a multicellular organism such as the *Drosophila* large intestine. In this work we propose a mathematical model for the core part of the Delta-Notch dependent boundary formation in the *Drosophila* large intestine (see Fig. 4 for a one-cell model) based on the work of [13]. That is, we only consider two key species, Delta (D) and Notch, and Notch can be in the inactive state (N) or active state (A). Inactive Notch can be converted into the active state when coupled with Delta from neighboring cells. Active Notch can inhibit the production of Delta of the same cell through target genes. Besides, Delta also inhibits the production of Notch of the same cell.

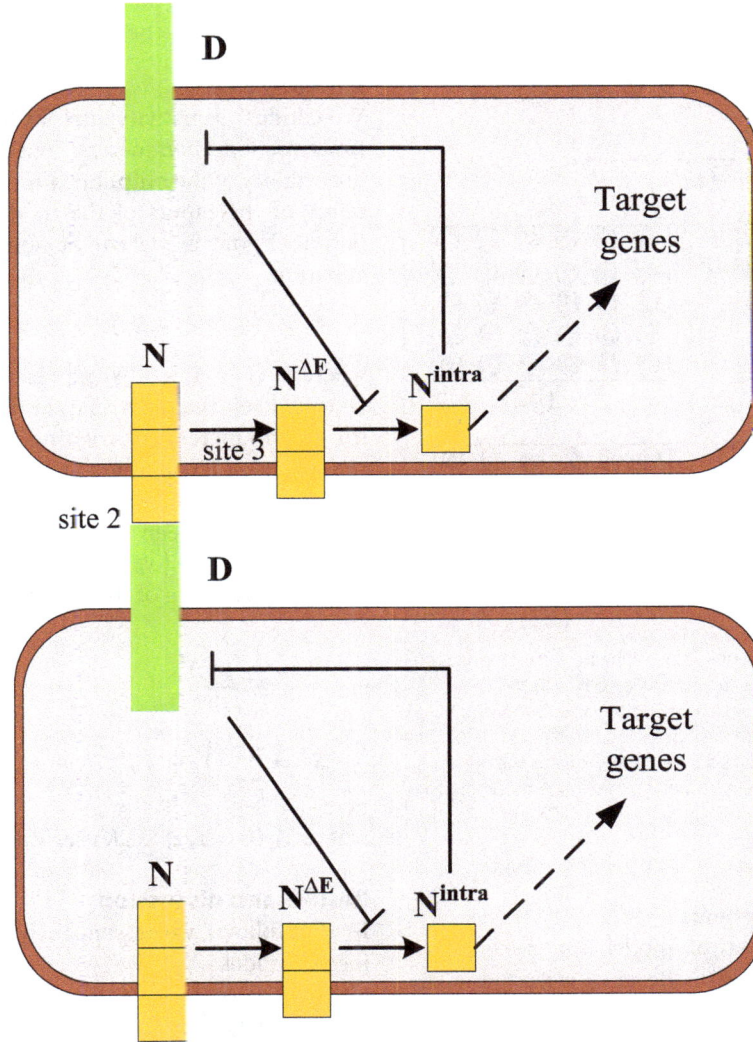

Fig. 2 A diagram of the Delta-Notch signaling pathway in boundary cell formation of the large intestine. This pathway shows the interaction of two neighboring cells, which is adapted from [1]

We arrange all cells in a regular $M \times N$ lattice (see Fig. 5 for an illustration), each site (cell) being a hexagon and having at most six neighboring cells. That is, we have $NC = M \times N/2$ cells for an $M \times N$ lattice.

The mathematical model is then given as follows:

$$\begin{cases} \dfrac{dD_i}{dt} = \dfrac{\lambda}{1 + \Delta \cdot A_i} - d_1 D_i - \sum_{NG(i)} f_1 \cdot D_i, \\[2em] \dfrac{dN_i}{dt} = \lambda_N - d_2 N_i + \sum_{j \in NG(i)} f_2 \cdot D_j - \dfrac{a N_i}{b D_i + N_i}, \quad (1) \\[2em] \dfrac{dA_i}{dt} = -d_3 A_i + \dfrac{a N_i}{b D_i + N_i}. \end{cases}$$

Here, $1 \leq i \leq NC$. D_i, N_i and A_i denote the concentration of Delta proteins, inactive and active Notch proteins,

respectively, in the ith cell. In $\frac{\lambda}{1 + \Delta \cdot A_i}$, λ represents the production rate of Delta proteins, and Δ the inhibition coefficient of active Notch proteins A_i. d_1 represents the degradation rate of Delta proteins. $\sum_{NG(i)} D_i$ represents the concentration of the Delta proteins (in the ith cell) that bind to the Notch proteins of the contacting cells of the ith cell, and $NG(i)$ denotes all the neighbors of the ith cell. f_1 represents the binding rate of Delta ligands to Notch receptors. λ_N and d_2 represent the production rate and the degradation rate of inactive Notch proteins, respectively. f_2 represents the binding rate of Delta ligands to Notch receptors. In $\frac{a N_i}{b D_i + N_i}$, a represents the conversion rate of Notch proteins from the inactive state to the active state, while b describes the inhibition effect of Delta on Notch in the same cell i. d_3 is the degradation rate of active Notch proteins.

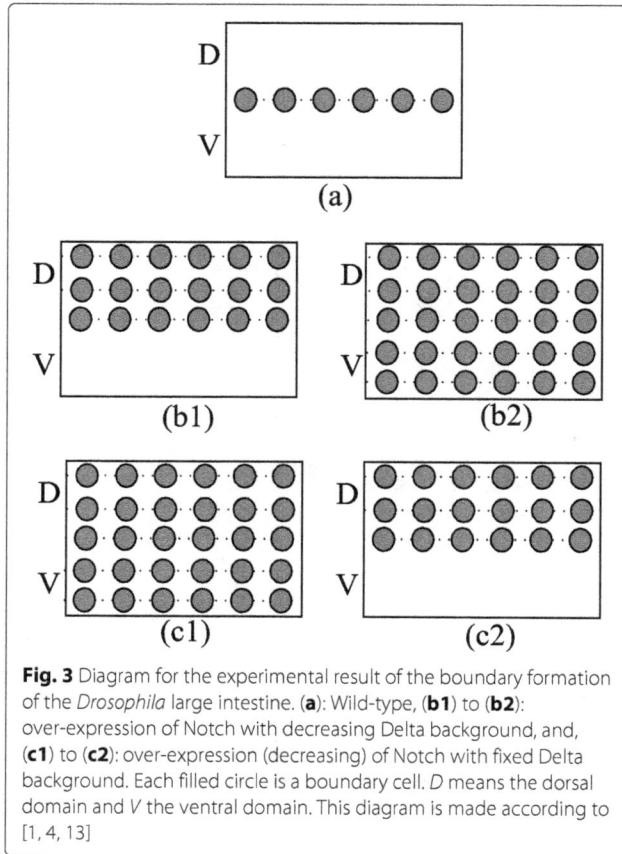

Fig. 3 Diagram for the experimental result of the boundary formation of the *Drosophila* large intestine. (**a**): Wild-type, (**b1**) to (**b2**): over-expression of Notch with decreasing Delta background, and, (**c1**) to (**c2**): over-expression (decreasing) of Notch with fixed Delta background. Each filled circle is a boundary cell. *D* means the dorsal domain and *V* the ventral domain. This diagram is made according to [1, 4, 13]

Simulation and analysis methods

We encode our mathematical model and perform all simulations with Matlab 2016. Besides, we employ the following analysis methods for our model.

Local stability analysis

We conduct local stability analysis to explore stability conditions of our model , with which we further determine values of parameters.

A set of autonomous ordinary differential equations can be written in the following vector form:

$$\dot{x} = f(x)$$

where $x = (x_1, x_2, \ldots, x_n)$ is the state vector and $f = (f_1, f_2, \ldots, f_n)$. The Jacobian matrix is:

$$J = \begin{bmatrix} \frac{\partial f_1}{\partial x_1} & \frac{\partial f_1}{\partial x_2} & \cdots & \frac{\partial f_1}{\partial x_n} \\ \frac{\partial f_2}{\partial x_1} & \frac{\partial f_2}{\partial x_2} & \cdots & \frac{\partial f_2}{\partial x_n} \\ \vdots & \vdots & & \vdots \\ \frac{\partial f_n}{\partial x_1} & \frac{\partial f_n}{\partial x_2} & \cdots & \frac{\partial f_n}{\partial x_n} \end{bmatrix}.$$

Assume x^* is an equilibrium point, i.e., $f(x^*) = 0$. J^* is the Jacobian matrix evaluated at x^*. The equilibrium

point x^* is stable if all the eigenvalues of the characteristic equation of J^* have negative real parts [22].

Sensitivity analysis

We conduct sensitivity analysis to investigate the significance of model parameters. Sensitivity analysis studies the uncertainty of the output of a model caused by the uncertainty of the inputs of the model [23, 24]. Assume the output (Y) of a model can be represented as the following equation:

$$Y = f(x_1, x_2, \ldots, x_n)$$

where x_i ($i = 1, 2, \ldots n$) represent the input variables or factors. Each time, we analyze the sensitivity of one factor (or parameter) by assigning a perturbation term $\varepsilon\beta$ to the factor, e.g., x_i, where ε denotes the disturbance intensity and β is a random number that satisfies the uniform distribution between [-1, 1]. We randomly sample values from the interval $[x_i - \varepsilon, x_i + \varepsilon]$ and then compute the mean and variance of the output as follows:

$$\bar{Y} = \frac{1}{K} \sum_{k=1}^{K} Y_k,$$

$$S^2 = \sum_{k=1}^{K} (Y_k - \bar{Y})^2 / (K - 1),$$

where Y_k ($k = 1, 2, \ldots, K$) are values of the output Y.

Results and discussion

In what follows, we present analysis and simulation results for the model.

Local stability analysis

System (1) gives the dynamic equations of NC cells, which is impossible to be analyzed for a big NC in theory. However, due to the locality of the Delta-Notch interaction, we can still obtain valuable insight into the whole system by only considering two cells to explore their dynamic behavior such as equilibria and stability [25–27] from the theoretical point of view. The mathematical model of two cells is given as follows:

$$\begin{cases} \dfrac{dD_1}{dt} = \dfrac{\lambda}{1 + \Delta \cdot A_1} - d_1 D_1 - f_1 \cdot D_1, \\[2mm] \dfrac{dN_1}{dt} = \lambda_N - d_2 N_1 + f_2 \cdot D_2 - \dfrac{aN_1}{bD_1 + N_1}, \\[2mm] \dfrac{dA_1}{dt} = -d_3 A_1 + \dfrac{aN_1}{bD_1 + N_1}, \\[2mm] \dfrac{dD_2}{dt} = \dfrac{\lambda}{1 + \Delta \cdot A_2} - d_1 D_2 - f_1 \cdot D_2, \\[2mm] \dfrac{dN_2}{dt} = \lambda_N - d_2 N_2 + f_2 \cdot D_1 - \dfrac{aN_2}{bD_2 + N_2}, \\[2mm] \dfrac{dA_2}{dt} = -d_3 A_2 + \dfrac{aN_2}{bD_2 + N_2}. \end{cases} \quad (2)$$

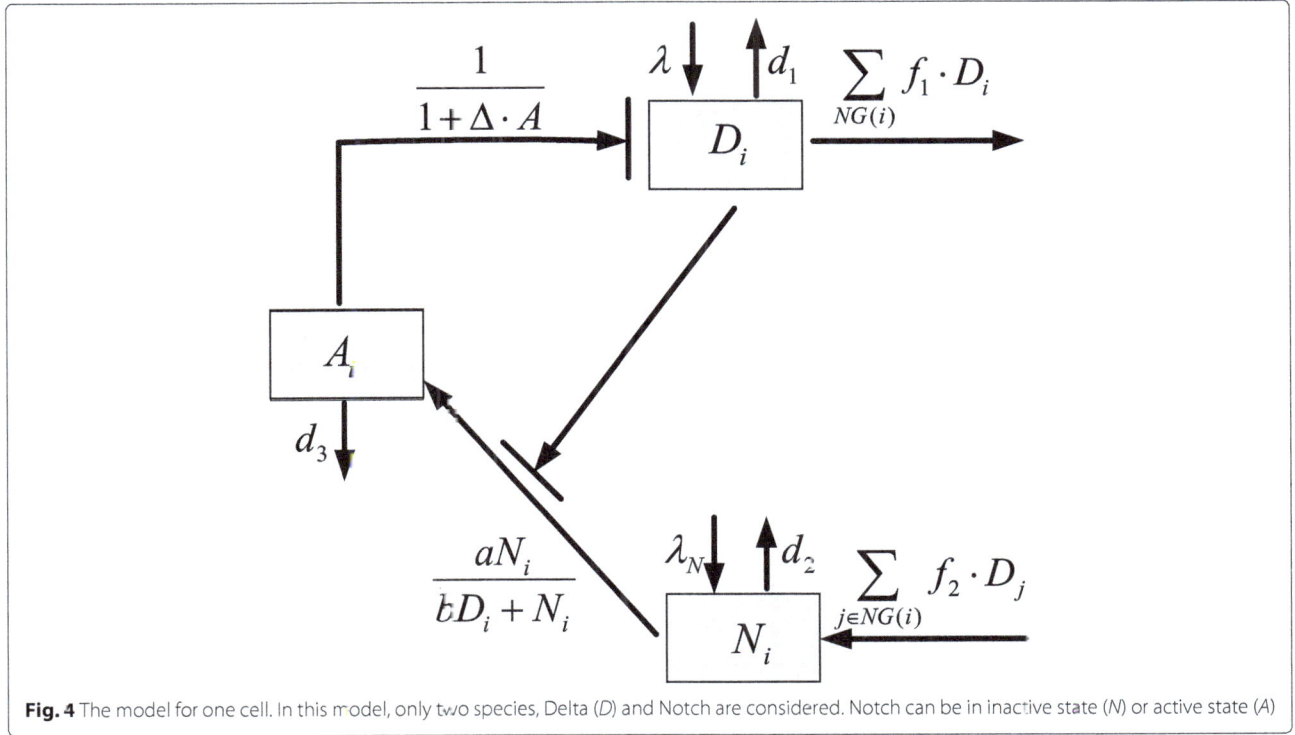

Fig. 4 The model for one cell. In this model, only two species, Delta (*D*) and Notch are considered. Notch can be in inactive state (*N*) or active state (*A*)

In the following, we will take λ (the production rate of Delta proteins) as an example to study the equilibria and stability of System (2).

(1) When $\lambda = 0$, which corresponds to mis-expression of Delta, we obtain the equilibrium

$$E_0 = \left(D_1^0, N_1^0, A_1^0, D_2^0, N_2^0, A_2^0\right),$$

where $D_1^0 = 0$, $N_1^0 = \frac{\lambda_N - a}{d_2}$, $A_1^0 = \frac{a}{d_3}$, $D_2^0 = 0$, $N_2^0 = \frac{\lambda_N - a}{d_2}$ and $A_2^0 = \frac{a}{d_3}$.

(2) When $\lambda > 0$, which corresponds to normal or over-expression of Delta, we obtain the equilibrium

$$E_1 = \left(D_1^1, N_1^1, A_1^1, D_2^1, N_2^1, A_2^1\right),$$

where $D_1^1 = \frac{\lambda}{(1+\Delta A_1^1)(d_1+f_1)}$, $N_1^1 = \frac{bd_3\lambda A_1^1}{(1+\Delta A_1^1)(d_1+f_1)(a-d_3 A_1^1)}$, $D_2^1 = \frac{\lambda}{(1+\Delta A_1^1)(d_1+f_1)}$, $N_2^1 = \frac{bd_3\lambda A_1^1}{(1+\Delta A_1^1)(d_1+f_1)(a-d_3 A_1^1)}$ and $A_2^1 = A_1^1$. A_1^1 is the solution of

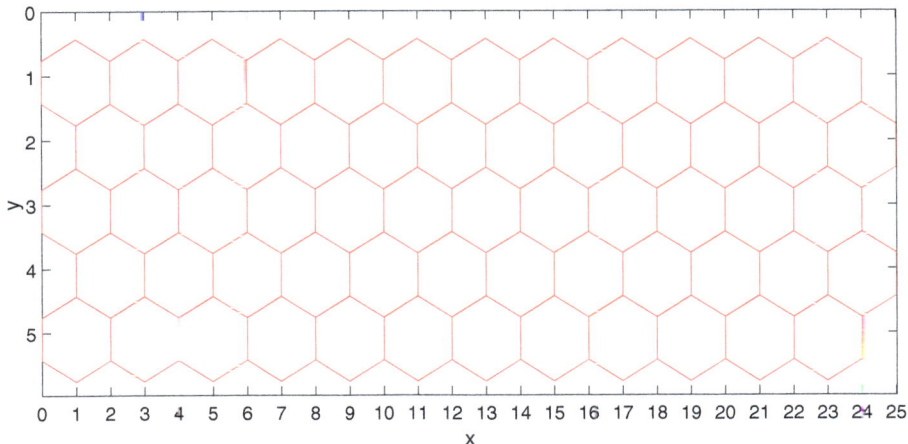

Fig. 5 A two-dimensional lattice for the model. The mathematical model proposed in this paper works on the two-dimensional lattice, which defines a patch of a tissue

$$-d_3^2(d_1+f_1)\Delta(A_1^1)^3+\left[d_3(d_1+f_1)(a\Delta-d_3)+\Delta d_3\lambda_N(d_1+f_1)\right](A_1^1)^2$$
$$+\left[ad_3(d_1+f_1)-\lambda_N(d_1+f_1)(a\Delta-d_3)+d_3f_2\lambda+bd_2d_3\lambda\right]A_1^1 \tag{3}$$
$$-a\lambda_N(d_1+f_1)-af_2\lambda=0.$$

Now, let

$$m_1=-d_3^2(d_1+f_1)\Delta,$$
$$m_2=d_3(d_1+f_1)(a\Delta-d_3)+\Delta d_3\lambda_N(d_1+f_1),$$
$$m_3=ad_3(d_1+f_1)-\lambda_N(d_1+f_1)(a\Delta-d_3)+d_3f_2\lambda+bd_2d_3\lambda,$$
$$m_4=-a\lambda_N(d_1+f_1)-af_2\lambda.$$

Then Eq. (3) becomes

$$m_1(A_1^1)^3+m_2(A_1^1)^2+m_3A_1^1+m_4=0.$$

Because $m_1=-d_3^2(d_1+f_1)\Delta\neq 0$, we further obtain

$$(A_1^1)^3+\frac{m_2}{m_1}(A_1^1)^2+\frac{m_3}{m_1}A_1^1+\frac{m_4}{m_1}=0.$$

Thus, we have

$$A_1^1=$$

$$\sqrt[3]{-\frac{\frac{2m_2^3}{27m_1^3}-\frac{m_2m_3}{3m_1^2}+\frac{m_4}{m_1}}{2}+\sqrt{\left(\frac{\frac{2m_2^3}{27m_1^3}-\frac{m_2m_3}{3m_1^2}+\frac{m_4}{m_1}}{2}\right)^2+\left(-\frac{m_2^2}{3m_1^2}+\frac{m_3}{m_1}\right)^3}}$$

$$+\sqrt[3]{-\frac{\frac{2m_2^3}{27m_1^3}-\frac{m_2m_3}{3m_1^2}+\frac{m_4}{m_1}}{2}-\sqrt{\left(\frac{\frac{2m_2^3}{27m_1^3}-\frac{m_2m_3}{3m_1^2}+\frac{m_4}{m_1}}{2}\right)^2+\left(-\frac{m_2^2}{3m_1^2}+\frac{m_3}{m_1}\right)^3}}$$

$$-\frac{m_2}{3m_1}.$$

Please note that the equilibria of System (2) we obtain above mean that the system will converge to a stable state after a period of time.

Now, we further study the stability of System (2) at equilibrium E_0. The Jacobi matrix [27] at E_0 is computed as follows:

$$J_{E_0}=\begin{bmatrix} -d_1-f_1 & 0 & 0 & 0 & 0 & 0 \\ \frac{abN_1^0}{(bD_1^0+N_1^0)^2} & -d_2-\frac{abD_1^0}{(bD_1^0+N_1^0)^2} & 0 & f_2 & 0 & 0 \\ \frac{-abN_1^0}{(bD_1^0+N_1^0)^2} & \frac{abD_1^0}{(bD_1^0+N_1^0)^2} & -d_3 & 0 & 0 & 0 \\ 0 & 0 & 0 & -d_1-f_1 & 0 & 0 \\ f_2 & 0 & 0 & \frac{abN_2^0}{(bD_2^0+N_2^0)^2} & -d_2-\frac{abD_2^0}{(bD_2^0+N_2^0)^2} & 0 \\ 0 & 0 & 0 & \frac{-abN_2^0}{(bD_2^0+N_2^0)^2} & \frac{abD_2^0}{(bD_2^0+N_2^0)^2} & -d_3 \end{bmatrix}.$$

Assume S is the eigenvalue of the characteristic equation $|J_{E_0}|$. Then the characteristic equation is calculated as follows [27]:

$$|J_{E_0}| = \begin{vmatrix} S+d_1+f_1 & 0 & 0 & 0 & 0 & 0 \\ \frac{-abN_1^0}{(bD_1^0+N_1^0)^2} & S+d_2+\frac{abD_1^0}{(bD_1^0-N_1^0)^2} & 0 & -f_2 & 0 & 0 \\ \frac{abN_1^0}{(bD_1^0+N_1^0)^2} & \frac{-abD_1^0}{(bD_1^0-N_1^0)^2} & S+d_3 & 0 & 0 & 0 \\ 0 & 0 & 0 & S+d_1+f_1 & 0 & 0 \\ -f_2 & 0 & 0 & \frac{-abN_2^0}{(bD_2^0+N_2^0)^2} & S+d_2+\frac{abD_2^0}{(bD_2^0+N_2^0)^2} & 0 \\ 0 & 0 & 0 & \frac{abN_2^0}{(bD_2^0+N_2^0)^2} & \frac{-abD_2^0}{(bD_2^0+N_2^0)^2} & S+d_3 \end{vmatrix}.$$

After some intermediate computation steps, we have

$$|J_{E_0}| = (S+d_1+f_1)^2(S+d_2)^2(S+d_3)^2 \tag{4}$$

It is obvious that all of the six eigenvalues of Eq. (4) are negative, which gives the following conclusion.

Lemma 1 *The equilibrium E_0 is locally asymptotically stable.*
For example, a simulation in this case is given in Fig. 6, with the following parameter values: $\lambda_N = 0.02$, $d_1 = d_2 = d_3 = 0.01$, $f_1 = f_2 = 0.665$, $a = 0.012$, $b = 69$, $\Delta = 10^6$ and $\lambda = 0$.

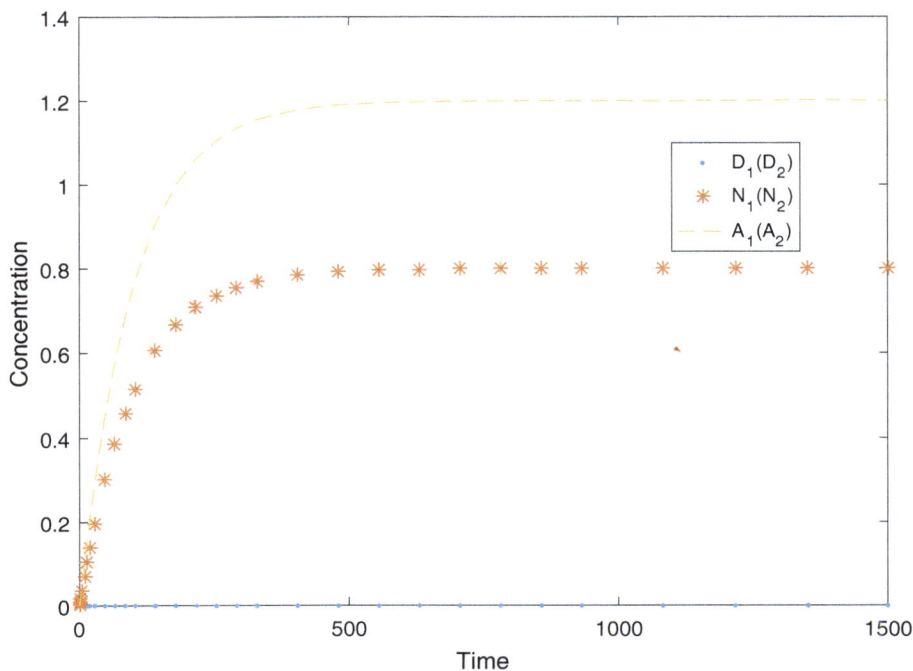

Fig. 6 A simulation plot at $\lambda = 0$. In this plot, the simulation traces of D_1 (N_1, A_1) and D_2 (N_2, A_2) overlap

Next, we investigate the stability of the equilibrium E_1. The Jacobi matrix at E_1 is given as follows:

$$J_{E_1} = \begin{bmatrix} -d_1 - f_1 & 0 & \frac{-\Delta\lambda}{(1+\Delta A_1^1)^2} & 0 & 0 & 0 \\ \frac{abN_1^1}{(bD_1^1+N_1^1)^2} & -d_2 - \frac{abD_1^1}{(bD_1^1+N_1^1)^2} & 0 & f_2 & 0 & 0 \\ \frac{-abN_1^1}{(bD_1^1+N_1^1)^2} & \frac{abD_1^0}{(bD_1^1+N_1^1)^2} & -d_3 & 0 & 0 & 0 \\ 0 & 0 & 0 & -d_1 - f_1 & 0 & \frac{-\Delta\lambda}{(1+\Delta A_2^1)^2} \\ f_2 & 0 & 0 & \frac{abN_2^1}{(bD_2^1+N_2^1)^2} & -d_2 - \frac{abD_2^1}{(bD_2^1+N_2^1)^2} & 0 \\ 0 & 0 & 0 & \frac{-abN_2^1}{(bD_2^1+N_2^1)^2} & \frac{abD_2^1}{(bD_2^1+N_2^1)^2} & -d_3 \end{bmatrix}.$$

The characteristic equation is:

$$|J_{E_1}| = \begin{vmatrix} S + d_1 + f_1 & 0 & \frac{\Delta\lambda}{(1+\Delta A_1^1)^2} & 0 & 0 & 0 \\ \frac{-abN_1^1}{(bD_1^1+N_1^1)^2} & S + d_2 + \frac{abD_1^1}{(bD_1^1+N_1^1)^2} & 0 & -f_2 & 0 & 0 \\ \frac{abN_1^1}{(bD_1^1+N_1^1)^2} & \frac{-abD_1^1}{(bD_1^1+N_1^1)^2} & S + d_3 & 0 & 0 & 0 \\ 0 & 0 & 0 & S + d_1 + f_1 & 0 & \frac{\Delta\lambda}{(1+\Delta A_2^1)^2} \\ -f_2 & 0 & 0 & \frac{-abN_2^1}{(bD_2^1+N_2^1)^2} & S + d_2 + \frac{abD_2^1}{(bD_2^1+N_2^1)^2} & 0 \\ 0 & 0 & 0 & \frac{abN_2^1}{(bD_2^1+N_2^1)^2} & \frac{-abD_2^1}{(bD_2^1+N_2^1)^2} & S + d_3 \end{vmatrix}.$$

Let

$$F_1 = \begin{bmatrix} S + d_1 + f_1 & 0 & \frac{\Delta\lambda}{(1+\Delta A_1^1)^2} \\ \frac{-abN_1^1}{(bD_1^1+N_1^1)^2} & S + d_2 + \frac{abD_1^1}{(bD_1^1+N_1^1)^2} & 0 \\ \frac{abN_1^1}{(bD_1^1+N_1^1)^2} & \frac{-abD_1^1}{(bD_1^1+N_1^1)^2} & S + d_3 \end{bmatrix}, F_2 = \begin{bmatrix} 0 & 0 & 0 \\ -f_2 & 0 & 0 \\ 0 & 0 & 0 \end{bmatrix}$$

and

$$F_3 = \begin{bmatrix} S + d_1 + f_1 & 0 & \frac{\Delta\lambda}{(1+\Delta A_2^1)^2} \\ \frac{-abN_1^1}{(bD_2^1+N_2^1)^2} & S + d_2 + \frac{abD_2^1}{(bD_2^1+N_2^1)^2} & 0 \\ \frac{abN_2^1}{(bD_2^1+N_2^1)^2} & \frac{-abD_1^1}{(bD_2^1+N_2^1)^2} & S + d_3 \end{bmatrix}.$$

Then we obtain

$$|J_{E_1}| = \begin{vmatrix} F_1 & F_2 \\ F_2 & F_3 \end{vmatrix}.$$

As $D_1^1 = D_2^1, N_1^1 = N_2^1, A_1^1 = A_2^1$, we have $F_1 = F_3$. We further have

$$|J_{E_1}| = \begin{vmatrix} F_1 & F_2 \\ F_2 & F_1 \end{vmatrix} = |F_1 + F_2| \cdot |F_1 - F_2|. \tag{5}$$

$$|F_1 + F_2| = \begin{vmatrix} S + d_1 + f_1 & 0 & \frac{\Delta\lambda}{(1+\Delta A_1^1)^2} \\ \frac{-abN_1^1}{(bD_1^1+N_1^1)^2} - f_2 & S + d_2 + \frac{abD_1^1}{(bD_1^1+N_1^1)^2} & 0 \\ \frac{abN_1^1}{(bD_1^1+N_1^1)^2} & \frac{-abD_1^1}{(bD_1^1+N_1^1)^2} & S + d_3 \end{vmatrix}$$

$$= S^3 + M_1 S^2 + M_2 S + M_3,$$

where

$$M_1 = d_2 + \frac{abD_1^1}{\left(bD_1^1 + N_1^1\right)^2} + d_1 + f_1 + d_3,$$

$$M_2 = (d_1 + f_1) d_3 + (d_1 + f_1 + d_3) \left(d_2 + \frac{abD_1^1}{\left(bD_1^1 + N_1^1\right)^2}\right)$$
$$+ \frac{\Delta\lambda}{\left(1 + \Delta A_1^1\right)^2} \cdot \frac{abN_1^1}{\left(bD_1^1 + N_1^1\right)^2},$$

$$M_3 = (d_1 + f_1) d_3 \left(d_2 + \frac{abD_1^1}{\left(bD_1^1 + N_1^1\right)^2}\right)$$
$$+ \frac{\Delta\lambda}{\left(1 + \Delta A_1^1\right)^2} \left[\frac{abN_1^1}{\left(bD_1^1 + N_1^1\right)^2} \left(d_2 + \frac{abD_1^1}{\left(bD_1^1 + N_1^1\right)^2}\right)\right.$$
$$\left. + \left(\frac{abN_1^1}{\left(bD_1^1 + N_1^1\right)^2} + f_2\right) \frac{abD_1^1}{\left(bD_1^1 + N_1^1\right)^2}\right].$$

Similarly, we have

$$|F_1 - F_2| = S^3 + M_1 S^2 + M_2 S + M'_3,$$

where

$$M'_3 = (d_1 + f_1) d_3 \left(d_2 + \frac{abD_1^1}{\left(bD_1^1 + N_1^1\right)^2}\right)$$
$$+ \frac{\Delta\lambda}{\left(1 + \Delta A_1^1\right)^2} \left[\frac{abN_1^1}{\left(bD_1^1 + N_1^1\right)^2} \left(d_2 + \frac{abD_1^1}{\left(bD_1^1 + N_1^1\right)^2}\right)\right.$$
$$\left. + \left(\frac{abN_1^1}{\left(bD_1^1 + N_1^1\right)^2} - f_2\right) \frac{abD_1^1}{\left(bD_1^1 + N_1^1\right)^2}\right].$$

Therefore, Eq. (5) becomes

$$|J_{E_1}| = |F_1 + F_2| \cdot |F_1 - F_2|$$
$$= \left(S^3 + M_1 S^2 + M_2 S + M_3\right)\left(S^3 + M_1 S^2 + M_2 S + M'_3\right). \tag{6}$$

Using the Routh-Hurwitz criterion for Eq. (6), we obtain

$$\Delta_1 \equiv 1 > 0, \quad \Delta_2 \equiv \begin{vmatrix} M_1 & 1 \\ M_3 & M_2 \end{vmatrix} > 0,$$

and

$$\Delta'_1 \equiv 1 > 0, \quad \Delta'_2 \equiv \begin{vmatrix} M_1 & 1 \\ M'_3 & M_2 \end{vmatrix} > 0.$$

Namely, $M_1 M_2 - M_3 > 0$, and $M_1 M_2 - M'_3 > 0$. Because $M_3 > M'_3$, we only need the condition $M_1 M_2 - M_3 > 0$ to be satisfied. That is, if $\frac{M_3}{M_1 M_2} < 1$, the eigenvalues of $|F_1 + F_2| = 0$ and $|F_1 - F_2| = 0$ are negative; thus, E_1 is locally stable.

Lemma 2 *If $\frac{M_3}{M_1 M_2} < 1$, the equilibrium E_1 is locally asymptotically stable; if $\frac{M_3}{M_1 M_2} > 1$, it is unstable.*

For example, a simulation in this case is given in Fig. 7 with the following parameter values: $\lambda_N = 0.027$, $d_1 = d_2 = d_3 = 0.01$, $f_1 = f_2 = 0.665$, $a = 0.012$, $b = 69$, $\Delta = 10^6$ and $\lambda = 35$.

Please note that the equilibria and stability analysis above show that the model would reach an equilibrium in either case of λ, which corresponds to that the system would result in a stable pattern in either normal or mutant conditions. If there are small disturbances of parameters (e.g., small environmental noises), the system would always converge to a stable state after a period of time when $\lambda = 0$ according to Lemma 1, but the system would converge to a stable state for $\lambda > 0$ when the stability conditions in Lemma 2 are satisfied. Besides, we will also use Lemma 2 to carefully choose appropriate parameter values for preserving the stability of the system. This is expected to be shown in the the boundary formation model of the *Drosophila* large intestine.

Deterministic simulation results

However, it is impossible to analyze the model above with a large number of cells in theory, e.g., a model with 60 cells would result in 180 dimensional ordinary differential equations. In this section, we will explore the model using numerical simulation.

The simulation starts from a prepattern of Delta expression in normal large intestines, that is, Delta is expressed only in the ventral region [13]. For this, we set the production rate of the Delta level as follows: 0 for cells at the first three rows (dorsal cells) and λ (greater than 0) for cells at the other rows (ventral cells). We set the time span to [0, 1500]. The parameter values are given in Table 1; except λ and λ_N, other parameter values are fixed throughout the paper. We use the same setting for all the following simulations.

Besides, we determine a boundary cell using the following rule. If the concentration of the active Notch proteins (A) in a cell is equal to or greater than 1.0, we regard the cell as a boundary cell.

In the following, we run the model given in System (1) with the parameter values given in Table 1 to obtain deterministic simulation results. We first validate the ability of our model to reproduce the wild-type result observed in the wet lab. We run simulation and obtain the result in the wild-type condition, illustrated in Fig. 8a. We can see that a single strand of boundary cells that are abutting the ventral region is produced, which is consistent with the experimental observation (corresponding to Scenario 1 and Fig. 3a). We also run the model in other conditions such as over-expression of Notch, and obtain simulation results (see Fig. 8b-c), which are also consistent with experimental observations (see Fig. 3b1-c2). That is,

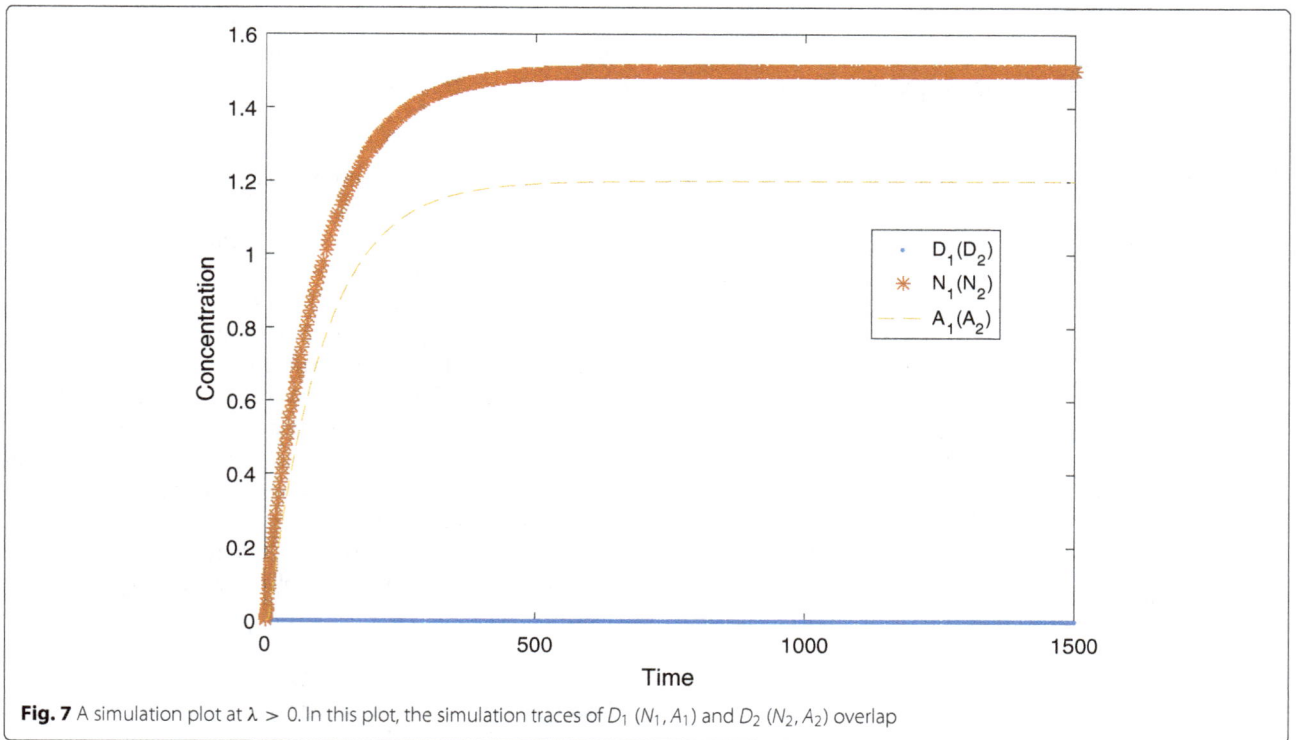

Fig. 7 A simulation plot at $\lambda > 0$. In this plot, the simulation traces of D_1 (N_1, A_1) and D_2 (N_2, A_2) overlap

our model has the ability to reproduce the experimental observations obtained in the wet lab.

Phenotype change due to the change of Delta or Notch expression

During the boundary formation in the *Drosophila* large intestine, different Delta or Notch expression (normal, mis-expression or over-expression) may result in different boundary cell distributions (phenotypes) [1]. In our model, we try to incorporate all these phenotypes

Table 1 Parameter values of the model

Parameter	Value
d_1	0.01
d_2	0.01
d_3	0.01
f_1	0.665
f_2	0.665
a	0.012
b	69
Δ	10^6
λ	[0, 12]
λ_N	[0, 0.1]

Except λ and λ_N, the other parameter values are used throughout the paper

reported so far. As shown in System (1), Delta or Notch expression is controlled by the rates λ and λ_N, respectively. Therefore, we need to map different Delta and Notch expression observed in experiments to different values of λ and λ_N in the model.

To do this, we explore the phenotype change due to the change of Delta or Notch expression in their parameter space and then determine what parameter value results in what kind of phenotype. As well, we cannot adopt a theoretical analysis due to the large number of equations involved, and instead, we still use numerical simulation. Besides, each time we tune only one parameter, either λ or λ_N, by fixing the other.

Phenotype change due to the change of Delta expression

By fixing $\lambda_N = 0.0005$, we gradually increase λ from 0 to 12 with a step 0.1 and then run simulation to obtain the simulation result, given in Table 2.

From the table above, we can see when $\lambda > 0.2$, the first three rows are stable boundary cells, while when $\lambda \leq 0.2$, all the five rows are boundary cells. This means there are only two phenotypes appearing due to the change of Delta expression.

Phenotype change due to the change of Notch expression

By fixing $\lambda = 9$, we gradually increase λ_N from 0 to 0.1 with a small step 0.0001 and then run simulation to obtain the simulation result, given in Table 3.

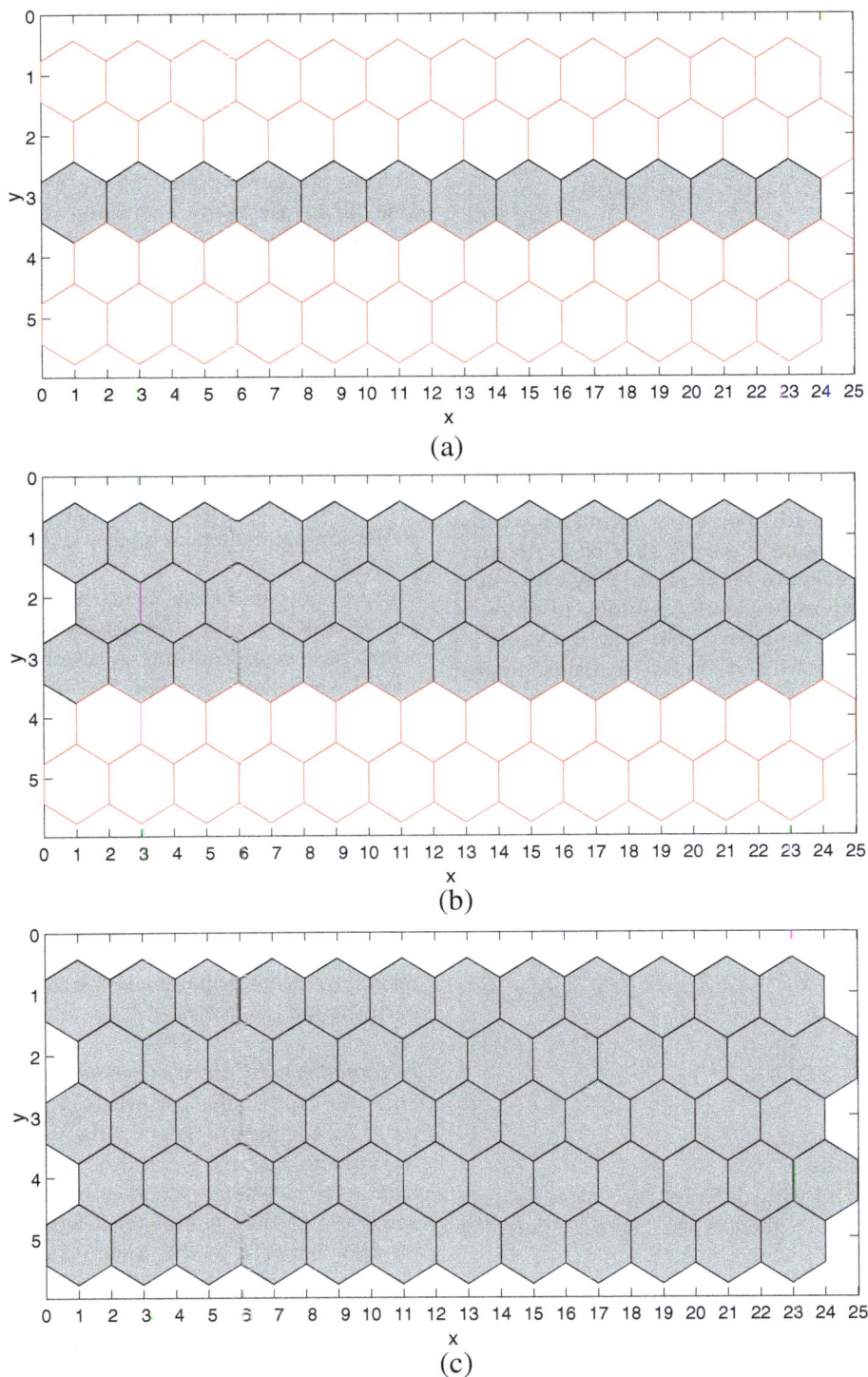

Fig. 8 Deterministic simulation results. **a** Wild-type: $\lambda_N = 0$ and $\lambda = 9$, **b** over-expression of Notch: $\lambda_N = 0.008$ and $\lambda = 9$, and, **c** over-expression of Notch: $\lambda_N = 0.1$ and $\lambda = 9$

Table 3 shows with the increase of λ_N, the number of boundary cells increases, which results in a couple of different phenotypes. Besides, we should notice that a tiny parameter change would result in a big change of the number of boundary cells, so we can deduce that parameter λ_N is very sensitive.

Table 2 Phenotype change due to the change of Delta expression

Expression	Phenotype
$12 \geq \lambda > 0.2$	At the first three rows are boundary cells
$0 \leq \lambda \leq 0.2$	At all five rows are boundary cells

By fixing $\lambda_N = 0.0005$, and gradually increasing λ from 0 to 12 with a step 0.1, we obtain the phenotype change with the change of Delta expression

According to the analysis result above, we finally map different Delta and Notch expression observed in experiments to different values of λ and λ_N in the model. Besides, we can reproduce all known biological phenotypes we have found by combining different values of λ and λ_N.

Simulation results with noises of Notch expression

As described in [1], the boundary formation of the *Drosophila* large intestine is greatly affected by temperature; the random effect of temperature may cause fluctuations of gene expression levels, resulting in different phenotypes, in which boundary cells may randomly be distributed in the dorsal and ventral domains. In this section, we will discuss how to use our model to obtain these random phenotypes.

From Tables 2 and 3, we know that only λ_N is sensitive to System (1), so we only add noises to λ_N. Then, System (1) with a random noise becomes the following form:

$$
\begin{cases}
\dfrac{dD_i}{dt} = \dfrac{\lambda}{1 + \Delta \cdot A_i} - d_1 D_i - \sum_{NG(i)} f_1 \cdot D_i, \\[3mm]
\dfrac{dN_i}{dt} = (\lambda_N + e) - d_2 N_i + \sum_{j \in NG(i)} f_2 \cdot D_j - \dfrac{a N_i}{b D_i + N_i}, \\[3mm]
\dfrac{dA_i}{dt} = -d_3 A_i + \dfrac{a N_i}{b D_i + N_i},
\end{cases}
\tag{7}
$$

where e is a small noise.

Running this model with different λ_N values, we can obtain different random phenotypes. Figure 9 illustrates

Table 3 Phenotype change due to the change of Notch expression

Expression	Phenotype
$0 \leq \lambda_N < 0.0001$	Only at the third row are boundary cells
$0.0001 \leq \lambda_N < 0.0098$	At the first three rows are boundary cells
$0.0098 \leq \lambda_N < 0.01$	At the first four rows are boundary cells
$0.01 \leq \lambda_N \leq 0.1$	At all five rows are boundary cells

By fixing $\lambda = 9$, and gradually increasing λ_N from 0 to 0.1 with a small step 0.0001, we obtain the phenotype change with the change of Notch expression

some phenotypes. So far, we have illustrated that our model reproduces all the phenotypes corresponding to those four scenarios we give above.

Perturbation influences of binding and inhibition parameters on boundary formation

During the development of the *Drosophila* large intestine, there are many environmental factors or noises, which may vary and even heavily affect the boundary formation. To make the model more realistic, we need to consider noises into the model, e.g., adding a random term to each parameter of interest, and analyze the model in noisy conditions. In the section above, we have discussed the random effects of Notch expression on the boundary formation. But in this section, we will further analyze the perturbation influences of the binding and inhibition parameters of the Delta-Notch pathway on the boundary formation. That is, we will explore the inhibition parameters, a and b, and binding parameters, f_1 and f_2.

For these parameters, which ones are of interest? To answer this, we may alternatively find those parameters which have significant influences on the dynamics of the model when they vary. For this, we use sensitivity analysis on the mathematical model to study the influence of parameter perturbation [28]. After that, we can design experiments by considering noises in vivo, and carefully add appropriate noise items to key parameters. We will explore this in the case of over-expression of Notch proteins. For the other cases, we can do the same exploration.

Next, with the parameter setting given in Fig. 8b (of course we can use any other setting), we will explore the influence of parameter perturbation on the boundary formation by changing parameters, a, b, f_1 and f_2, to different perturbation intensities.

Perturbation influence of parameter a

First, we explore the perturbation influence of parameter a. All the parameters take the values given in Fig. 8b. In order to do this, we add a perturbation term $\varepsilon\beta$ to parameter a, where ε denotes the disturbance intensity and β is a random number that satisfies the uniform distribution between [-1, 1]. Then, System (1) with a random disturbance becomes the following form:

$$
\begin{cases}
\dfrac{dD_i}{dt} = \dfrac{\lambda}{1 + \Delta \cdot A_i} - d_1 D_i - \sum_{NG(i)} f_1 \cdot D_i, \\[3mm]
\dfrac{dN_i}{dt} = \lambda_N - d_2 N_i + \sum_{j \in NG(i)} f_2 \cdot D_j - \dfrac{(a + \varepsilon\beta_i)N_i}{b D_i + N_i}, \\[3mm]
\dfrac{dA_i}{dt} = -d_3 A_i + \dfrac{(a + \varepsilon\beta_i)N_i}{b D_i + N_i}.
\end{cases}
\tag{8}
$$

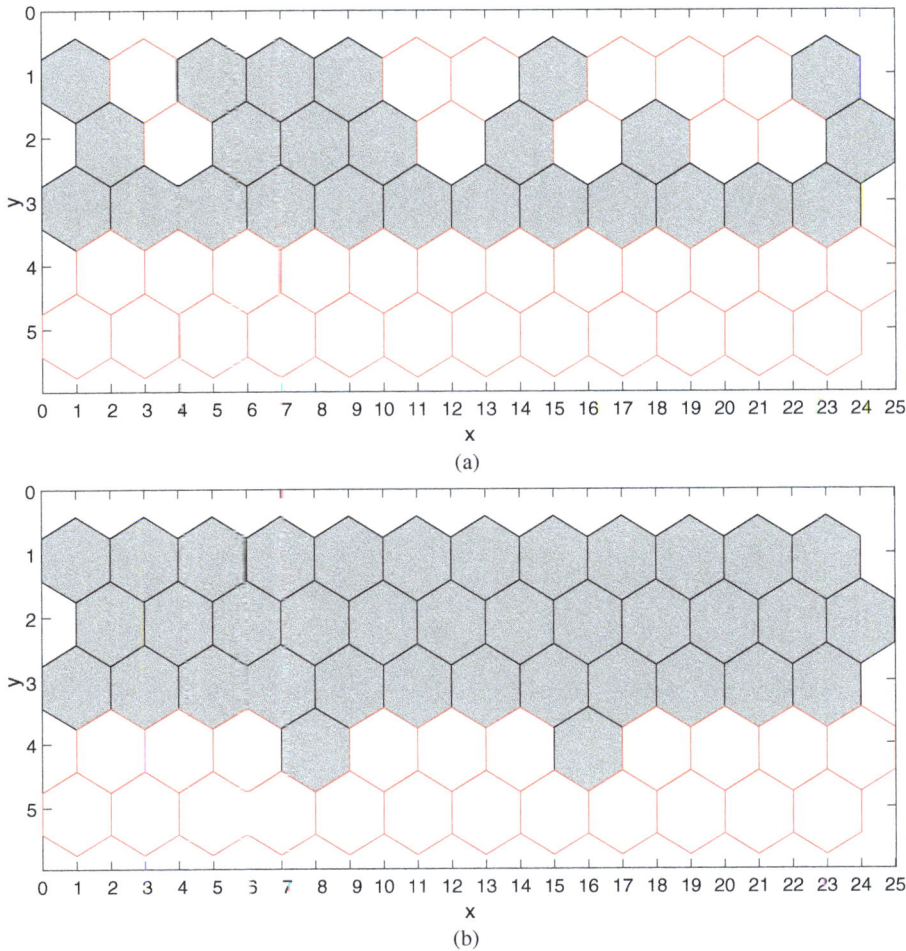

Fig. 9 Two random simulation runs at $\lambda_N = 0.0005$ and $\lambda = 0.3$. The noises are randomly sampled from $[-5 \cdot 10^{-4}, 5 \cdot 10^{-4}]$

At the beginning of each simulation, a random noise $\varepsilon\beta_i$ for each cell i is generated for a by sampling β_i. By varying ε between $[0, 0.001]$ and performing simulation for 50 times at each ε level, we compute the mean and standard error of the number of boundary cells. For example, Fig. 10 gives the result when $\varepsilon = 10^{-7}$, $\varepsilon = 10^{-6}$, $\varepsilon = 10^{-5}$ and $\varepsilon = 10^{-4}$, respectively.

From Fig. 10, we can see that a small perturbation of parameter a may cause a big change of the number of boundary cells (mean), i.e., the bigger the perturbation intensity is, the less the number of boundary cells is (from 36 boundary cells without perturbation to about 15 boundary cells with a perturbation intensity $\varepsilon = 10^{-4}$). Furthermore, from the error bars, we know that the bigger the perturbation intensity is, the bigger the standard error is, although the smaller the mean is.

Perturbation influence of parameter b
We further explore the perturbation influence of parameter b. Similarly we add a perturbation term $\varepsilon\beta$ to b, i.e.,

replacing all b with $b + \varepsilon\beta$ in System (1). We also simulate the model in the same way as for parameter a, and obtain the mean and standard error of the number of boundary cells when $\varepsilon = 0.01$, $\varepsilon = 0.05$, $\varepsilon = 0.1$ and $\varepsilon = 0.2$, respectively, illustrated in Fig. 11.

Figure 11 shows that a big perturbation causes a small change (e.g., there are 36 boundary cells without perturbation and around 34 boundary cells with the perturbation intensity $\varepsilon = 0.2$), which is quite different from the effect of parameter a. Therefore, the perturbation influence of parameter b is much smaller than that of a.

Perturbation influence of parameters f_1 and f_2
We further explore the perturbation of f_1 and f_2. For simplicity, we assume $f_1 = f_2 = f$. Similarly, we add a perturbation term $\varepsilon\beta$ to f_1 and f_2, i.e., replacing all f_1 and f_2 with $f + \varepsilon\beta$ in System (1). We also simulate the model in the same way as above, and obtain the mean and standard error of the number of boundary cells when $\varepsilon = 0.001$,

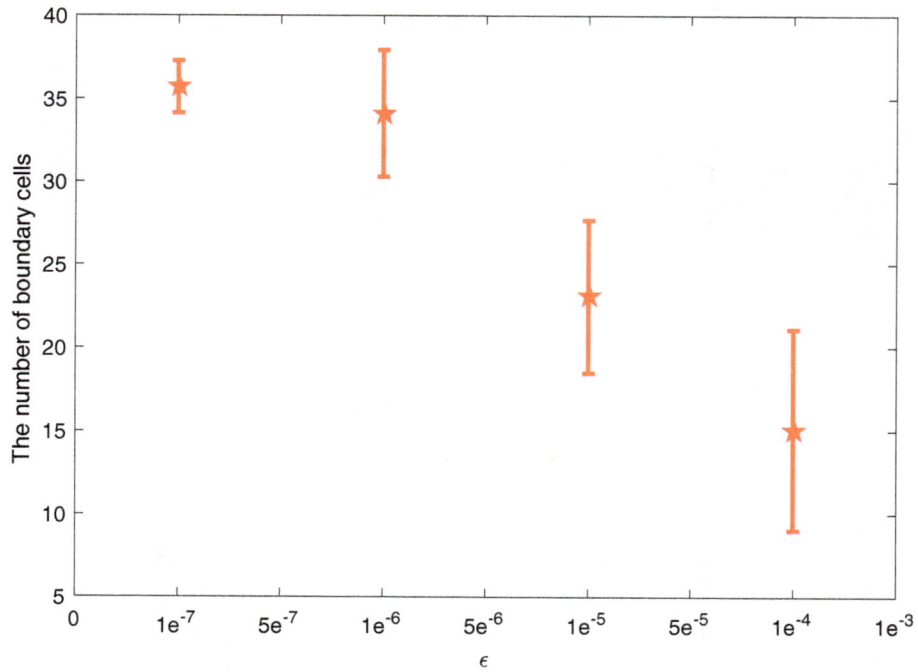

Fig. 10 Perturbation influence of parameter a. The considered disturbance intensities are: $\varepsilon = 10^{-7}$, $\varepsilon = 10^{-6}$, $\varepsilon = 10^{-5}$ and $\varepsilon = 10^{-4}$, respectively

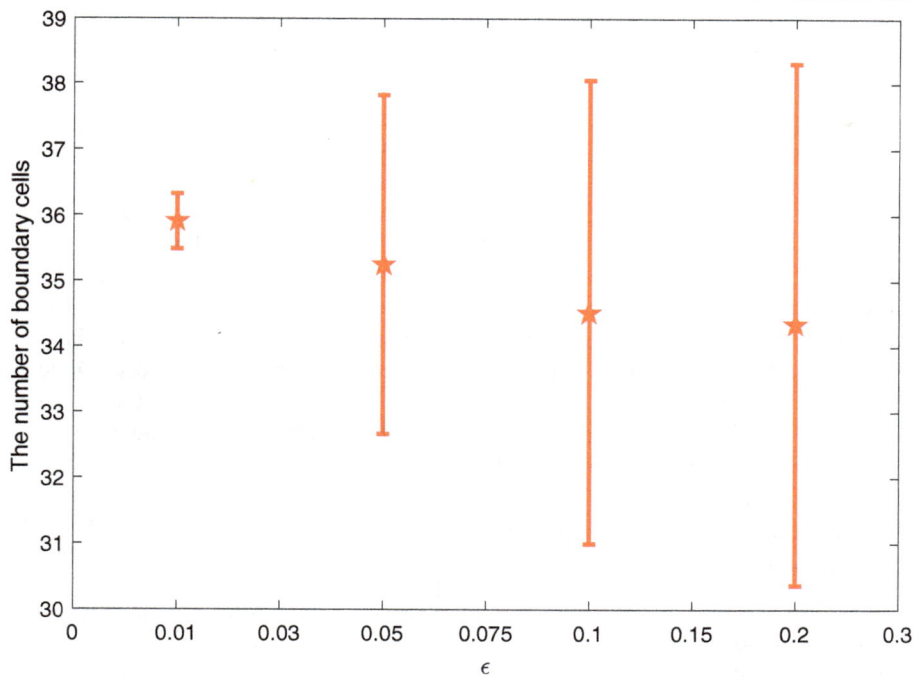

Fig. 11 Perturbation influence of parameter b. The considered disturbance intensities are: $\varepsilon = 0.01$, $\varepsilon = 0.05$, $\varepsilon = 0.1$ and $\varepsilon = 0.2$, respectively

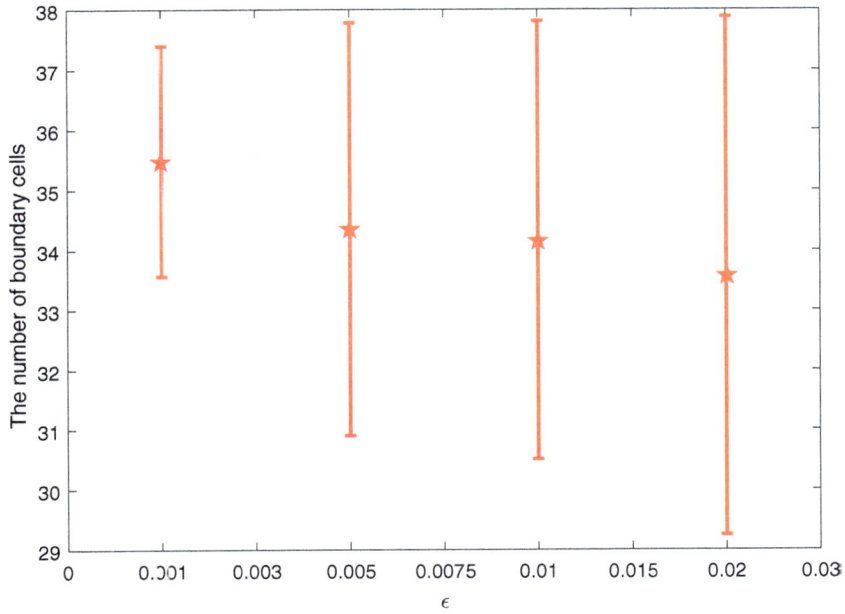

Fig. 12 Perturbation influence of parameters f_1 and f_2. The considered disturbance intensities are: $\varepsilon = 0.001$, $\varepsilon = 0.005$, $\varepsilon = 0.01$ and $\varepsilon = 0.02$, respectively

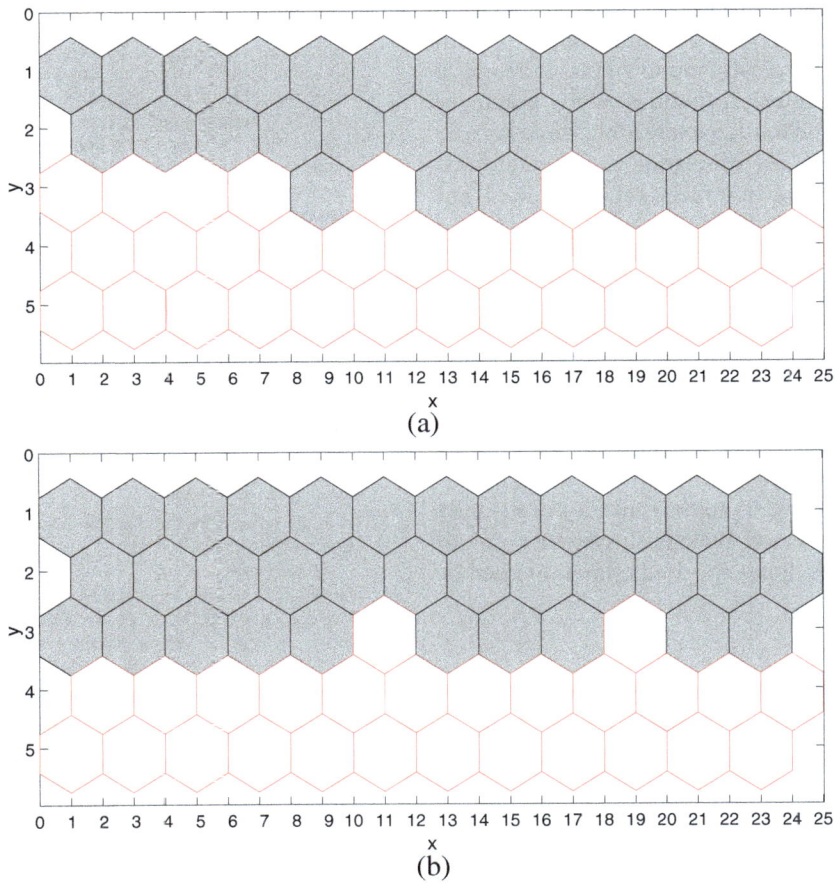

(a)

(b)

Fig. 13 Two simulation runs for parameter a with a perturbation intensity $\varepsilon = 10^{-6}$. All the parameters take the values given in Fig. 8b

$\varepsilon = 0.005$, $\varepsilon = 0.01$ and $\varepsilon = 0.02$, respectively, given in Fig. 12.

From Fig. 12, we can see, similar to parameter b, the perturbation of parameter f does not have an obvious influence on the number of boundary cells in the given perturbation intensities.

In conclusion, the analysis results above show that the perturbation of parameter a has an obvious influence on the boundary formation, while the other three parameters not. Therefore, when we consider noises in the model, we only add a random term to parameter a, and neglect others. That is, we finally obtain a model with random noises, which is given in System (8). With this model, we can make reasonable predictions in noisy conditions by setting the appropriate intensity of the noise according to the analysis result above (see, e.g., Fig. 13).

Conclusion

In this paper, we give a mathematical model for the Delta-Notch dependent boundary formation in the *Drosophila* large intestine. We aim to use this model to better interpret related biological phenomena and therefore we perform not only theoretical but also simulation analysis to achieve the goal. By combining different analysis techniques, we finally confirm that the model is both mathematically and biologically sound and is sufficient to interpret related experimental findings of the boundary formation in the *Drosophila* large intestine. Moreover, by modulating parameters, our simulation can generate various abberant patterns of boundary cells that have not been reported in biological studies so far. Biologically, these varieties of phenotypes are assumed to be results of variation of gene products in the Delta-Notch pathway. To further verify the validity of our simulation method, we are now trying to develop a biological experimental system to manipulate levels of forced-gene expression. In a next step, we will shift the purpose of the model from interpretation to prediction. That is, we will use this model to make different predications in different conditions and then help biologists to have an idea of detailed mechanisms of complicated biological systems, and to design experiments to validate the predictions obtained by our simulation.

Acknowledgment
We would like to thank Prof. Monika Heiner for fruitful discussions and good comments. We also would like to thank the anonymous reviewers for their constructive comments.

Funding
This work was mainly supported by National Natural Science Foundation of China (61273226), which also financed the article-processing charge. FL was supported by JSPS Invitation Fellowships for Research in Japan (L15542) for his stay in Japan during this research. None of the funding agencies had any role in the design of the study and collection, analysis, and interpretation of data and in writing the manuscript.

Authors' contributions
FL conceived the study and developed the model. DS performed the simulations and analyzed the model. HM and RM provided the biological knowledge. All authors read and approved the final manuscript.

About this supplement
This article has been published as part of BMC Systems Biology Volume 11 Supplement 4, 2017: Selected papers from the 10th International Conference on Systems Biology (ISB 2016). The full contents of the supplement are available online at https://bmcsystbiol.biomedcentral.com/articles/supplements/volume-11-supplement-4.

Competing interests
The authors declare that they have no competing interests.

Author details
[1]Control and Simulation Center, Harbin Institute of Technology, West Dazhi Street 92, 150001 Harbin, People's Republic of China. [2]School of Software Engineering, South China University of Technology, Building B7, 510006 Guangzhou, People's Republic of China. [3]Faculty of Science, Yamaguchi University, Yoshida 1677-1, 753-8512 Yamaguchi, Japan.

References
1. Takashima S, Yoshimori H, Yamasaki N, Matsuno K, Murakami R. Cell-fate choice and boundary formation by combined action of Notch and engrailed in the *Drosophila* hindgut. Dev Genes Evol. 2002;212:534–41.
2. Murakami R, Shiotsuki Y. Ultrastructure of the hindgut of *Drosophila* larva, with special reference to the domains identified by specific gene expression patterns. J Morphol. 2001;248:144–50.
3. Takashima S, Murakami R. Regulation of pattern formation in the *Drosophila* hindgut by wg, hh, dpp, and en. Mech Dev. 2001;101:79–90.
4. Hamaguchi T, Takashima S, Okamoto A, Imaoka M, Okumura T, Murakami R. Dorsoventral patterning of the *Drosophila* hindgut is determined by interaction of genes under the control of two independent gene regulatory systems, the dorsal and terminal systems. Mechanisms of Development. 2012;29:236–43.
5. Kumichel A, Knust E. Apical localisation of crumbs in the boundary cells of the *Drosophila* hindgut is independent of its canonical interaction partner stardust. PLoS ONE. 2014;9(5):94038.
6. Wang Y, Zhang SX, Chen L. Modelling biological systems from molecules to dynamical networks. BMC Systems Biology. 2012;6(Suppl 1):1.
7. Grimm O, Coppey M, Wieschaus E. Modelling the bicoid gradient. Development. 2010;137:2253–64.
8. Murray JD. Mathematical Biology. Berlin: Springer; 2003.
9. Nakamasu A, Takahashi G, Kanbe A, Kondo S. Interactions between zebrafish pigment cells responsible for the generation of Turing patterns. PNAS. 2009;106(21):8029–34.
10. Liu F, Blätke M, Heiner M, Yang M. Modelling and simulating reaction-diffusion systems using coloured Petri nets. Comput Biol Med. 2014;53:297–308.
11. Morelli LG, Uriu K, Ares S, Oates AC. Computational approaches to developmental patterning. Science. 2012;36:187–91.
12. Iwaki D, Lengyel JA. A Delta-Notch signaling border regulated by engrailed/invected repression specifies boundary cells in the *Drosophila* hindgut. Mech Dev. 2002;114:71–84.
13. Matsuno H, Murakami R, Yamane R, Yamasaki N, Fujita S, Yoshimori H, Miyano S. Boundary formation by notch signaling in *Drosophila* multicellular systems: experimental observations and gene network modeling by genomic object net. In: Proceedings of Pac Symp Biocomput: 3-7 January 2003; Kauai, Hawaii. World Scientific Press; 2003. p. 152–63.
14. Hayward P, Kalmar T, Arias AM. Wnt/notch signalling and information processing during development. Development. 2008;135:411.
15. Muskavitch MA. Delta-Notch signaling and *Drosophila* cell fate choice. Dev Biol. 1994;166(2):415–30.

16. Bray SJ. Notch signalling: a simple pathway becomes complex. Nat Rev Mol Cell Biol. 2006;7:678–89.

17. Kopan R, Ilagan MXG. The canonical Notch signaling pathway: unfolding the activation mechanism. Cell. 2009;137 216–33.

18. Baker NE. Notch signaling in the nervous system: pieces still missing from the puzzle. BioEssays. 2000;22:264–73.

19. Collier JE, Monk NA, Maini PK, Lewis JH. Pattern formation by lateral inhibition with feedback: a mathematical mode of Delta-Notch intercellular signaling. J Theor Biol. 1996;183(4):429–46.

20. Boareto M, Jolly MK, Lu M, Onuchic JN, Clementi C, Ben-Jacob E. Jagged-Delta asymmetry in Notch signaling can give rise to a sender/receiver hybrid phenotype. PNAS. 2015;22(5):402–9.

21. Sprinzak D, Lakhanpal A, LeBon L, Santat LA, Fontes M, Anderson GA, Garcia-Ojalvo J, Elowitz MB. Cis-interactions between Notch and Delta generate mutually exclusive signalling states. Nature. 2010;465:86–90.

22. Boyce WE. Elementary Differential Equations and Boundary Value Problems, Binder Ready Version, (10th Edition). USA: Wiley; 2012.

23. Hamby DM. A review of techniques for parameter sensitivity analysis of environmental models. Environ Monit Assess. 1994;32(2):135–54.

24. Turányi T. Sensitivity analysis of complex kinetic systems: Tools and applications. J Math Chem. 1990;5(3):203–48.

25. Seydel R. Practical Bifurcation and Stability Analysis. New York: Springer; 2010.

26. Lakshmikantham V, Leela S, Martynyuk AA. Stability Analysis of Nonlinear Systems. Switzerland: Springer; 2015.

27. Khanh NH, Huy NB. Stability analysis of a computer virus propagation model with antidote in vulnerable system Acta Math Sci. 2016;36B(1): 49–61.

28. Shivamoggi BK. Perturbation Methods for Differential Equations. New York: Springer; 2003.

10

Parameter identifiability-based optimal observation remedy for biological networks

Yulin Wang[1] and Hongyu Miao[2*]

Abstract

Background: To systematically understand the interactions between numerous biological components, a variety of biological networks on different levels and scales have been constructed and made available in public databases or knowledge repositories. Graphical models such as structural equation models have long been used to describe biological networks for various quantitative analysis tasks, especially key biological parameter estimation. However, limited by resources or technical capacities, partial observation is a common problem in experimental observations of biological networks, and it thus becomes an important problem how to select unobserved nodes for additional measurements such that all unknown model parameters become identifiable. To the best knowledge of our authors, a solution to this problem does not exist until this study.

Results: The identifiability-based observation problem for biological networks is mathematically formulated for the first time based on linear recursive structural equation models, and then a dynamic programming strategy is developed to obtain the optimal observation strategies. The efficiency of the dynamic programming algorithm is achieved by avoiding both symbolic computation and matrix operations as used in other studies. We also provided necessary theoretical justifications to the proposed method. Finally, we verified the algorithm using synthetic network structures and illustrated the application of the proposed method in practice using a real biological network related to influenza A virus infection.

Conclusions: The proposed approach is the first solution to the structural identifiability-based optimal observation remedy problem. It is applicable to an arbitrary directed acyclic biological network (recursive SEMs) without bidirectional edges, and it is a computerizable method. Observation remedy is an important issue in experiment design for biological networks, and we believe that this study provides a solid basis for dealing with more challenging design issues (e.g., feedback loops, dynamic or nonlinear networks) in the future. We implemented our method in R, which is freely accessible at https://github.com/Hongyu-Miao/SIOOR.

Keywords: Biological network, Graphical model, Structural identifiability analysis, Structural equation model, Observation strategy

Background

The emergence of young research fields such as systems biology and network medicine [1, 2] reflects some exciting changes in biomedical investigators' view of biology and practice. Particularly, it has been increasingly recognized that thinking in networks may lead to novel scientific insights and findings [3] that the traditional reductionism approaches cannot grant [4]. The recent development of experimental techniques (e.g., a variety of high-throughput omics approaches) also provides unprecedented opportunities for biomedical investigators to construct numerous biological networks at different levels and scales; for instance, protein-protein interaction networks [5, 6], gene regulatory networks [7–10], functional RNA networks [11–13], and metabolic networks [14, 15] can be found in a number of databases or knowledge repositories nowadays [9, 16, 17]. All such previous efforts provide a solid basis for further advancing our understanding of biological systems and the associated outcomes qualitatively or quantitatively.

* Correspondence: Hongyu.Miao@uth.tmc.edu
[2]Department of Biostatistics, School of Public Health, University of Texas Health Science Center at Houston, Houston, TX 77030, USA
Full list of author information is available at the end of the article

Graphical models have long been considered as a natural mathematical representation of biological network for various quantitative analysis tasks such as parameter inference [18–21]. Specifically, given a biological network structure and experimental observations of certain variables associated with network nodes, it is often of significant interest to determine the unknown coefficients associated with network edges. For instance, to understand the responses of a biological network (e.g., activation or inhibition) to different environmental signals (e.g., different signaling molecules or different doses of the same signaling molecule), edge coefficients are likely to vary under different conditions and thus need to be estimated under each condition for the same given network structure [18]. In such a scenario, although the structure of the corresponding graphical model is known and fixed, concerns about the accuracy and reliability of parameter estimates often raise due to, e.g., the existence of unobserved node variables (i.e., latent variables). In practice, latent variables are not uncommon due to various technical limitations, ethic issues, financial affordability, and so on [18, 20]. Therefore, a natural question to ask is: what is the remedy that enables us to obtain reliable parameter estimates for a given graphical model structure with partially observed variables?

To the best knowledge of our authors, the aforementioned important question has rarely been tackled before in the context of quantifying unknown model parameters of biological networks; and in this study, we make the very first attempt to address this question from the structural identifiability point of view. By the definition in Miao et al. [18], an unknown model parameter is structurally identifiable if it can be uniquely determined for a given model structure under the assumptions that sample size is sufficiently large and data quality is not of concern. Of course, one can also take the effects of sample size and data noise into consideration and conduct the so-called practical identifiability analysis [18]; however, this is out of the scope of this study as practical identifiability analysis is not feasible at certain experimental design stage when real data are not available. On the contrary, structural identifiability analysis allows us to detect flaws in model structure and observation scheme before data collection, and thus should be investigated first. Our solution to the question mentioned at the end of the previous paragraph is thus a strategy that identifies a minimum number of unobserved nodes, for which the associated node variables should be observed in experiments such that all unknown parameters become structurally identifiable. This is a useful and cost-effective remedy if some of the model parameters are not identifiable given the

original observation scheme, and we thus name it the structural identifiability-based optimal observation remedy (SIOOR).

Since biological networks can be represented by many different types of mathematical or statistical models, it is impossible to devise the SIOOR strategy for every different model type in one study. Therefore, we consider a linear structural equation model [22] here because it is a representative graphical model type and has been widely applied in various disciplines including systems biology [23–27]. A number of previous studies have investigated the parameter identifiability problem of SEMs, but the majority of these studies only derived theoretical criteria or conditions for identifiability verification, including Pearl's back door and front door criteria [28], Brito and Pearl's generalized instrumental variable criterion [29], Tian's accessory set approach [30]. Only a few studies proposed computerizable identifiability analysis approaches, including Drton's condition [31] and Foygel's half-trek criterion [32] (implemented in R package SEMID), Sullivant's computer algebra method and the more recent Wang's identifiability matrix method [33, 34]. More importantly, all such criterions and methods assume that the observation strategy is given (i.e., it is pre-specified which variables are observed and which are not), and none of them considered the remedy strategy if a given observation strategy does not grant identifiability to all unknown model parameters. The focus of this study is thus to investigate how to choose a minimum number of nodes that are not observed in the original observation strategy for additional experimental measurements such that all unknown model parameters become identifiable. This study leads to a general and computerizable solution to the SIOOR problem for the first time.

More specifically, in the case that a given observation strategy of a biological network cannot grant identifiability of all unknown parameters in the corresponding SEM due to the existence of unobserved variables, we propose a dynamic programming (DP) approach to search for all possible SIOOR strategies. The proposed approach is a generic and computerizable method that can deal with recursive SEMs. It should be stressed that SIOOR strategy does not involve any power or sample size calculation and thus cannot be compared with the traditional experimental design approaches [35, 36]. Also, it should be stressed that the observability problem in control theory is different from the SIOOR problem because the aim of observability analysis is to determine the internal states of a system from its external outputs [37]. For clarification purpose, we also compare Liu's graphic approach for observability analysis [38] with our SIOOR strategy in this study.

This article is organized as follows. In the Methods Section, the structural identifiability-based optimal observation remedy problem is mathematically formulated. We then propose a dynamic programming approach with theoretical justification to solve the problem for recursive SEMs. In the Results and Discussion Section, we describe our algorithm implementation and validate the proposed method using selected benchmark networks. Also, a real substructure from the influenza virus A [39] KEEG pathway is chosen as an example to illustrate the application of the proposed method in practice.

Methods

In this section, several key concepts and definitions are introduced for solving the SIOOR problem, including Observation Strategy (OS), Cardinality of Observation Strategy [4], and Identifiability Gain (IG). The design of the dynamical programming algorithm is also described. In addition, we provide the necessary theoretical justification for the proposed method.

Problem formulation

A directed biological network can be denoted by $G = (\mathbf{V}, \mathbf{E})$, where $\mathbf{V}$ denotes the node set and $\mathbf{E}$ denotes the edge set. Let V_i ($i = 1, 2, \ldots, n$) denote the i-th node, and Y_i denote the variable associated with V_i. If Y_i is a linear function of the remaining node variables, the corresponding SEM can be specified as follows,

$$Y_i = \sum_{j \neq i} c_{ij} Y_j + \varepsilon_i, \quad i, j = 1, \cdots, n,$$

where c_{ij} denotes the coefficient associated with the directed edge $V_j \rightarrow V_i$, and ε_i denotes the disturbance error term that follows a certain distribution (Gaussian or non-Gaussian [40, 41]) with mean zero. For simplicity, all disturbance error terms are assumed to be independent. By definition, $\mathbf{E}$ specifies the structure of the coefficient matrix $\mathbf{C} = [c_{ij}]$, i.e., $c_{ij} = 0$ if no edge exists in $\mathbf{E}$ from V_j to V_i for $i \neq j$. When a network structure contains one or more loops, G is a directed cyclic graph (DCG) and the corresponding SEM is called a non-recursive model; otherwise, G is a directed acyclic graph [42] and the corresponding SEM is called recursive. Although Drton's condition [31] and Foygel's half-trek criterion [32] are applicable to the identifiability analysis of non-recursive SEMs, the identifiability of parameters on a loop may be still inconclusive. Due to the lack of mature structural identifiability analysis techniques for examining every unknown parameter of a non-recursive SEM, this study focuses on recursive SEMs (i.e., DAGs) only.

Definition 1 (*observation strategy*). Given a graph G $= (\mathbf{V}, \mathbf{E})$, its observation strategy can be denoted by a binary vector $O = (O_{V_1}, \cdots, O_{V_n})^T$, where $O_{V_i} = 1$ if node V_i is observed and $O_{V_i} = 0$ if V_i is unobserved.

Observation strategy is important to parameter identifiability. In general, for a given network structure, the more observed nodes an observation strategy contains, the more likely all model parameters are identifiable. However, more observed nodes are usually associated with a higher experiment cost, so it is also desirable to reduce any unnecessary cost. The goal of SIOOR is thus to improve a given observation strategy by observing a minimum number of originally unobserved nodes such that all nonzero parameters in $\mathbf{C}$ become identifiable. For this purpose, let $\mathbf{P}$ denote the vector of all nonzero parameters in $\mathbf{C}$, and let $\mathbf{D}$ denote the vector of identifiability status of every element in $\mathbf{P}$. That is, if P_i is locally or globally identifiable (i.e., P_i has a finite number of possible values or a unique value within the parameter space, see [18]), $D_i = 1$; otherwise, $D_i = 1$. When all the parameters in a model are locally or globally identifiable, this model is called identifiable. Consequently, the SIOOR problem can be formulated as follows

$$\min_{\text{observed } V_i} \sum_{i=1}^{n} O_{V_i}, \text{ subject to } \mathbf{D} = \mathbf{1}, \quad (1)$$

where $\sum_{i=1}^{n} O_{V_i}$ is the total number of observed nodes in an observation strategy O, and $\mathbf{1}$ denotes a vector of ones. For clarification, we stress that the observation measurements are for the random variables associated with network nodes, and we assume $(n - m)$ of them are observed in the original observation strategy, where n denotes the total number of nodes and $0 < m \leq n$.

The objective function above is minimized with respect to the originally unobserved nodes, subject to the constraint $\mathbf{D} = \mathbf{1}$. During the minimization process, it needs to be repeatedly verified whether all parameters have become identifiable (i.e., $\mathbf{D} = \mathbf{1}$). For this purpose, an efficient algorithm for structural identifiability analysis of SEMs is needed. Here we consider the identifiability matrix method proposed by Wang et al. [34]. Briefly, structural identifiability of parameters can be verified by examining the number of solutions to the symbolic polynomial identifiability equations generated by Wright's path coefficient method [43, 44]. To avoid the expensive symbolic computation involved in reducing such identifiability equations, the identifiability matrix method proposes to derive binary matrices from symbolic polynomials and thus enable us to determine the number of solutions via several simple matrix operations. It is noteworthy that Wang's work [34] does not explicitly handle colliders involving bidirectional arcs when generating identifiability equations with Wright's method, however, the identifiability matrix method is still applicable here as we do not consider bidirectional arcs in DAGs.

Identifiability gain and must-be-observed nodes

The optimization problem in the previous section is combinatorial in nature. Therefore, if the number of the originally unobserved nodes (denoted by m) is not small, enumerating all the 2^m different possible observation strategies over these nodes will be computationally expensive. We thus need an efficient algorithm such as dynamic programming to obtain the solutions. For this purpose, a few more definitions need to be introduced first.

Definition 2 (*redundant identifiability equation*). Given a set of identifiability equations, an identifiability equation $IE(V_i,V_j)$ is redundant with respect to that set if it can be expressed as a linear combination of the equations in that set.

Definition 3 (*cardinality of observation strategy*). Given an observation strategy O for a network G, one symbolic polynomial identifiability equation can be generated for each pair of d-connected [28] observed nodes using, e.g., Wright's path coefficient method. Then the total number of non-redundant identifiability equations is called the cardinality of O, denoted by $f(O)$.

The Wright's path coefficient method generates identifiability equations for recursive SEMs by calculating the covariance between two node variables, which is equal to the sum of the products of edge coefficients along each d-connected path, i.e., $IE(V_i, V_j) : Cov$

$(V_i, V_j) = \sum_{path_k} \prod_{edge_l} \theta_l$. After removing all redundant iden-

tifiability equations and redundant monomials, the identifiability result of each parameter can be determined by Theorem 1 in [34]. That is, if the number of non-redundant identifiability equations is less than the number of unknown parameters, then the parameters have an infinite number of possible values within the parameter space and are thus unidentifiable; otherwise, the parameters have a limited number of solutions or even a unique solution and are thus at least locally identifiable [45]. Let N_u denote the total number of unknown parameters in $\mathbf{P}$. For every parameter in $\mathbf{P}$ being locally or globally identifiable, the inequality $f(O) \geq N_u$ should hold according to Theorem 1 in [34]. Therefore, the optimization problem can also be formulated as follows

$$\min_{\text{observed } V_i} \sum_{i=1}^{n} O_{V_i}, \text{ subject to } f(O) \geq N_u, \qquad (2)$$

where the calculation of $f(O)$ is a key challenge because it depends on specific network structure and observation strategy and thus has no closed-form solution. We thus introduce the following definition.

Definition 4 (*identifiability gain*). Given a network $G = (\mathbf{V,E})$, let $O^{(k)}$ and $f(O^{(k)})$ denote an observation strategy and its cardinality, respectively. Let V_i be an unobserved node in $O^{(k)}$, and only V_i becomes observed in a new observation strategy $O^{(k+1)}$ with the observation statuses of other nodes remaining unchanged. Let $f(O^{(k+1)})$ denote the cardinality of $O^{(k+1)}$. Then the identifiability gain of observing V_i, denoted by $g(V_i, O^{(k)})$, is calculated as $g(V_i, O^{(k)}) = f(O^{(k+1)}) - f(O^{(k)})$.

By definition, $g(V_i, O^{(k)})$ is the difference in cardinality between two consecutive observation strategies $O^{(k)}$ and $O^{(k+1)}$. That is, after V_i becomes observed in $O^{(k+1)}$, we need to find out the number of newly added non-redundant identifiability equations. First, if another node V_j ($i \neq j$) is observed in both $O^{(k)}$ and $O^{(k+1)}$ and there exists a Wright's path [46] of length 1 connecting V_i and V_j, it can be shown that the newly added identifiability equation, denoted by $IE(V_i,V_j)$, is non-redundant (see Lemma 1 and Additional file 1 for theoretical justification). However, if the length of every Wright's path between V_i and V_j is greater than 1, the identifiability equation $IE(V_i,V_j)$ is not always redundant, and it depends on both the node's observation status and the structure of the network. Here we introduce the concept of detour-path before we further elucidate the redundancy issue. Consider a DAG $G = (\mathbf{V,E})$ and two d-connected observed nodes V_i and V_j. Assume that there exists a Wright's path P_{ji} between V_i and V_j as well as an observed node $V_k (k \neq i, j)$ on P_{ji}, and the direction of P_{ji} is from V_i to V_k and then to V_j. Now let P_{ki} and P_{jk} denote the two segments of P_{ji}, then P_{ki} entering node V_k has an arrow pointing into V_k while P_{jk} exiting node V_k has an arrow pointing away from V_k. However, if there exists another Wright's path between V_k and V_j, denoted by $\tilde{P}_{kj}$, which has no any other observed nodes besides V_k and V_j and has an arrow pointing into V_k, then V_k is a collider with respect to P_{ki} and $\tilde{P}_{kj}$. Thus, we call the Wright's path segment P_{jk} the detour-path, and call V_i, V_j and V_k the upstream node, the downstream node, and the collider node of the detour-path P_{jk}, respectively. By definition, a detour-path can have only one downstream node and one collider node but may have one or more upstream nodes. Moreover, multiple detour-paths can share the same upstream node, the same downstream node or the same collider node. Several examples are given in Fig. 1 to illustrate the concept of detour-path.

In addition, when an upstream node V_i is shared by two or more detour-paths that have the same downstream node, V_i is called a shared upstream node; otherwise, V_i is called an exclusive upstream node. Note that a detour-path can have both exclusive and shared upstream nodes in the same time, and the collider node of one detour-path can be an upstream node of another detour-path. Consider two detour-paths that have no exclusive upstream nodes, if they share the same downstream node and at least one upstream node, or one upstream node of one detour path is the collider node of the other detour-path, then two detour-paths are intersecting. One can tell that if P_{jk_1} intersects with P_{jk_2} and P_{jk_2} intersects with P_{jk_3}, then P_{jk_1} also intersects with P_{jk_3}. Then we consider a downstream node V_j, let S_IDP denote all the intersecting detour-paths, and let S_SUN denote all the shared upstream nodes of S_IDP. Similar to a single unknown parameter, the coefficient product $WP = \prod_{edge_l} \theta_l$ of a Wright's path P can be deemed as a single parameter and one can tell its structural identifiability based on identifiability equations. If a detour-path P has at least one exclusive upstream node, then the Wright's coefficient WP of P is globally identifiable (see Lemma 2 and Additional file 1 for theoretical justification). Also, for a group of intersecting detour-paths, if the node number of S_SUN is equal to or greater than the number of intersecting detour-paths in S_IDP,

then the Wright's coefficient of each detour-path in S_IDP is globally identifiable (see Lemma 3 and Additional file 1 for theoretical justification).

Given a DAG $G = (\mathbf{V}, \mathbf{E})$, consider two observed nodes V_i, V_j and an unobserved node V_u. V_u may not be on any Wright's paths between V_i and V_j. For this case, if only V_u becomes observed in $O^{(k+1)}$, then for each observed node V_i in $O^{(k)}$, one can check whether the identifiability equation $IE(V_i, V_u)$ is redundant according to Lemma 4 (see Additional file 1 for theoretical justification). That is, when none of the Wright's paths between V_i and V_u contains detour-paths, $IE(V_i, V_u)$ is redundant if and only if each Wright's path between V_i and V_u passes at least one observed node other than V_i and V_u; otherwise, $IE(V_i, V_u)$ is redundant if and only if the Wright's coefficient of each detour-path between V_i and V_u is globally identifiable in $O^{(k)}$ and each Wright's path between V_i and V_u passes at least one observed node other than V_i and V_u. If V_u is on a Wright's path between V_i and V_j, and the sufficient and necessary condition for one of the identifiability equations $IE(V_i, V_u)$ and $IE(V_j, V_u)$ being redundant is similar to Lemma 4 and given in Lemma 5 (see Additional file 1 for theoretical justification). Note that it can be determined whether the Wright's coefficient of a detour-path is globally identifiable according to Lemma 2 and Lemma 3.

Based on Lemma 4 and Lemma 5, we propose a novel graphic method to calculate the identifiability gain $g(V_i, O^{(k)})$. Let des_i denote the descendant node set of V_i, anc_i denote the ancestor node set of V_i, rel_i denote the set of nodes that are not included in des_i

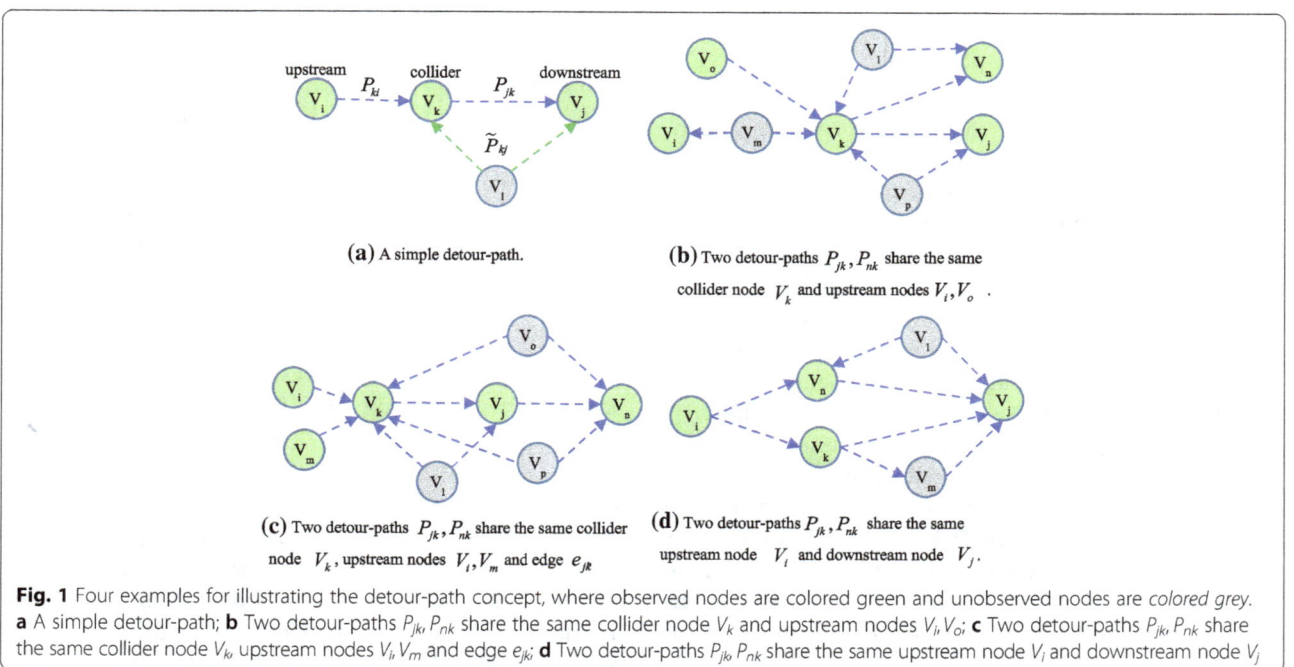

(a) A simple detour-path.

(b) Two detour-paths P_{jk}, P_{nk} share the same collider node V_k and upstream nodes V_i, V_o.

(c) Two detour-paths P_{jk}, P_{nk} share the same collider node V_k, upstream nodes V_i, V_m and edge e_{jk}.

(d) Two detour-paths P_{jk}, P_{nk} share the same upstream node V_i and downstream node V_j.

Fig. 1 Four examples for illustrating the detour-path concept, where observed nodes are colored green and unobserved nodes are *colored grey*. **a** A simple detour-path; **b** Two detour-paths P_{jk}, P_{nk} share the same collider node V_k and upstream nodes V_i, V_o; **c** Two detour-paths P_{jk}, P_{nk} share the same collider node V_k, upstream nodes V_i, V_m and edge e_{jk}; **d** Two detour-paths P_{jk}, P_{nk} share the same upstream node V_i and downstream node V_j

or anc_i. Moreover, let $bound_i \subset anc_i$ denote the boundary node set, in which every node has at least one outgoing edge to a node in rel_i. Then we can calculate $g(V_i, O^{(k)})$ by removing the following edges from the original graph **G**: i) all the incoming edges to the observed nodes that are not collider nodes of detour-paths in anc_i; ii) all the outgoing edges from some observed nodes in des_i and rel_i and these observed nodes are not the collider nodes of the detour-paths whose Wright's coefficients are unidentifiable in $O^{(k)}$; and iii) all the outgoing edges from the observed nodes in $bound_i$ to nodes in rel_i, and then we get a new graph denoted by G'. Let N_w denote the total number of the observed nodes that are connected with V_i via any Wright's path in graph G'. Furthermore, one can tell from the edge-removal operation that there still exist some redundant identifiability equations in G', because the following two types of redundancy cases have not been considered in the edge-removal operation: V_i is the downstream node of an arbitrary detour-path, and V_i is on a Wright's path between two observed nodes in G'. Let N_r denote the number of redundant identifiability equations in G'. According to the topological structure of G' and the node's observation status, we can obtain N_r based on Lemma 4 and Lemma 5 (see the details in Implementation and Verification Section). It can be shown that the identifiability gain is $g(V_i, O^{(k)}) = N_v - N_r$ (see Theorem 1 and Additional file 1 for theoretical justification).

For a given DAG **G** and an observation strategy $O^{(k)}$, different unobserved nodes may associate with different identifiability gains. Naturally, our strategy is to choose the unobserved node in $O^{(k)}$ with the maximum identifiability gain if it becomes observed in $O^{(k+1)}$. However, we also recognize that, to assure that all model parameters are at least locally identifiable, certain nodes of a DAG must be observed if they are unobserved in an observation strategy (see Lemma 6 and Additional file 1 for theoretical justification). For convenience, we call such nodes the must-be-observed [14] nodes, and let $O^{(0)}{}_M$ denote the observation strategy, in which only the MBO nodes are observed.

Lemma 1 *Given a DAG* G = (V,E), *an observed node V_i, and an unobserved node V_u in $O^{(k)}$, if only V_u becomes observed in $O^{(k+1)}$, the identifiability equation $IE(V_i, V_u)$ is non-redundant if there exists a Wright's path of length 1 connecting V_i and V_u.*

Lemma 2 *If a detour-path P has one or more exclusive upstream node, the Wright's coefficient WP of P is globally identifiable.*

Lemma 3 *For a group of intersecting detour-paths, if the number of the shared upstream nodes in S_SUN is equal to or greater than the number of intersecting*

detour-paths in S_IDP, then the Wright's coefficient of each detour-path in S_IDP is globally identifiable.

Lemma 4 *Given a DAG* G = (V,E), *an observed node V_i, and an unobserved node V_u in $O^{(k)}$, if only V_u becomes observed in $O^{(k+1)}$, there exist two cases:*

1) each Wright's path between V_i and V_u passes at least one observed node other than V_i and V_u when none of the Wright's paths between V_i and V_u contains detour-paths;

2) each Wright's path between V_i and V_u passes at least one observed node other than V_i and V_u, and the Wright's coefficient of each detour-path between V_i and V_u is globally identifiable in $O^{(k)}$ when certain Wright's paths between V_i and V_u contain detour-paths.

Then the identifiability equation $IE(V_i, V_u)$ is redundant if and only if one of the above conditions holds.

Lemma 5 *Given a DAG* G = (V,E), *two d-connected observed nodes V_i and V_j, and an unobserved node V_u in $O^{(k)}$, if V_u is on a Wright's path between V_i and V_j and only V_u becomes observed in $O^{(k+1)}$, there exist two cases:*

1) each Wright's path between V_i and V_j passes at least one observed node other than V_i and V_j when none of the Wright's paths between V_i and V_j contains detour-paths;

2) each Wright's path between V_i and V_j passes at least one observed node other than V_i and V_j, and the Wright's coefficient of each detour-path between V_i and V_j is globally identifiable in $O^{(k)}$ when certain Wright's paths between V_i and V_j contain detour-paths.

Then one of the two identifiability equations $IE(V_i, V_u)$ and $IE(V_j, V_u)$ is redundant if and only if one of the above conditions holds.

Theorem 1 *Given a DAG* G = (V, E) *and an unobserved node V_i in an observation strategy O, let G' denote the sub-graph after the edge-removal operation. Then the identifiability gain is $g(V_i, O) = N_w - N_r$, where N_w denotes the total number of the observed nodes that are connected with V_i via any Wright's path in graph G', and N_r denotes the number of redundant identifiability equations in graph G'.*

Lemma 6 *For a given DAG* G = (V, E), *the following nodes must be observed to assure that all the parameters of the corresponding SEM are at least locally identifiable*

1) The nodes with an out-degree 0;
2) The nodes with an out-degree 1;
3) The nodes with an in-degree 0 and an out-degree less than 3.

Dynamic programming strategy

Let $O^{(0)}{}_G$ denote a given observation strategy. If some of the MBO nodes are not observed in $O^{(0)}{}_G$, $O^{(0)}{}_M$ should be incorporated into $O^{(0)}{}_G$ according to Lemma 6. Therefore, the initial observation strategy, denoted by $O^{(0)}$, should always be $O^{(0)} = \left(O^{(0)}{}_M | O^{(0)}{}_G\right)$, where the OR operator is an element-wise operation. For example, for a DAG with six nodes, if $O^{(0)}{}_M = \begin{bmatrix} 1 & 0 & 1 & 0 & 0 & 0 \end{bmatrix}^T$ and $O^{(0)}{}_G = \begin{bmatrix} 0 & 1 & 1 & 0 & 0 & 0 \end{bmatrix}^T$, then $O^{(0)} = \begin{bmatrix} 1 & 1 & 1 & 0 & 0 & 0 \end{bmatrix}^T$.

The dynamic programming strategy starts with the calculation of the cardinality of $O^{(0)}$ (that is, $f(O^{(0)})$) based on Theorem 1. Specifically, let R be the number of observed nodes in $O^{(0)}$, $V_{o\_r}(r = 1, 2, \cdots, R)$ be the r-th observed node in $O^{(0)}$, and $O^{(0)}\{V_{o\_1}, ..., V_{o\_r}\}$ be the observation strategy in which only the first r observed nodes in $O^{(0)}$ are observed. Then $f(O^{(0)}) = \sum_{r=1}^{R-1} g\left(V_{o\_(r+1)}, O^{(0)}\{V_{o\_1}, ...,, V_{o\_r}\}\right)$ can be calculated according to Theorem 1. Note that the order at which V_{o_r} is selected into $O^{(0)}\{V_{o\_1}, ..., V_{o\_R}\}$ will not change the observation strategy (e.g., $O^{(0)}\{V_{o\_1}, V_{o\_2}\} = O^{(0)}\{V_{o\_2}, V_{o\_1}\}$) and thus have no effect on the value of $f(O^{(0)})$.

The second step of our dynamic programming strategy is to define stages and their associated states. Let S denote the number of unobserved nodes in $O^{(0)}$, and let V_{u_s} ($s = 1, 2, \cdots, S$) denote the s-th unobserved node in $O^{(0)}$, then the dynamic programming procedure can be divided into $S + 1$ stages. For illustration purpose, we consider a simple example with 5 unobserved nodes, as shown in Fig. 2. The 0-th stage is actually the initialization step as described in the previous paragraph, and it has only one state, i.e., $O^{(0)}$. At the first stage, there are $S = 5$ different states; that is, only one of the unobserved nodes $\{V_{u\_1}, V_{u\_2}, \cdots, V_{u\_5}\}$ in $O^{(0)}$ will be selected to observe. At the second stage, since one of the five unobserved nodes has been selected at the

previous stage, there are only four unobserved nodes for selection and thus four states exist (that is, $\{V_{u\_2}, V_{u\_3}, V_{u\_4}, V_{u\_5}\}$). Therefore, as shown in Fig. 2, except for stages 0 and 1, each subsequent stage has one less states than its previous stage; also, the upper triangular region (see the area above the labels of stages 1–5 in Fig. 2) is empty because the selection order of unobserved nodes does not affect the eventual observation strategy so the inclusion of such states in the upper triangular region is redundant. One can tell that the proposed stage and state definitions satisfy the optimality principle of dynamic programming [47–49].

The third step is to compute the state transition costs for searching the optimal state transition path(s). According to the definitions of stages and states, there may exist several different states at the s-th stage that can transit to the same state at the $(s + 1)$-th stage. For instance, four states V_{u_1}, V_{u_2}, V_{u_3} and V_{u_4} at the first stage can transit to V_{u_5} at the second stage, as shown in Fig. 2. The state transition cost from state V_{u_i} to state V_{u_j} ($i \neq j$) between two consecutive stages is just the identifiability gain $g(V_{u\_j}, O^{(k)}\{..., V_{u\_i}, ...\})$, where $O^{(k)}\{..., V_{u\_i}, ...\}$ means that V_{u_i} is observed in $O^{(k)}$. Then the cardinality of an observation strategy can be computed by adding $f(O^{(0)})$ and all the state transition costs along the state transition path. Since the goal of the dynamic programming strategy is to search for the optimal observation strategies, when there exist multiple transition paths from state V_{u_i} in $O^{(k)}$ to state V_{u_j} in $O^{(k+1)}$ ($i \neq j$), the transition path associated with the maximum identifiability gain will be chosen; that is, $f\left(O^{(k+1)}_{V_{u\_j}}\right) = \max_{V_{u\_i}, i \neq j} \left(g\left(V_{u\_j}, O^{(k)}_{V_{u\_i}}\right) + f\left(O^{(k)}_{V_{u\_i}}\right)\right)$, where $O^{(k)}_{V_{u\_i}}$ is a convenient notation for $O^{(k)}\{..., V_{u\_i}, ...\}$.

The dynamic programming strategy above can be mathematically described in Eq. (3), and we have implemented this strategy in R (see the "Implementation and verification" Section),

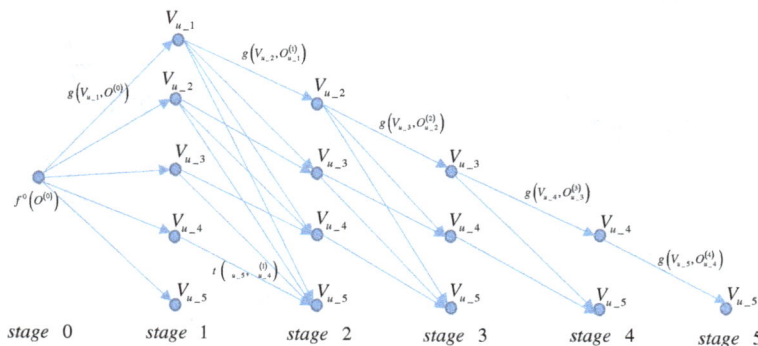

Fig. 2 Schematic illustration of the stages, states and state transition costs in the proposed dynamic programming strategy

$$\begin{cases} f\left(O_{V_{u\_s}}^{(1)}\right) = f\left(O^{(0)}\right) + g\left(V_{u\_s}, O^{(C)}\right), \quad s = 1, 2, \dots, S, \\ f\left(O_{V_{u\_j}}^{(k+1)}\right) = \max\limits_{V_{u\_i}, i \neq j}\left(g\left(V_{u\_j}, C_{V_{u\_i}}^{(k)}\right) + f\left(O_{V_{u\_i}}^{(k)}\right)\right), \quad k \\ \quad = 1, 2, \cdots k \leq S-1. \end{cases}$$

(3)

It should be stressed that it is not necessary to finish all the S iterations as shown in Eq. (3). Once the cardinality $f\left(O_{V_{u\_i}}^{(k)}\right)$ at the k-th stage becomes equal to or greater than the number of unknown parameters N_u, the dynamic programming process will stop and we get the SIOOR strategies.

Results and Discussion
Overview of the framework
Observation strategy design is an under-investigated problem for biological networks, despite the fact that a variety of biological networks have been actively constructed and used in numerous benchside or bedside studies. However, the existence of latent variables is a common problem due to cost, technical or other limitations, and has significantly hampered our capability to quantitatively investigate and understand such networks via, e.g., key network parameter estimation from experimental data. Identifiability analysis has long been recognized as a powerful tool to assure the accuracy and reliability of parameter estimation techniques; however, identifiability-based observation strategy design for biological networks turns out to be an unexplored field despite its substantial importance to biological network studies like structure identification.

To the best knowledge of our authors, this is the first study that tackles the problem of identifiability-based observation strategy design for biological networks described by linear SEMs. First, we introduce several new concepts such as cardinality of observation strategy and identifiability gain and mathematically formulate the identifiability-based optimal observation problem. Second, for a given network structure, the key idea is to turn a minimum number of unobserved nodes in the original observation strategy into observed such that the number of non-redundant identifiability equations becomes greater than or equal to the number of unknown model parameters (i.e., the whole system is at least locally identifiable). By counting the number of non-redundant identifiability equations, we avoid performing actual identifiability analysis on SEM and the proposed method is thus computationally efficient. Third, by defining the concepts of stage division and state transition, a dynamic programming strategy is proposed to solve the maximization problem without involving any time-consuming symbolic computation or matrix operations [33, 34]. Fourth, an efficient computing algorithm is proposed to calculate the identifiability gain of each unobserved node in a given observation strategy. More specifically, the computing process is significantly simplified by counting the number of observed nodes that connect with the node of concern via Wright's paths after removing certain edges from the original graph.

It takes a non-constant time to compute the node identifiability gain in each iteration, and the algorithm complexity depends on the number of observed nodes. Furthermore, the number of iterations of the dynamic programming algorithm does not depend on the total number of nodes, but the number of unobserved nodes in the original observation strategy. Let S denote the number of unobserved nodes and T denote the number of observed nodes in the original observation strategy, then the computation complexity of the dynamic programing strategy is $O(S^2 \cdot T)$.

Implementation and verification
The flowchart of the proposed algorithm for searching the structural identifiability-based optimal observation remedy is shown in Fig. 3. We have implemented the dynamic programming algorithm in R, and all the source codes and examples are freely accessible at https://github.com/Hongyu-Miao/SIOOR.

Here we describe several important technical details of the implementation. First, at the state transition step, i.e., $f\left(O_{V_{u\_i}}^{(k+1)}\right) = \max\limits_{V_{u\_i}, i \neq j}\left(g\left(V_{u\_j}, O_{V_{A\_i}}^{(k)}\right) + f\left(O_{V_{u\_i}}^{(k)}\right)\right)$, if there exist multiple transitions that produce the same $f(O^{(k+1)})$, our current implementation chooses only one such transition to update the next-stage observation strategy. If needed, the R code can be slightly modified to enumerate all optimal observation strategies. Second, since the boundary node set is just a subset of the ancestor node set for a given node, the processing of the boundary nodes is incorporated into the processing of the ancestor nodes in the current implementation.

In order to verify the implementation, synthetic DAGs can be generated for this purpose, like the two DAG examples in Fig. 4. The first DAG contains 8 nodes and 13 edges, and it has only a single input node and a single output node. Moreover, the first example considers a special initial observation strategy (i.e., all nodes are unobserved) to illustrate the capability of the proposed method to design optimal observation strategy from scratch. The second DAG has multiple input and output nodes, and it considers a more general situation, that is, there exists both observed and unobserved nodes in the initial observation strategy. We analyzed the two examples using the proposed algorithm, and used the identifiability matrix method [34] to verify that the obtained observation remedies do grant (local) identifiability to all model parameters.

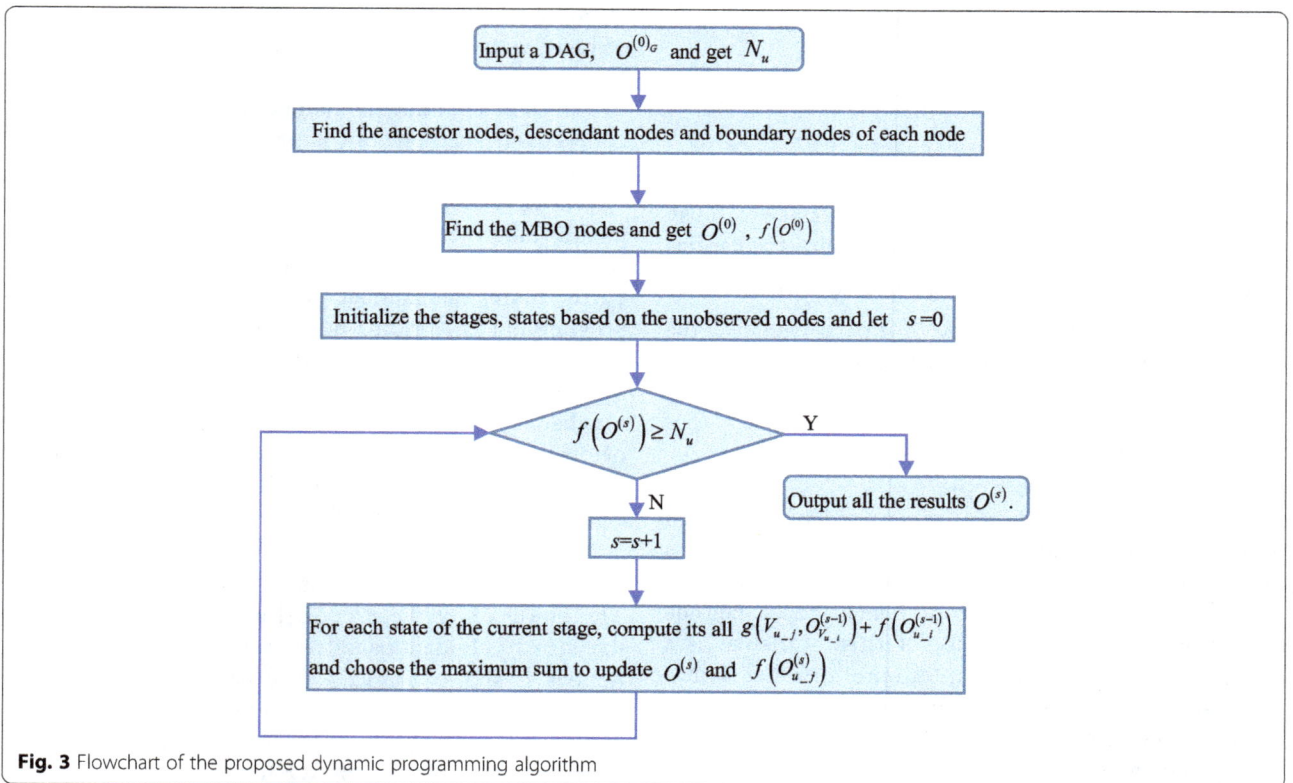

Fig. 3 Flowchart of the proposed dynamic programming algorithm

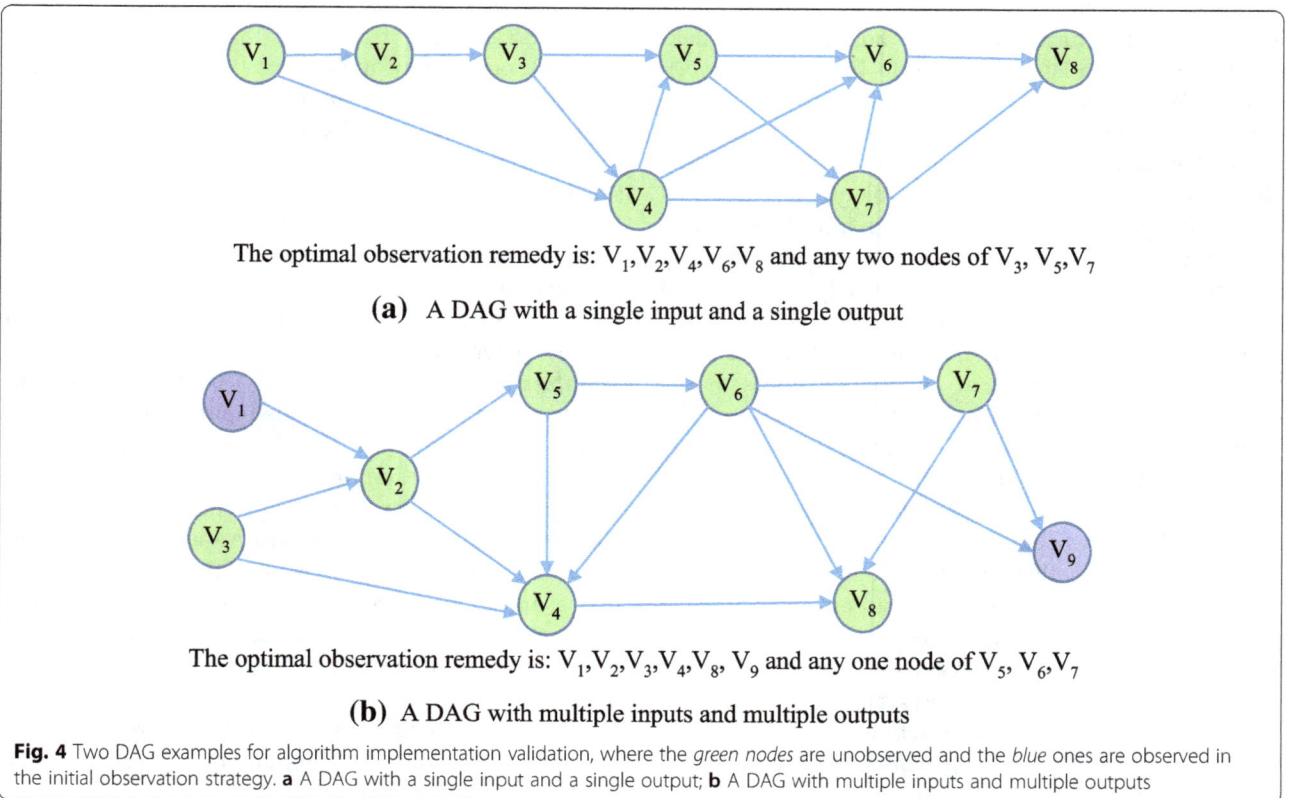

The optimal observation remedy is: V_1, V_2, V_4, V_6, V_8 and any two nodes of V_3, V_5, V_7

(a) A DAG with a single input and a single output

The optimal observation remedy is: $V_1, V_2, V_3, V_4, V_8, V_9$ and any one node of V_5, V_6, V_7

(b) A DAG with multiple inputs and multiple outputs

Fig. 4 Two DAG examples for algorithm implementation validation, where the *green nodes* are unobserved and the *blue* ones are observed in the initial observation strategy. **a** A DAG with a single input and a single output; **b** A DAG with multiple inputs and multiple outputs

Applications to real biological networks

Since it is impossible to cover all the biological networks in various databases and knowledge repositories [16, 17] in one study, we choose the biological network associated with influenza A virus [39] as an application example for illustration purpose. IAV can infect birds as well as mammals including human, and it has been one of the major infectious pathogens that have caused millions of human deaths. It is thus of great scientific significance to systematically understand IAV infection and immune response mechanisms. Therefore, Matsuoka et al. [50] manually curated a comprehensive database, called FluMap, for depicting the influenza virus life cycle at the molecular level from over 500 previous publications. There are mainly five modules in FluMap: virus entry, virus replication and transcription, post-translational processing, transportation of virus proteins, and packaging and budding. Given the critical role of virus replication in influenza virus life cycle, numerous experimental studies (e.g., [42, 51–53]) have made attempts to understand virus replication mechanisms and their clinical implications. Thus, we choose to focus on the IAV replication module and analyze its observation strategy.

Since IAV replication involves many different biomolecules and complex interactions, it is usually infeasible to observe all such components and their interactions in one study. The question of concern here is how to choose a minimal number of nodes in Fig. 5 to observe such that all the model parameters become at least locally identifiable. Note that Fig. 5 is derived from Matsuoka's work [50], and consists of 22 nodes and 26 edges; for simplicity, the catalyzers and inhibitors in this network are treated as reactants.

A relevant concept, called observability, has been previously investigated by Liu et al. [38] for complex dynamic systems. Although observability analysis also deals with observation strategies considering the existence of latent variables, it is very different from identifiability analysis in two aspects: 1) the focus of observability analysis is not model parameters but how to infer the unobserved state variables from experimentally measured outputs of a system; 2) the graphical approach proposed by Liu et al. was developed for the so-called balance equations based on mass-action kinetics, the model structures of which are very different from static linear SEMs. However, it is of interest to compare the identifiability-based observation results with those of the observability-based method. For this purpose, we assume that all the nodes in Fig. 5 are initially unobserved. After applying the proposed dynamic programing method, we get the optimal observation strategy shown in Fig. 6(a) for achieving parameter identifiability. The optimal observation strategy produced by Liu's observability approach is shown in Fig. 6(b). According to Fig. 6(a) and (b), one can tell that the identifiability-based observation strategy contains 20 observed nodes and 2 unobserved nodes, while the observability-based strategy contains 3 observed nodes and 19 unobserved nodes. That is, for the IAV replication module, the system internal states can be inferred from a few observed output nodes if a balance equation model is used; however, it needs much more observed nodes to achieve parameter identifiability if a linear SEM is used. Such an observation is not only due to the different goals of observability and identifiability analyses, but also the differences in the underlying model structures used in observability or identifiability analyses.

Moreover, besides the nodes with an out-degree 0 or 1 as mentioned in Lemma 3, the identifiability-based observation strategy is also likely to select the nodes with a high out-degree as unobserved nodes; for instance, the two unobserved nodes viral_RNA and NP(ub) in Fig. 6(a) have the highest out-degrees 2 and 3, respectively. This is because, if an unobserved node has

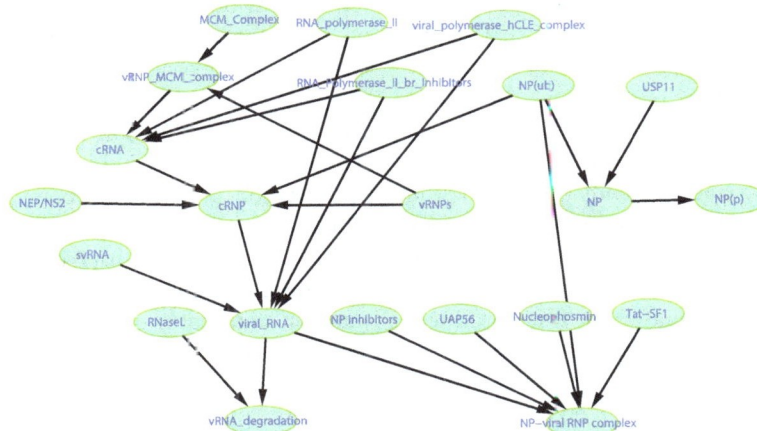

Fig. 5 An application example based on the influenza A virus replication module, where all nodes are initially unobserved and in *green color*

(a) Identifiability-based optimal observation strategy.

(b) Observability-based optimal observation strategy.

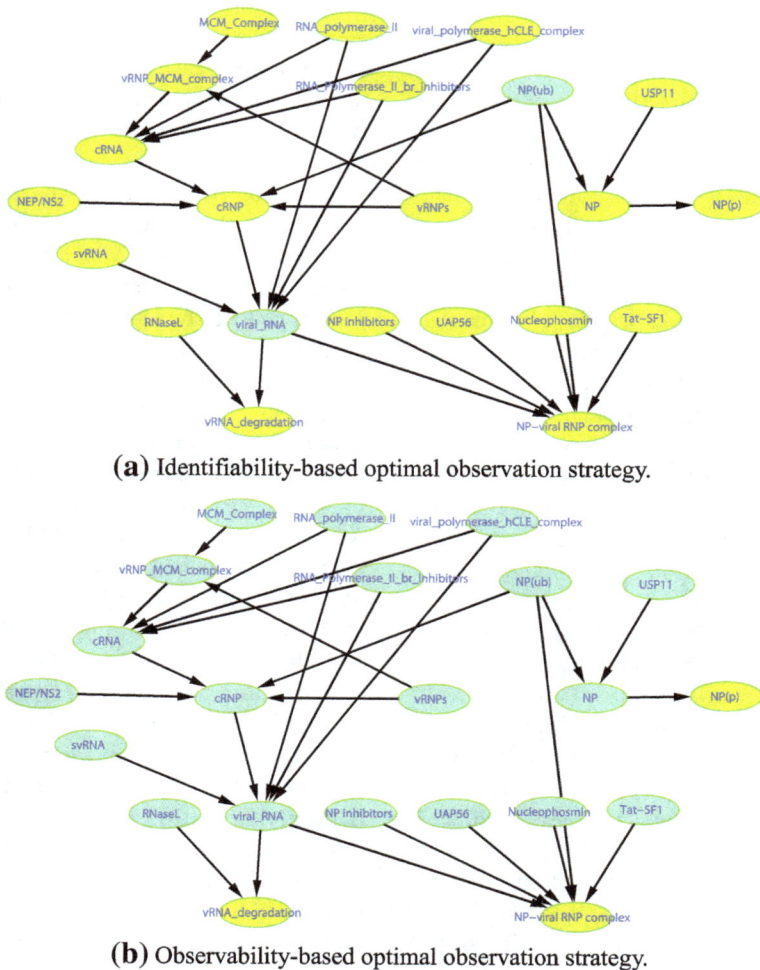

Fig. 6 The optimal observation strategies for the influenza A virus replication module based on **a** identifiability and **b** observability, where the *yellow nodes* are observed and the *green nodes* are unobserved

a high out-degree, this node is connected with many out-neighbor nodes; when its out-neighbor nodes are observed, there will exist multiple Wright's paths that connect such out-neighbor nodes and pass this unobserved node, and the corresponding identifiability equations thus contain the parameters associated with the out-edges of this unobserved node such that these parameters can be identifiable. Interestingly, the observability-based strategy tends to select the nodes with a low out-degree as observed nodes, for example, all the nodes with 0 out-degree are observed in Fig. 6(b). It is because the nodes with an out-degree 0 in a DAG are usually the final products of chemical reactions, instead of reactants, and thus the internal states associated with other nodes can be easily inferred based on the balance equations if all the final products of chemical reactions are measured.

Conclusions

In this study, we address an important problem for biological networks: the design of observation strategies for all edge coefficients being identifiable. Linear SEMs are used as the mathematical representation of biological networks, which allows us to formulate the problem as a constrained optimization problem. A dynamic programming strategy was then developed to solve the constrained optimization problem to obtain the optimal observation strategies at the cost of turning a minimal number of unobserved nodes into observed. The proposed solution is novel and efficient because it avoids both symbolic computation and matrix operations as used in other studies, and we provided necessary theoretical justifications for the proposed algorithm. As verified by multiple examples (synthetic or real networks), the proposed solution is generic and can be applied to an arbitrary DAG (recursive SEMs) without bidirectional edges.

We also recognize that many real biological networks are dynamic, nonlinear, or have feedback loops, which are beyond the capability of the method developed in this study. However, this study provides a basis for determining the identifiability-based optimal observation remedy for more complex biological networks, and we expect to tackle the more challenging problems in the future.

Abbreviations
SEM: Structure equation model; DAG: Directed acyclic graph; SIOOR: Structural identifiability-based optimal observation remedy; IAV: Influenza A virus; DP: Dynamic programming; OS: Observation strategy; G: Identifiability gain; DCG: Directed cyclic graph

Acknowledgements
The authors thank Dr. Yu Luo and Ms. Lijie Wang for useful suggestions and discussions.

Funding
This work was partially supported by the Fundamental Research Funds for the Central Universities of China ZYGX2014J064 (YW) and NSF grant 1620957 (HM).

Authors' contributions
YW contributed to method development, computational analyses, real network analyses and manuscript writing. HM proposed the idea, oversaw the study, and significantly contributed to manuscript preparation. All authors have read and approved the final version of the manuscript.

Authors' information
YW is Assistant Professor at School of Computer Science and Engineering, University of Electronic Science and Technology of China. HM is Associate Professor at the Department of Biostatistics, School of Public Health, University of Texas Health Science Center at Houston, USA.

Competing interests
The authors declare that they have no competing interests.

Author details
[1]School of Computer Science and Engineering, University of Electronic Science and Technology of China, Chengdu, Sichuan, China. [2]Department of Biostatistics, School of Public Health, University of Texas Health Science Center at Houston, Houston, TX 77030, USA.

References
1. Butcher EC, Berg EL, Kunkel EJ. Systems biology in drug discovery. Nat Biotech. 2004;22(10):1253–9.
2. Barabasi A-L, Gulbahce N, Loscalzo J. Network medicine: a network-based approach to human disease. Nat Rev Genet. 2011;12(1):56–68.
3. Seiple IB, Zhang Z, Jakubec P, Langlois-Mercie A, Wright PM, Hog DT, et al. A platform for the discovery of new macrolide antibiotics. Nature. 2016;533(7603):338–45.
4. Bansal M, Yang J, Karan C, Menden MP, Costelo JC, Tang H, et al. A community computational challenge to predict the activity of pairs of compounds. Nat Biotech. 2014;32(12):1213–22. doi:10.1038/nbt.3052.
5. Rua JF, Venkatesan K, Hao T, Hirozane-Kishikawa T, Dricot A, Li N. Towards a proteome-scale map of the human protein-protein interaction network. Nature. 2005;437. doi:10.1038/nature04209
6. Stelzl U, Worm U, Lalowski M, Haenig C, Brembeck FH, Goehler H, et al. A human protein-protein interaction network: a resource for annotating the proteome. Cell. 2005;122(6):957–68. doi:10.1016/j.cell.2005.08.029.

7. Carninci P, Kasukawa T, Katayama S, Gough J, Frith M, Maeda N. The transcriptional landscape of the mammalian genome. Science. 2005; 309(5740):1559–63. doi:10.1126/science.1112014.
8. Minguez P, Parca L, Diella F, Mende DR, Kumar RD, Helmercitterich M, et al. Deciphering a global network of functionally associated post-translational modifications. Mol Syst Biol. 2012;8(1):599.
9. Minguez P, Letunic I, Parca L, Bork P. PTMcode: a database of known and predicted functional associations between post-translational modifications in proteins. Nucleic Acids Res. 2013;41. doi:10.1093/nar/gks1230.
10. Liu Z, Wu H, Zhu J, Miao H. Systematic identification of transcriptional and post-transcriptional regulations in human respiratory epithelial cells during influenza A virus infection. BMC Bioinformatics. 2014;15(1):336.
11. Reynolds A, Leake D, Boese Q, Scaringe S, Marshall W, Khvorova A. Rational siRNA design for RNA interference. Nat Biotechnol. 2004;22(3):326–30.
12. Lewis BP, Burge CB, Bartel DP. Conserved seed pairing, often flanked by adenosines, indicates that thousands of human genes are microRNA targets. Cell. 2005;120(1):15–20.
13. Ponting CP, Oliver PL, Reik W. Evolution and functions of long noncoding RNAs. Cell. 2009;136(4):629–41.
14. Jeong H, Tombor B, Albert R, Oltvai ZN, Barabási AL. The large-scale organization of metabolic networks. Nature. 2000;407. doi:10.1038/35036627.
15. Duarte NC, Becker SA, Jamshidi N, Thiele I, Mo ML, Vo TD, et al. Global reconstruction of the human metabolic network based on genomic and bibliomic data. Proc Natl Acad Sci U S A. 2007;104(6):1777–82.
16. Gerstein MB, Kundaje A, Hariharan M, Landt SG, Yan K-K, Cheng C, et al. Architecture of the human regulatory network derived from ENCODE data. Nature. 2012;489(7414):91–100.
17. Kanehisa M, Goto S, Sato Y, Kawashima M, Furumichi M, Tanabe M. Data, information, knowledge and principle: back to metabolism in KEGG. Nucleic Acids Res. 2014;42(D1):D199–205.
18. Miao H, Xia X, Perelson AS, Wu H. On identifiability of nonlinear ODE models and applications in viral dynamics. SIAM Rev. 2011;53(1):3–39.
19. Giraud C, Tsybakov A. Discussion: latent variable graphical model selection via convex optimization. Ann Stat. 2012;40(4):1984–8.
20. Shamaiah M, Lee SH, Vikalo H. Graphical models and inference on graphs in genomics: challenges of high-throughput data analysis. IEEE Signal Process Mag. 2012;29(1):51–65. doi:10.1109/MSP.2011.943012.
21. Domke J. Learning graphical model parameters with approximate marginal inference. IEEE Trans Pattern Anal Mach Intell. 2013;35(10):2454–67.
22. Mazman SG, Usluel YK. Modeling educational usage of Facebook. Comput Educ. 2010;55(2):444–53.
23. Lewis BP, Burge CB, Bartel DP. Conserved seed pairing, often flanked by adenosines, indicates that thousands of human genes are microRNA targets. Cell. 2005;120. doi:10.1016/j.cell.2004.12.035.
24. Duarte NC, Eecker SA, Jamshidi N, Thiele I, Mo ML, Vo TD. Global reconstruction of the human metabolic network based on genomic and bibliomic data. Proc Natl Acad Sci. 2007;104. doi:10.1073/pnas.0610772104.
25. Ponting CP, Oliver PL, Reik W. Evolution and functions of long noncoding RNAs. Cell. 2009;136. doi:10.1016/j.cell.2009.02.006.
26. Minguez P, Parca L, Diella F, Mende DR, Kumar R, Helmer-Citterich M. Deciphering a global network of functionally associated post-translational modifications. Mol Syst Biol. 2012;8.
27. Cai XBJ, Giannrakis GB. Inference of gene regulatory networks with sparse structural equation models exploiting genetic perturbations. PLoS Comput Biol. 2013;9. doi:10.1371/journal.pcbi.1003058.
28. Pearl J. Causality: models, reasoning, and inference. 2nd ed. Cambridge: Cambridge University Press; 2009.
29. Brito C, Pearl J. Generalized instrumental variables. Uncertainty in artificial intelligence. 2002. p. 85–93.
30. Tian J. A criterion for parameter identification in structural equation models. arXiv preprint arXiv:12065289. 2012.
31. Drton M, Foygel R, Sullivant S. Global identifiability of linear structural equation models. Ann Stat. 2011;39(2):865–86.
32. Foygel R, Draisma J, Drton M. Half-trek criterion for generic identifiability of linear structural equation models. Ann Stat. 2012;40(3):1682–713.
33. Sullivant S, Garcia-Puente LD, Spielvogel S. Identifying causal effects with computer algebra. Proceedings of the Twenty-Sixth Conference on Uncertainty in Artificial Intelligence (UAI). Arlington: AUAI Press; 2010.

34. Wang Y, Lu N, Miao H. Structural identifiability of cyclic graphical models of biological networks with latent variables. BMC Syst Biol. 2016;10(1):1–15. doi:10.1186/s12918-016-0287-y.

35. Kreutz C, Timmer J. Systems biology: experimental design. FEBS J. 2009; 276(4):923–42.

36. Marvel S, Williams CM. Set membership experimental design for biological systems. BMC Syst Biol. 2012;6(1):21.

37. Liu AR, Bitmead RR. Stochastic observability in network state estimation and control. Automatica. 2011;47(1):65–78.

38. Liu Y, Slotine JE, Barabasi A. Observability of complex systems. Proc Natl Acad Sci U S A. 2013;110(7):2460–5.

39. Pirsiavash H, Ramanan D, Fowlkes CC. Globally-optimal greedy algorithms for tracking a variable number of objects. Computer vision and pattern recognition. 2011.

40. Shimizu S, Hoyer PO, Hyvärinen A, Kerminen A. A linear non-Gaussian acyclic model for causal discovery. J Mach Learn Res. 2006;7:2003–30.

41. Hoyer PO, Hyvarinen A, Scheines R, Spirtes PL, Ramsey J, Lacerda G, et al. Causal discovery of linear acyclic models with arbitrary distributions. arXiv preprint arXiv:12063260. 2012.

42. Watanabe T, Kiso M, Fukuyama S, Nakajima N, Imai M, Yamada S, et al. Characterization of H7N9 influenza A viruses isolated from humans. Nature. 2013;501(7468):551–5. doi:10.1038/nature12392.

43. Wright S. The method of path coefficients. Ann Math Stat. 1934;5(3):161–215.

44. Wright S. Path coefficients and path regressions: alternative or complementary concepts? Biometrics. 1960;16. doi:10.2307/2527551.

45. Garcia C, Li T. On the number of solutions to polynomial systems of equations. SIAM J Numer Anal. 1979.

46. Sullivant S, Talaska K, Draisma J. Trek separation for Gaussian graphical models. Ann Stat. 2010;38(3):1665–85.

47. Felzenszwalb PF, Zabih R. Dynamic programming and graph algorithms in computer vision. IEEE Trans Pattern Anal Mach Intell. 2011;33(4):721–40. doi:10.1109/TPAMI.2010.135.

48. Tran D, Yuan J, Forsyth D. Video event detection: from subvolume localization to spatiotemporal path search. IEEE Trans Pattern Anal Mach Intell. 2014;36(2):404–16. doi:10.1109/TPAMI.2013.137.

49. Jiang H, Tian T, Sclaroff S. Scale and rotation invariant matching using linearly augmented trees. IEEE Trans Pattern Anal Mach Intell. 2015;37(12):2558–72.

50. Matsuoka Y, Matsumae H, Katoh M, Eisfeld AJ, Neumann G, Hase T, et al. A comprehensive map of the influenza A virus replication cycle. BMC Syst Biol. 2013;7(1):97.

51. Honda A, Mizumoto K, Ishihama A. Minimum molecular architectures for transcription and replication of the influenza virus. Proc Natl Acad Sci U S A. 2002;99(20):13166–71.

52. Konig R, Stertz S, Zhou Y, Inoue A, Hoffmann HH, Bhattacharyya S, et al. Human host factors required for influenza virus replication. Nature. 2010; 463(7282):813–7.

53. York A, Hutchinson E, Fodor E. Interactome analysis of the influenza A virus transcription/replication machinery identifies protein phosphatase 6 as a cellular factor required for efficient virus replication. J Virol. 2014; 88(22):13284–99.

Finding low-conductance sets with dense interactions (FLCD) for better protein complex prediction

Yijie Wang and Xiaoning Qian[*]

Abstract

Background: Intuitively, proteins in the same protein complexes should highly interact with each other but rarely interact with the other proteins in protein-protein interaction (PPI) networks. Surprisingly, many existing computational algorithms do not directly detect protein complexes based on both of these topological properties. Most of them, depending on mathematical definitions of either "modularity" or "conductance", have their own limitations: Modularity has the inherent resolution problem ignoring small protein complexes; and conductance characterizes the separability of complexes but fails to capture the interaction density within complexes.

Results: In this paper, we propose a two-step algorithm FLCD (**F**inding **L**ow-**C**onductance sets with **D**ense interactions) to predict overlapping protein complexes with the desired topological structure, which is densely connected inside and well separated from the rest of the networks. First, FLCD detects well-separated subnetworks based on approximating a potential low-conductance set through a personalized PageRank vector from a protein and then solving a mixed integer programming (MIP) problem to find the minimum-conductance set within the identified low-conductance set. At the second step, the densely connected parts in those subnetworks are discovered as the protein complexes by solving another MIP problem that aims to find the dense subnetwork in the minimum-conductance set.

Conclusion: Experiments on four large-scale yeast PPI networks from different public databases demonstrate that the complexes predicted by FLCD have better correspondence with the yeast protein complex gold standards than other three state-of-the-art algorithms (ClusterONE, LinkComm, and SR-MCL). Additionally, results of FLCD show higher biological relevance with respect to Gene Ontology (GO) terms by GO enrichment analysis.

Keywords: Protein complex identification, Low conductance set, Dense subnetwork, Mixed integer programming

Background

Recent developments of high-throughput profiling techniques, such as yeast two-hybrid (Y2H) and tandem affinity purification (TAP) with mass spectrometry (MS), allow scientists to generate large-scale protein-protein interaction (PPI) datasets for different species [1–5]. These interactome data have enabled us to discover biological insights from a systematic point of view through PPI networks, where nodes represent proteins and edges denote biological relationships (either physical binding or statistical association) between two proteins. In this paper, we focus on predicting protein complexes in derived PPI networks from high-throughput profiling.

Based on the inherent topological structures of protein complexes [6], prediction of protein complexes can be formulated as searching for subnetworks that are densely connected inside and well separated from the rest of the

---

*Correspondence: xqian@ece.tamu.edu
Department of Electrical & Computer Engineering. Texas A and M University,
MS 3128, TAMU, College Station, TX, USA

PPI networks. Many algorithms have been developed and applied for this purpose of detecting protein complexes.

These existing algorithms can be grouped into three categories. The first category includes the algorithms that mimic Markovian random walk on graphs, pioneered by MCL [7]. MCL does not have explicit mathematical definitions for the desired properties of subnetworks to detect as protein complexes. Similar to random walk, it iteratively implements "Expand" and "Inflation" operations to generate non-overlapping complexes. R-MCL [8] and SR-MCL [9] are improved versions of MCL. R-MCL penalizes the large complexes at each iteration in order to obtain more size-balanced complexes with a similar number of nodes within them. SR-MCL executes R-MCL many times to yield overlapping complexes. All those algorithms have shown good empirical performance, despite the mystery of parameter tuning and the lack of theoretic understanding of their working mechanisms.

Algorithms in the second category do not directly predict complexes according to the topological structure of subnetworks but resemble traditional clustering methods based on derived similarity measures between nodes or edges. For example, MCODE [1], CFinder [10], and RRW [11] grow complexes from single nodes by iteratively adding similar nodes in terms of different similarity criteria that help form local dense subnetworks. However, they only concentrate on the internal connectivity of the subnetworks and neglect the connectivity between the subnetworks and the rest of the networks. LinkComm [12] represents networks with edge graphs, whose nodes are interactions and edges reflect the similarity between interactions, and derives potential complexes by hierarchical clustering to partition the edge graphs.

Algorithms in the third category detect complexes based on explicit topological definitions of protein complexes. For example, modularity [13] and conductance [6, 14] are two widely used definitions. Algorithms based on modularity [15] aim to detect subnetworks that have higher than expected internal connections. And algorithms, such as ClusterONE [6], based on finding low-conductance sets, focus on the separability of the subnetworks, which can be quantified by the ratios between the external connections of subnetworks and the total number of interactions of the proteins within the subnetworks. However, these methods have their own limitations. Modularity-based methods have the inherent resolution problem [16], which leads to ignorance of small-size protein complexes. Algorithms based on conductance minimization [6, 17] consider the relationships between the internal connections and the external connections of subnetworks, but neglect the density of the interactions within the subnetworks.

In this paper, we propose a two-step algorithm FLCD (**F**inding **L**ow-**C**onductance sets with **D**ense interactions) to detect protein complexes that have dense interactions inside and sparse interactions outside in a given PPI network. FLCD explicitly takes care of both the internal and external connectivity of protein complexes in two steps. FLCD first identifies a low-conductance set around a protein, which is locally well separated from the rest of the network. Then a densely connected subnetwork within the low-conductance set is detected based on the definition of the edge density of a subnetwork proposed in [18]. We compare our FLCD with three state-of-the-art overlapping complex prediction algorithms, which are ClusterONE [6], LinkComm [12], and SR-MCL [9], respectively. Experimental results on four different yeast PPI networks from different publicly accessible databases demonstrate that our FLCD outperforms all competing algorithms for biological significance in terms of yeast protein complex gold standards and Gene Ontology (GO) term annotations [19].

Results and discussion

We first introduce the implementation details of the algorithms that we take for comparison; the information of the PPI networks, the reference protein complex datasets as our gold standards, and the GO terms we use for evaluation; and the criteria for the performance comparison. In order to demonstrate the robust performance of FLCD, we then compare predicted protein complexes from *three* selected state-of-the-art protein complex prediction algorithms based on *two* golden standard protein complex datasets on *four* public yeast PPI networks. What's more, we apply GO enrichment analysis to the entire set of detected complexes by all the competing algorithms. At the end, we illustrate differences between protein complexes predicted by all competing algorithms corresponding to specific reference complexes to further demonstrate the superiority of our FLCD.

Algorithms, data, and evaluation metrics
Algorithms
We compare our FLCD algorithm with other three state-of-the-art overlapping complex prediction algorithms, which are ClusterONE [6], LinkComm [12], and SR-MCL [9]. The JAVA implementation of ClusterONE does not require any tuning parameters. For LinkComm, we set the tuning parameter t (the threshold to cut the dendrogram for hierarchical clustering) to 0.2 that achieves the best performance empirically in our experiments. For SR-MCL, we set the inflation parameter $I = 3$ and other parameters to their default settings since they yield the best results in our experiments. We set the only parameter k of our FLCD, the size of local neighbors based on personalized PageRank computation, to 20.

Data

We take four yeast PPI networks for performance evaluation: SceDIP, SceBG, SceIntAct, and SceMINT, extracted respectively from the Database of Interacting Proteins (DIP) [2], the Biological General Repository for Interaction Datasets (BioGRID) [3], the IntAct Molecular Interaction Database (IntAct) [4], and the Molecular INTeraction database (MINT) [5]. We note that we only consider protein-protein interactions by removing all genetic interactions from SceBG. We download the protein complex gold standards from the supplementary data in [6], which are obtained from the Saccharomyces Genome Database (SGD) [20] and the Munich Information Center for Protein Sequences (MIPS) [21] databases. For each PPI network, we remove reference protein complexes if their size smaller than 3 or half of the proteins of them are not in the network. The detailed information of four PPI networks and the gold standard reference complex datasets are provided in Table 1.

Due to the possible incompleteness of the reference protein complexes, we further examine the biological relevance of every predicted complex by GO enrichment analysis. We download the mappings of yeast genes and proteins to GO terms according to [20] (version 20150411).

Evaluation metrics for protein complex prediction

For the protein complex prediction, we assess the performance of all competing algorithms by a composite score consisting of three quality measures: F-measure [9, 14]; the geometric accuracy (Acc) score [14]; and the maximum matching ratio (MMR) [6]. For fair comparison, we remove predicted complexes of two or fewer proteins by all competing algorithms.

For a gold standard reference protein complex set $C = \{c_1, c_2, \ldots, c_n\}$ and a set of predicted complexes $S = \{s_1, s_2, \ldots, s_m\}$, the F-measure is defined as the harmonic mean of precision and recall defined as follows:

$$\text{precision} = \frac{|N_{cs}|}{|S|}; \quad \text{recall} = \frac{|N_{cp}|}{|C|}, \quad (1)$$

in which $N_{cs} = \{s_i \in S | NA(c_j, s_i) \geq 0.25, \exists c_j \in C\}$ is the set of the complexes that match to one or more reference protein complexes; $|N_{cs}|$ is the size of the set N_{cs}.

$N_{cp} = \{c_i \in C | NA(c_i, s_j) \geq 0.25, \exists s_j \in S\}$ is a set of reference protein complexes that are matched by predicted complexes. We consider a reference protein complex c_j is matched by a predicted complex s_j if $NA(c_i, s_j) \geq 0.25$ [9, 22], where $NA(c_i, s_j) = \frac{|c_i \cap s_j|^2}{|c_i| \times |s_j|}$ is called neighborhood affinity. Finally, the F-measure is

$$\text{F-measure} = 2 \times \frac{\text{precision} * \text{recall}}{\text{precision} + \text{recall}}. \quad (2)$$

The geometric accuracy (Acc) score is the geometric mean of two other measures — the cluster-wise sensitivity (Sn) and cluster-wise positive predictive value (PPV) [6]. Given m predicted and n reference complexes, let t_{ij} denote the number of proteins that exist in both predicted complex s_i and reference complex c_j, and w_j represent the number of proteins in reference complex c_j. Then Sn and PPV can be computed as

$$\text{Sn} = \frac{\sum_{j=1}^{n} \max_{i=1,\ldots,m} t_{ij}}{\sum_{j=1}^{n} w_j}; \quad \text{PPV} = \frac{\sum_{i=1}^{m} \max_{j=1,\ldots,n} t_{ij}}{\sum_{i=1}^{m} \sum_{j=1}^{n} t_{ij}}. \quad (3)$$

The Acc score provides a balanced measure of Sn and PPV: $\text{Acc} = \sqrt{\text{Sn} \times \text{PPV}}$.

The maximum matching ratio (MMR) is the ratio of the weight of maximum weight matching to the size of the reference set.

GO enrichment analysis

Suppose that a given PPI network has N proteins with M proteins annotated with one GO term and the predicted complex has n proteins with m proteins annotated with the same GO term. The p-value of the complex enriched with that GO term can be calculated as similarly done in [23]:

$$p\text{-value} = \sum_{i=m}^{n} \frac{\binom{m}{i}\binom{N-M}{N-i}}{\binom{N}{n}}. \quad (4)$$

We choose the lowest p-value of all its enriched GO terms for a predicted complex as its final p-value. A GO term is statistically significantly enriched when the p-value of any complex corresponding to this GO term is lower than $1e-3$.

Comparison on protein complex prediction

We apply all competing algorithms to search for potential protein complexes in four yeast PPI networks and compare them in terms of the composite score, consisting of F-measure, Acc score and MMR based on both the SGD and MIPS reference protein complex datasets.

We note that the different sizes and different numbers of detected complexes would affect the scores for the metrics that we have employed. However, in the context of complex prediction, there is no universal gold-standard

Table 1 The detailed information of four yeast PPI networks and the numbers of covered SGD and MIPS reference complexes

Network	#. proteins	#. interactions	SGD	MIPS
SceDIP	5136	22491	224	184
SceBG	6438	80577	234	189
SceIntAct	5453	54134	231	187
SceMINT	5414	27315	230	188

metric. Hence, we apply three aforementioned metrics that have been commonly adopted in many other related works [6, 9]. We also note that the average sizes of the complexes generated by FLCD in our experiments are from 6 to 8 for four networks under study. The average complex sizes are indeed comparable to the average sizes of detected complexes by other algorithms. For example, the average sizes of complexes produced by LinkCommunity are from 5 to 6; The average sizes of complexes produced by ClusterONE are from 7 to 9; The average sizes of complexes produced by SR-MCL are from 8 to 10. Furthermore, the total numbers of predicted complexes yielded by FLCD, LinkCommunity and SR-MCL are much larger than that of ClusterONE. The reason is that the post-processing procedure of ClusterONE filters out complexes with lower scores but FLCD and LinkCommunity output all complexes without filtering.

As shown in Figs. 1 and 2, FLCD clearly outperforms other state-of-the-art algorithms for all four networks on both SGD and MIPS reference datasets. Therefore, the complexes detected by FLCD have the best correspondence with the reference datasets. The detailed evaluation scores in Figs. 1 and 2 are displayed in Tables 2 and 3, respectively.

When we take SGD reference dataset as our gold standard protein complexes, from Table 2, we find that FLCD consistently achieves the best MMR scores among all competing algorithms because FLCD is the only algorithm that can capture the desired network structure of protein complexes. In the table, we also compare F-measure and the precision and recall scores that are used to compute F-measure. We observe that for all four PPI networks, FLCD predicts the largest number of matched reference protein complexes, and therefore FLCD attains the best recall scores for all PPI networks. With respect to the precision score, FLCD is the best for SceMINT but ClusterONE performs the best for the rest. However, since the post-processing step in ClusterONE only keeps the dense complexes, ClusterONE has low coverage. Based on the precision and recall scores, we find that FLCD attains the best F-measures for SceDIP and SceMINT PPI networks and ClusterONE obtains the best scores for SceBG and SceIntAct PPI networks. In addition to MMR and F-measure, we show comparison on the cluster-wise sensitivity (Sn), the cluster-wise positive predictive value (PPV) and the Acc score. We notice that FLCD has the best Acc scores for SceBG and SceIntAct. LinkComm obtains the best Acc scores for SceDIP and SceMINT, since LinkComm detects several large-size and many small-size complexes, which favors both the Sn and PPV scores [6]. We also compare the coverage of the competing algorithms and notice that SR-MCL has the largest coverage and FLCD has competitive coverage to SR-MCL. Here, the coverage is defined as the number of proteins covered by all predicted complexes, which is typically used to evaluate whether complex prediction algorithms can help comprehensively predict functionalities for all the proteins in a given network.

For MIPS reference dataset, we notice the similar trend for the evaluation scores in Table 3. FLCD finds the largest number of matched reference complexes in MIPS and attains the best recall scores, F-measures and MMR scores for all four PPI networks. The Acc scores of FLCD are competitive to LinkComm, which achieves the best Acc scores for all four yeast PPI networks. FLCD covers the competitive number of proteins to SR-MCL, which covers

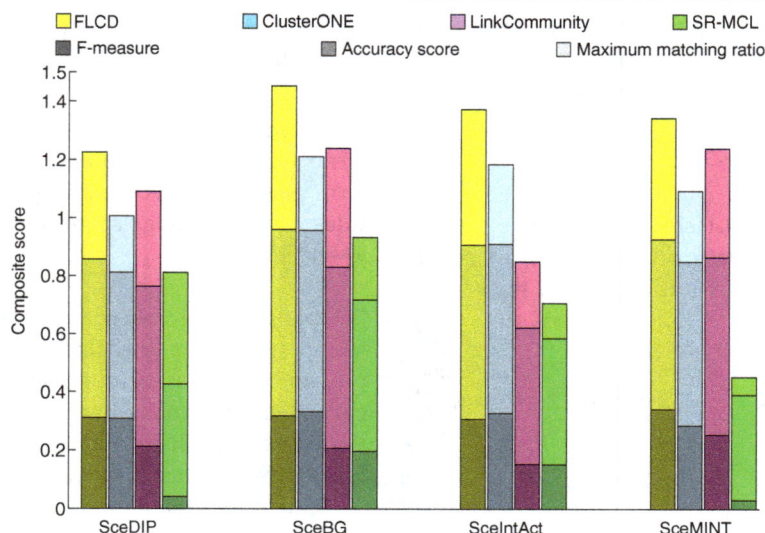

Fig. 1 Comparison of all competing algorithms by SGD reference dataset in terms of the composite scores. *Shades of the same color* indicate different evaluating scores. Each bar height reflects the value of the composite score

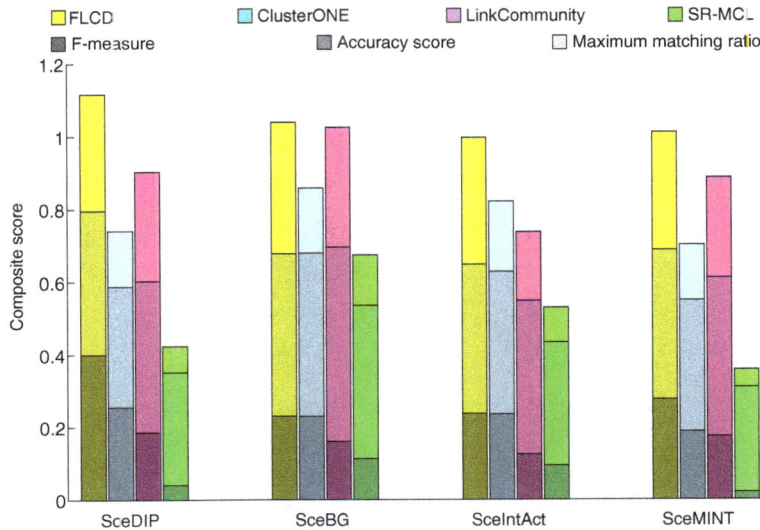

Fig. 2 Comparison of all competing algorithms by MIPS reference dataset in terms of the composite scores. *Shades of the same color* indicate different evaluating scores. Each bar height reflects the value of the composite score

the largest number of proteins in all four yeast PPI networks. However, by the overall performance, which is represented by the composite score, FLCD is superior to other competing algorithms as shown in Fig. 2.

In summary, considering the composite score based on three metrics, our FLCD outperforms the other algorithms. To further validate all competing algorithms, we perform GO enrichment analysis in the next section to see whether all predicted complexes by different algorithms have significant biological meaning.

Comparison on GO enrichment analysis

We perform GO enrichment analysis for all protein complexes predicted by the competing algorithms and report

Table 2 Comparison of protein complex prediction by SGD reference dataset

Network	Method	# complex	#. matched	coverage	Recall	Precision	F-measure	Sn	PPV	Acc	MMR
SceDIP	FLCD	2134	**152**	3921	0.6786	0.2020	**0.3113**	0.5964	0.5003	0.5462	**0.3685**
	CONE	380	86	1503	0.3839	0.2579	0.3085	0.4082	0.6203	0.5032	0.1950
	LinkC	1839	137	3735	0.6116	0.1289	0.2130	0.6290	0.4820	**0.5506**	0.3276
	SR-MCL	3216	44	4678	0.2228	0.0221	0.0412	0.5120	0.2893	0.3489	0.0708
SceBG	FLCD	4027	**183**	5836	0.7821	0.2000	0.3181	0.7363	0.5621	**0.6433**	**0.4920**
	CONE	522	122	2735	0.5214	0.2433	**0.3318**	0.6488	0.6035	0.6257	0.2542
	LinkC	5382	164	6076	0.7008	0.1217	0.2072	0.8880	0.4373	0.6231	0.4100
	SR-MCL	1862	108	5889	0.4615	0.1245	0.1961	0.8999	0.3034	0.5225	0.2151
SceIntAct	FLCD	3394	**172**	4678	0.7446	0.1933	0.3069	0.6699	0.5391	**0.6009**	**0.4661**
	CONE	496	117	1994	0.5065	0.2419	**0.3275**	0.5742	0.5944	0.5842	0.2742
	LinkC	1297	93	5290	0.4026	0.0941	0.1525	0.9223	0.2393	0.4698	0.2285
	SR-MCL	1079	68	5342	0.2294	0.0437	0.1517	0.7784	0.2402	0.4341	0.1213
SceMINT	FLCD	2483	**157**	4210	0.6826	0.2280	**0.3418**	0.6524	0.5284	0.5871	**0.4163**
	CONE	513	110	2335	0.4783	0.2027	0.2848	0.5370	0.5954	0.5654	0.2442
	LinkC	2201	144	4068	0.6261	0.1595	0.2542	0.6757	0.5540	**0.6119**	0.3743
	SR-MCL	3698	33	4976	0.1435	0.0169	0.0302	0.5013	0.2597	0.3608	0.0609

CONE and LinkC are short for ClusterONE and LinkComm, respectively
Bold values denote the best scores corresponding to specific criteria

Table 3 Comparison of protein complex prediction by MIPS reference dataset

Network	Method	# complex	#. matched	Coverage	Recall	Precision	F-measure	Sn	PPV	Acc	MMR
SceDIP	FLCD	2134	**120**	3921	0.6522	0.1603	**0.2573**	0.4001	0.3901	0.3951	**0.3206**
	CONE	380	74	1503	0.4022	0.1868	0.2551	0.2749	0.4015	0.3322	0.1533
	LinkC	1839	109	3735	0.5924	0.1104	0.1862	0.4775	0.3646	**0.4173**	0.2993
	SR-MCL	2851	41	4687	0.1964	0.0230	0.0402	0.4592	0.2104	0.3108	0.0726
SceBG	FLCD	4027	**124**	5836	0.6561	0.1393	**0.2298**	0.4643	0.4315	0.4476	**0.3611**
	CONE	522	86	2735	0.4450	0.1533	0.2293	0.4537	0.4452	0.4494	0.1795
	LinkC	5382	109	6076	0.6349	0.0918	0.1604	0.8179	0.3504	**0.5354**	0.3285
	SR-MCL	1862	65	5889	0.3439	0.0673	0.1126	0.7360	0.2436	0.4234	0.1384
SceIntAct	FLCD	3394	**120**	4678	0.6417	0.1452	**0.2368**	0.4183	0.4034	0.4108	**0.3482**
	CONE	496	79	1994	0.4225	0.1633	0.2356	0.3587	0.4296	0.3925	0.1927
	LinkC	1297	80	5290	0.4278	0.0732	0.1251	0.9028	0.1986	**0.4234**	0.1886
	SR-MCL	1079	45	5342	0.1337	0.0190	0.0941	0.6246	0.1850	0.3399	0.0960
SceMINT	FLCD	2483	**111**	4210	0.5904	0.1800	**0.2759**	0.4147	0.4086	0.4116	**0.3231**
	CONE	513	67	2335	0.3564	0.1267	0.1869	0.3274	0.4017	0.3626	0.1519
	LinkC	2201	100	4068	0.5319	0.1040	0.1740	0.4744	0.4038	**0.4377**	0.2744
	SR-MCL	3698	24	4976	0.1277	0.0112	0.0205	0.4192	0.1999	0.2894	0.0481

CONE and LinkC are short for ClusterONE and LinkComm, respectively
Bold values denote the best scores corresponding to specific criteria

the percentages of the predicted protein complexes that are significantly enriched with at least one GO term and the total number of GO terms that are enriched in the predicted complexes in Table 4. We find that our FLCD achieves the best percentages of the enriched predicted protein complexes in SceDIP and SceIntAct PPI networks. ClusterONE obtains the best percentages for SceBG and SceMINT PPI networks but with the smaller number of GO terms enriched in the detected complexes because ClusterONE may remove meaningful functional modules in its post-processing step. Furthermore, the protein complexes detected by FLCD are significantly associated with the largest number of GO terms over all competing algorithms on all four PPI networks.

To further examine the statistical significance of the complexes detected by the competing algorithms, we compare the p-values of the complexes under GO terms of biological process, molecular function, and cellular component domains. We use the lowest p-value for each predicted complex and show the comparison of the statistical significance of the complexes detected by all competing algorithms in Fig. 3. The y-axis of Fig. 3 represents the negative log-p-values while the x-axis is the ordered list of the complexes detected by all competing algorithms in terms of their negative log-p-values. Since complexes with significant biological relevance have lower p-values, higher values in Fig. 3 represent the higher quality of the detected complexes. As shown in Fig. 3, for all four

Table 4 Comparison by GO enrichment analysis

Network	Method	# complex	% enriched	# GO
SceDIP	FLCD	2134	**72.2**	**1442**
	CONE	380	71.8	852
	LinkC	1839	67.4	1273
	SR-MCL	2851	23.5	957
SceBG	FLCD	4027	72.4	**1800**
	CONE	522	**77.4**	1282
	LinkC	5382	39.8	1554
	SR-MCL	1862	56.4	1702
SceIntAct	FLCD	3394	62.4	**1414**
	CONE	496	**65.6**	1031
	LinkC	1297	46.5	1129
	SR-MCL	1079	44.7	888
SceMINT	FLCD	2483	**62.3**	**1416**
	CONE	513	59.4	954
	LinkC	2201	32.1	1123
	SR-MCL	3698	19.7	856

"% enriched" presents the percentage of complexes that are enriched with at least one GO term.
"# GO" denotes the number of enriched GO terms
Bold values denote the best scores corresponding to specific criteria

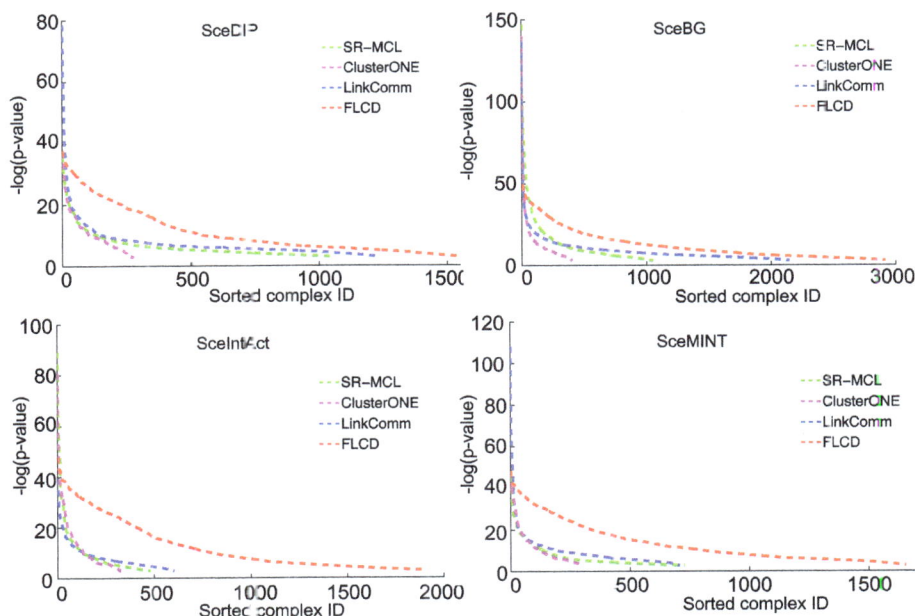

Fig. 3 Statistical significance of the predicted complexes of all competing algorithms

yeast PPI networks, in addition to the fact that FLCD detects significantly more GO-enriched complexes, FLCD clearly outperforms other competing algorithms because the curves of FLCD are consistently on top of the others. The outperformance of FLCD further demonstrates that network structure that has dense internal connectivity and sparse external connectivity can better depict complexes of biological significance and FLCD provides an effective way to predict complexes with the desired network structure through explicitly taking care of internal and external connectivity of potential subnetworks.

Examples of predicted complexes

We further show the differences between the competing algorithms by illustrating the predicted complexes corresponding to two specific reference protein complexes. The first reference protein complex is the Smc5-Smc6 complex. In Fig. 4, the Smc5-Smc6 complexes predicted by FLCD, ClusterONE, LinkComm, and SR-MCL are displayed from (a.1) to (a.4), respectively. We notice that FLCD successfully identifies the Smc5-Smc6 complex as shown in Fig. 4(a.1). ClusterONE fails to detect the protein annotated as NSE4, probably due to the inaccuracy of the greedy algorithm used in ClusterONE. Also, we find that the protein annotated as GEX1 only interacts with the protein NSE3 but it is falsely added to the Smc5-Smc6 complex by ClusterONE. Because ClusterONE focuses on the separability of a complex but does not directly consider the internal density of the complex, it may mistakenly add proteins with small degrees into the final

result. The complex in Fig. 4(a.3) predicted by LinkComm contains false positives and false negatives since the similarities between interactions used in LinkComm can not describe the topological structure of protein complexes. In Fig 4(a.4), we find out that the Smc5-Smc6 complex predicted by SR-MCL consists of many false positives. However, it is hard to explain the performance of SR-MCL on predicting the Smc5-Smc6 complex due to the unclear working mechanism of SR-MCL.

Similarly, we show the predicted RNase complexes by all competing algorithms in Fig. 4 from (b.1) to (b.4). In (b.1), we observe that FLCD detects all proteins in the reference RNase complex but mistakenly includes the protein SKI7 due to the existence of false positive interactions between SKI7 and proteins in RNase complex. In addition to SKI7, the predicted complex by ClusterONE (shown in Fig. 4(b.2)) contains two false positive proteins with very small degrees due to the ignorance of the internal density. Because LinkComm does not explicitly characterize the separability of the complexes, it also recruits some false positive proteins as clearly shown in Fig. 4(b.3). For the complex obtained by SR-MCL, we note that it has lots of false positive proteins and the topological property of the predicted complex is not clear.

Conclusions

We propose an algorithm FLCD to predict protein complexes in protein-protein interaction networks. FLCD can better characterize the topological structure of a protein complex, which is densely connected inside and well

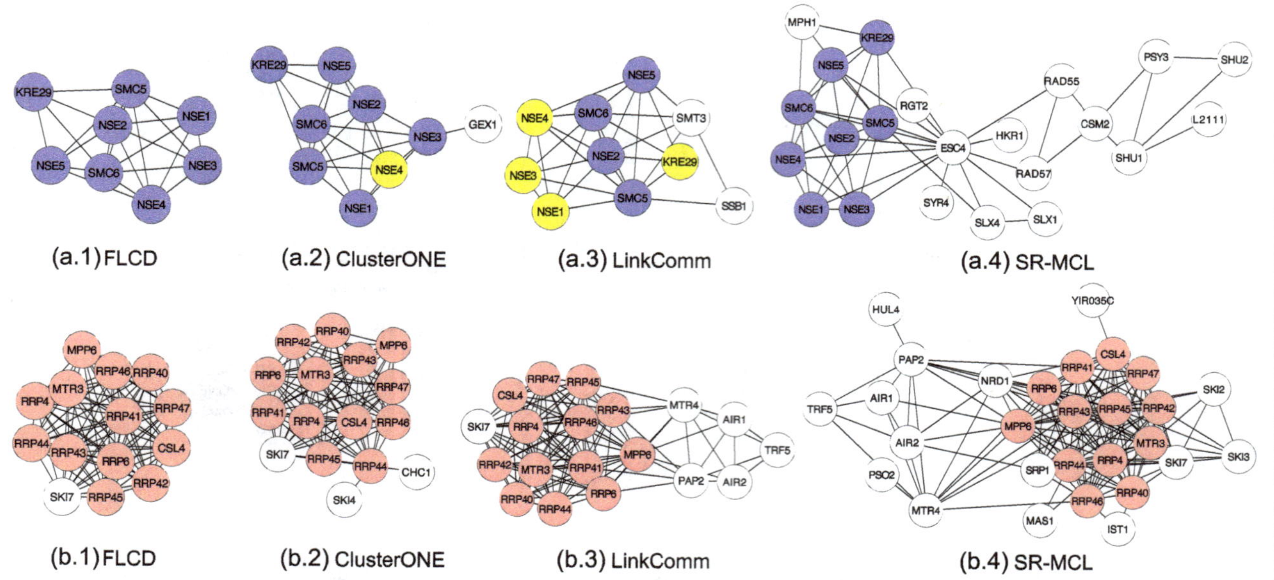

Fig. 4 Illustrations of predicted complexes in SceBG network. *a.1* to *a.4* are Smc5-Smc6 complexes predicted by FLCD, ClusterONE, LinkComm, and SR-MCL, respectively. *Nodes in blue* are proteins in the reference Smc5-Smc6 complex and nodes in white are proteins outside the reference Smc5-Smc6 complex. *Nodes in yellow* are proteins failed to be detected by the corresponding algorithms. *b.1* to *b.4* are RNase complexes predicted by FLCD, ClusterONE, LinkComm, and SR-MCL, respectively. *Nodes in red* are proteins in the reference RNase complex and *nodes in white* are proteins outside the reference RNase complex

separated from the rest of the networks. We compare FLCD with other three state-of-the-art algorithms on protein complex prediction. The comparison results show that FLCD achieves superior performances. Furthermore, GO enrichment analysis of the results of the competing algorithms demonstrates that FLCD finds more biologically meaningful complexes, within which proteins tend to be in the same cellular components and have similar functions and/or participate in the same biological processes.

Methods
Terminologies and definitions
Let an undirected graph $G = (V, E)$ represent a PPI network, where V denotes the set of proteins in G and E is the interaction set. A is the adjacency matrix of G with $A_{ij} = A_{ji}$ and $A_{ij} = 1$ denoting node i interacts with node j and $A_{ij} = 0$ otherwise. The degree matrix D of G is a diagonal matrix with $D_{ii} = d_i$, where $d_i = \sum_j A_{ij}$ is the number of interactions connecting to protein i.

For a set S of proteins, the conductance of S in G is defined as [17]

$$\phi(S) = \frac{|E(S, \bar{S})|}{\min\{vol(S), vol(\bar{S})\}}, \quad S \cup \bar{S} = V, \qquad (5)$$

where $E(S, \bar{S})$ denotes the edge cut, the set of edges between the set S and its complement set $\bar{S}$, $|\cdot|$ denotes the set size, and $vol(T) = \sum_{i \in T} d_i$ is the number of all

incident interactions of the set T. Here we make a mild assumption that $vol(S) \ll vol(V)$ for a small protein complex S in the large-scale PPI network G, which means $vol(S) = \min\{vol(S), vol(\bar{S})\}$. Hence, we have

$$\phi(S) = \frac{|E(S, \bar{S})|}{vol(S)} = \frac{\sum_i \left(D_{ii}^S - \sum_j A_{ij}^S\right)}{\sum_i D_{ii}^S}, \qquad (6)$$

where A^S is the adjacency matrix of the induced subnetwork with respect to set S and D^S is the degree matrix for the nodes in S, where $D_{ii}^S = \sum_j A_{ij} = d_i$ for $i \in S$. For the same set S, the density of S is defined as [18]

$$\mathcal{D}(S) = \frac{|E(S, S)|}{|S|} = \frac{1}{2} \frac{\sum_{ij} A_{ij}^S}{\sum_i \mathbf{1}_{i \in S}}, \qquad (7)$$

where $\mathbf{1}_{i \in S}$ is the indicator function depending on whether $i \in S$.

Motivation
FLCD is motivated by conductance minimization to identify well separated subnetworks in a given network. However, FLCD can overcome the problem of conductance minimization, which pays no attention to the internal connectivity within subnetworks as potential protein complexes. Figure 5 shows a motivating example: We can find two complexes enclosed in the red dotted lines in the network based on conductance minimization. The conductances of the complexes within red dotted lines are $\frac{2}{11}$

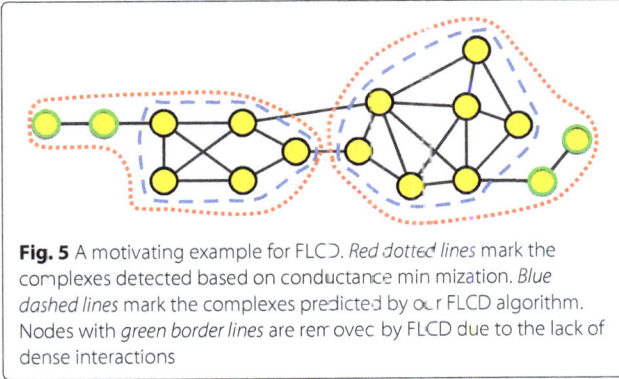

Fig. 5 A motivating example for FLCD. *Red dotted lines* mark the complexes detected based on conductance minimization. *Blue dashed lines* mark the complexes predicted by our FLCD algorithm. Nodes with *green border lines* are removed by FLCD due to the lack of dense interactions

and $\frac{2}{17}$ and the conductances of complexes within blue dashed lines are $\frac{3}{10}$ and $\frac{3}{16}$. Obviously, the conductances of the complexes within red dotted lines are lower than the complexes within blue dashed lines, indicating that the complexes within red dotted lines are topologically more separable than the complexes within blue dashed lines. However, the complexes within the blue dashed lines are more likely to be the desired complexes since the nodes with green border lines can not be confidently grouped into potential protein complexes due to their low degrees.

FLCD explicitly considers both the separability and internal edge density of complexes in two steps respectively. At the first step, it takes care of the separability of complexes by ensuring low conductance to hope for the complexes to have unique biological functions. At the second step, FLCD preserves the densely connected parts of the complexes identified in the first step. Because PPI networks are noisy and typically sparse, instead of finding cliques, we use the definition of internal density in (7) to search for dense subnetworks as final predicted complexes.

Searching for a low-conductance set H_v^*

Given a starting protein v, our goal is to find a protein set H_v^* with low conductance including v. We first apply the algorithm proposed in [17] to find a potential set H with low conductance, then the minimum-conductance set H_v^* in H is identified through solving a mixed integer programming (MIP) problem exactly.

Following [17], a low-conductance set including v can be efficiently approximated via the personalized PageRank vector of v. The personalized PageRank vector $p(\alpha, v)$ of v on G is the stationary distribution of the random walk on G, in which at every step, the random walker has the probability of α to restart the random walk at v and otherwise performs a lazy random walk. Mathematically, $p(\alpha, v)$ is the unique solution to

$$p(\alpha, v) = \alpha e_v + (1 - \alpha)p(\alpha, v)W, \tag{8}$$

where $\alpha \in (0, 1]$ is the "teleportation" constant, e_v is the indicator vector of v and $W = \frac{1}{2}(I + D^{-1}A)$ is the underlying probability transition matrix of the lazy random walk. We apply the local algorithm in [17] to efficiently approximate $\hat{p} \approx p(\alpha, v)$. Then we sort the nodes based on $\hat{p}$ and attain an ordered set $\mathcal{H} = \{v_1, v_2, \dots, v_n\}$, whose elements satisfy $\hat{p}(v_i) > \hat{p}(v_{i+1})$. Inspired by PageRank-Nibble [17] that sweeps the ordered set $\mathcal{H}$ to get the low-conductance set, we propose to find the minimum low-conductance set within a subnetwork of size k, which consists of the top k elements in $\mathcal{H}$, by solving a MIP problem. We take the top k elements out of $\mathcal{H}$, which are more likely to comprise a low-conductance set with v, and put them in H. The minimum-conductance set H_v^* in H can be derived by solving the following optimization problem based on (6):

$$\text{min:} \quad \frac{x^T (D^H - A^H) x}{x^T d^H} \tag{9}$$
$$s.t. \; x_v = 1, \; x_i \in \{0, 1\},$$

where x is a binary vector with $x_i = 1$ indicating that node i in H is assigned into H_v^* and $x_i = 0$ otherwise; and d^H is a vector containing the degrees of every node in H. We force node v to be in the low-conductance set by setting $x_v = 1$. By algebraic manipulations, (9) can be transformed into the following equivalent formulation:

$$\text{min:} \quad z$$
$$s.t. \; z\sum_i x_i d_i^H - \sum_i \sum_j \left(D_{ij}^H - A_{ij}^H\right) x_i x_j \geq 0, \tag{10}$$
$$x_v = 1, \; x_i \in \{0, 1\}.$$

After using standard techniques [24] to linearize zx_i and $x_i x_j$, the optimization problem can be solved by any MIP solver, such as Gurobi [25]. Because the size of $|H| = k$ is much smaller than $|V| = n$ and we only focus on identifying one low-conductance set, we can efficiently obtain the minimum-conductance set H_v^* in H by solving (10) exactly.

If node v is in a connected component of size k' and we set $k > k'$, then we might have a trivial solution that the low-conductance set is the connected component with conductance 0. To avoid this, we apply the following procedure. We check every derived low-conductance set of size k' to see whether it has exactly 0 conductance, which implies that it is a connected component with size k'. If that is the case, we then set $k = k' - 1$, and re-solve the MIP to get a non-trivial solution.

Conservation of the densest subnetwork C_v^* in H_v^*

The induced subnetwork G_v with respect to the protein set H_v^* is well separated from the rest of the network; however, there may exist nodes with low degrees in H_v^*. As illustrated in Fig. 5, to remove low-degree nodes (nodes with green border lines) as well as reserve densely connected subnetworks, we apply the definition of the internal density (7) to find the densest subnetwork in H_v^*. Because the problem size is small for such a local optimization problem, we can again take the full advantages of the power of MIP solvers. The node set $C_v^* \in H_v^*$ corresponding to the densest subnetwork can be identified based on (7) by deriving the exactly optimal solution to the following MIP problem:

$$\text{max:} \quad \frac{r^T A_{ij}^{H_v^*} r}{r^T \mathbf{1}} \tag{11}$$
$$s.t. \quad r_i \in \{0, 1\},$$

where $\mathbf{1}$ is an all-one vector and r is the binary vector indicating the memberships of the nodes from H_v^* in the densest subnetwork. This optimization problem explicitly searches for the subnetwork with the highest internal density and it can be transformed into the equivalent problem, as similarly done in (10):

$$\text{max:} \quad w$$
$$s.t. \quad w \sum_i r_i - \sum_i \sum_j A_{ij}^{H_v^*} r_i r_j \leq 0, \tag{12}$$
$$r_i \in \{0, 1\},$$

which can also be cast into the MIP framework with the exactly optimal solution obtained by using standard MIP solvers after linearization [24].

The FLCD algorithm

The step-by-step procedure of FLCD algorithm is given in Table 5. The FLCD algorithm screens every protein with degree higher than two. For each selected protein, the FLCD algorithm first searches for the minimum-conductance set around it and then finds the densest subnetwork in the minimum-conductance set, which is

Table 5 The FLCD algorithm

Algorithm: The FLCD Algorithm
Input: $\mathcal{S} = V$ and $k = 20$.
Output: A set of predicted complexes R.
1 While ($\exists v \in \mathcal{S}$ and $d_v \geq 3$)
2   Estimate $\hat{p} \approx p(\alpha, v)$.
3   Sort nodes in V based on $\hat{p}$ and collect the top k nodes in H_v.
4   Finding the lowest-conductance set $H_v^* \in H_v$ based on (10).
5   Identifying the node set C_v^* of the densest subnetwork in H_v^* based on (12).
6   Considering C_v^* as one predicted complex, let $R = \{R, C_v^*\}$ and $\mathcal{S} = \mathcal{S} - v$.
7 EndWhile
8 Remove duplicated complexes and complexes with size smaller than three in R.

considered as a predicted complex. After screening every possible proteins, we remove the duplicated complexes and complexes with size smaller than three. There is only one parameter k for the FLCD algorithm, where k can be considered as the upper bound of the sizes of the desired protein complexes. Also, the MIP problems (10) and (12) are both NP hard. The actual computational complexity of solving these MIP problems depends on the problem size of these local problems determined by k. The smaller k is, the less time it takes the FLCD algorithm to search for subnetworks as potential protein complexes. Throughout the experiments in this paper, we set $k = 20$.

Acknowledgements
XQ was partially supported by Awards CCF-1447235, CCF-1553281, IOS-1547557 from the National Science Foundation (NSF) and AFRI-2015-67013-22816 from the United States Department of Agriculture (USDA).

Funding
The publication costs of this article was funded by Award CCF-1447235 from the National Science Foundation.

Authors' contributions
Conceived the algorithm: YW, XQ. Implemented the algorithm and performed the experiments: YW. Analyzed the results: YW, XQ. Wrote the paper: YW, XQ. Both authors have read and approved the final manuscript.

Competing interests
The authors declare that they have no competing interests.

About this supplement
This article has been published as part of BMC Systems Biology Volume 11 Supplement 3, 2017: Selected original research articles from the Third International Workshop on Computational Network Biology: Modeling, Analysis, and Control (CNB-MAC 2016): systems biology. The full contents of the supplement are available online at http://bmcsystbiol.biomedcentral.com/articles/supplements/volume-11-supplement-3.

References
1. Bader GD, Hogue CW. An automated method for finding molecular complexes in large protein interaction networks. BMC Bioinformatics. 2003;4:2.
2. Salwinski L, Miller CS, Smith AJ, Pettit FK, Bowie JU, Eisenberg D. The Database of Interacting Proteins: 2004 update. Nucleic Acids Res. 2004;32: 449–51.
3. Stark C, Breitkreutz BJ, Reguly T, Boucher L, Breitkreutz A, Tyers M. BioGRID: A general repository for interaction datasets. Nucleic Acids Res. 2006;34:535–9.
4. Kerrien S, Aranda B, Breuza L, et al. The IntAct molecular interaction database in 2012. Nucleic Acids Res. 2012;40(D1):841–6.
5. Licata L, Briganti L, Peluso D, et al. MINT, the molecular interaction database: 2012 update. Nucleic Acids Res. 2012;40:857–61.
6. Nepusz T, Yu H, Paccanaro A. Detecting overlapping protein complexes in protein-protein interaction networks. Nat Methods. 2012;9:471–2.

7. van Dongen S. A cluster algorithm for graphs. 2000. Technical Report INS-R0010, National Research Institute for Mathematics and Computer Science in the Netherlands, Amsterdam.

8. Satuluri V, Parthasarathy S. Scalable Graph Clustering Using Stochastic Flows: Applications to Community Discovery. In: 15th ACM SIGKDD International Conference on Knowledge Discovery and Data Mining (KDD'09). Paris; 2009. p. 737–46.

9. Shih YK, Parthasarathy S. Identifying functional modules in interaction networks through overlapping Markov clustering. Bioinformatics. 2012;28(18):1473–9.

10. Palla G, Derényi I, Farkas I, Vicsek T. Uncovering the overlapping community structure of complex networks in nature and society. Nature. 2005;435:814–8.

11. Macropol K, Can T, Singh AK. RRW: Repeated random walks on genome-scale protein networks for local cluster discovery. BMC Bioinformatics. 2009;10:283.

12. Ahn YY, Bagrow JP, Lehmann S. Link communities reveal multiscale complexity in networks. Nature. 2010;466:761–4.

13. Gavin AC, Aloy P, Grandi P, et al. Proteome survey reveals modularity of the yeast cell machinery. Nature. 2006;440:631–6.

14. Wang Y, Qian X. Functional module identification in protein interaction networks by interaction patterns. Bioinformatics. 2014;30(1):81–93.

15. Newman MEJ. Modularity and community structure in networks. Proc Nat Acad Sci USA. 2006;103:8577–82.

16. Fortunato S, Barthélemy M. Resolution limit in community detection. Proc Nat Acad Sci USA. 2007;104(1):36–41.

17. Andersen R, Chung F, Lang K. Local Graph Partitioning Using PageRank Vectors. In: 2006 47th Annual IEEE Symposium on Foundations of Computer Science (FOCS'06). Berkeley; 2006. p. 475–86.

18. Corneil DG, Perl Y. Clustering and domination in perfect graphs. Discrete Appl Math. 1984;9(1):27–39.

19. Ashburner M, Ball CA, Blake JA, et al. Gene ontology: Tool for the unification of biology. The Gene Ontology Consortium. Nat Genet. 2000;25(1):25–9.

20. Hong EL, et al. Gene ontology annotations at SGD: New data sources and annotation methods. Nucleic Acids Res. 2008;36:577–81.

21. Mewes HW, et al. MIPS: Analysis and annotation of proteins from whole genomes. Nucleic Acids Res. 2004;32(Database issue):D41–44.

22. Wang Y, Qian X. Joint clustering of protein interaction networks through Markov random walk. BMC Syst Biol. 2014;8(suppl 1):9.

23. Shih Y, Parthasarathy S. Scalable global alignment for multiple biological networks. BMC Bioinformatics. 2012;13(Suppl 3):11.

24. Fan N, Pardalos P. Multi-way clustering and biclustering by the ratio cut and normalized cut in graphs. J Combinatorial Optimization. 2010;23(2):224–51.

25. Gurobi Optimization Inc. Gurobi Optimizer Reference Manual. 2016. http://www.gurobi.com.

12

Hybrid method to solve HP model on 3D lattice and to probe protein stability upon amino acid mutations

Yuzhen Guo[1][*][†], Fengying Tao[1][†], Zikai Wu[4,5] and Yong Wang[2,3]

Abstract

Background: Predicting protein structure from amino acid sequence is a prominent problem in computational biology. The long range interactions (or non-local interactions) are known as the main source of complexity for protein folding and dynamics and play the dominant role in the compact architecture. Some simple but exact model, such as HP model, captures the pain point for this difficult problem and has important implications to understand the mapping between protein sequence and structure.

Results: In this paper, we formulate the biological problem into optimization model to study the hydrophobic-hydrophilic model on 3D square lattice. This is a combinatorial optimization problem and known as NP-hard. Particle swarm optimization is utilized as the heuristic framework to solve the hard problem. To avoid premature in computation, we incorporated the Tabu search strategy. In addition, a pulling strategy was designed to accelerate the convergence of algorithm based on the characteristic of native protein structure. Together a novel hybrid method combining particle swarm optimization, Tabu strategy, and pulling strategy can fold the amino acid sequences on 3D square lattice efficiently. Promising results are reported in several examples by comparing with existing methods. This allows us to use this tool to study the protein stability upon amino acid mutation on 3D lattice. In particular, we evaluate the effect of single amino acid mutation and double amino acids mutation via 3D HP lattice model and some useful insights are derived.

Conclusion: We propose a novel hybrid method to combine several heuristic strategies to study HP model on 3D lattice. The results indicate that our hybrid method can predict protein structure more accurately and efficiently. Furthermore, it serves as a useful tools to probe the protein stability on 3D lattice and provides some biological insights.

Keywords: Protein structure prediction, HP model, 3D lattice, Particle swarm optimization, Protein stability

Background

Protein is the substantial basis of biological activity. The function of protein is determined by its structure which is believed to be decided by the amino acid sequence according to Anfinsen's experiments. So the research on protein structure prediction (also called protein folding

*Correspondence: guoyuzhen@nuaa.edu.cn
[†]Equal contributors
[1]Department of Mathematics, Nanjing University of Aeronautics and Astronautics, 210000 Nanjing, People's Republic of China
Full list of author information is available at the end of the article

problem) is very significant and fundamental in exploring the fundamental principle to map sequence, structure, and function.

To capture the backbone of protein structure prediction, Dill and his collaborators introduced HP lattice model to simplify real world complexity in 1995 [1]. HP lattice model is an abstracted scaffold, and eventually convert the protein structure prediction problem to an optimization problem on lattice. The aim is to find the optimal structure with the lowest energy. Computationally, solving this problem is NP-hard. For this reason many researchers have been attracted to study this problem by proposing many heuristic algorithms. In recent years, for 2D HP protein folding problem, many methods have been proposed, e.g. , PSO (Particle Swarm Optimization) [2], ACO (Ant Colony Algorithm) [3], ABO (Artificial Bee Colony) [4] and SOM (Self-Organizing Mapping) [5] etc.

One issue for 2D lattice model is that it's too simplified to constrain the amino acid sequence on a 2D plane. One step forward is to fold the sequence on 3D lattice and make it a better and native approximation. So far, several algorithms have been applied for 3D HP protein structure prediction problem, such as UEGO (Universal Evolutionary Global Optimization) [6], GA (Genetic Algorithms) [7], TS (Tabu Search) [8], EA (Evolutionary Algorithm) [9] and so on. Each method has its advantage to capture some special structure in the problem. In this paper, we aim to propose a hybrid method and improve the efficiency to solve the 3D HP protein structure prediction problem.

PSO was introduced by Kennedy and Eberhart [10]. It is a swarm intelligence optimization algorithm which imitates the foraging behaviors of birds and fish. As a simple meta-heuristic, it has been used to solve optimization problem with nonlinear, non-differentiable, and multi-modal function. Originally, this algorithm was designed for solving continuous optimization problem. Here, we started from the basic PSO framework and firstly extend the algorithm to the combinatorial optimization, into which we formally formulate the HP model on 3D lattice. In addition, we improved PSO as follows: a) redefined velocity for discrete model; b) employed modified Tabu search strategy to avoid premature convergence; c) designed pulling strategy to speed up convergence.

We showed that our hybrid algorithm can predict structures of amino acid sequences with different length efficiently. With this useful tool, we simulated the effects after single amino acid mutation and double amino acids mutation, respectively. Some biological insights are obtained.

The remainder of this paper is organized as follows. Firstly, a mathematical model was established for 3D HP problem. Secondly, we explained the PSO algorithm and proposed modified Tabu search method and pulling strategy. Thirdly, the performance of our algorithm was validated. Fourthly, the amino acid mutation result was obtained and analyzed. Finally, conclusions were presented.

Methods

Combinatorial optimization formulation for 3D HP lattice model

In HP model, every amino acid sequence is abstracted as an alphabetic string with H (hydrophobic amino acid) and P (hydrophilic amino acid). The protein conformation is a self-avoiding path on a 2D lattice. It is assumed that the main driving forces of the formation of the tertiary structure are the interactions among hydrophobic amino acids which are adjacent on lattice but not adjacent in the sequence, denoted as H-H interactions. The free energy of a protein conformation (X) is expressed by the number of H-H interactions. Based on Anfinsen's assumption [11], the configuration tends to form a core in the spatial structure shield from the surrounding solvent by hydrophilic amino acids with the minimal free energy. So the more H-H interactions, the lower the free energy. We assumed that the free energy equals to the minus number of H-H interactions. HP lattice model has been used for solving protein structure prediction problem on 2D and 3D lattices widely. In this paper, we focused on the 3D HP square lattice model.

At present, relative coordinates and space coordinates have been used to denote the protein conformation. For a sequence S with L amino acids, X is a string of length $L-1$ over the symbols $\{r(ight), l(eft), f(orward), d(own), u(p)\}$ in relative coordinates, these five symbols reflect the relative location of contiguous amino acids on lattice. In space coordinates, X records the 3D coordinates of L amino acids, namely, $X = (X(1), X(2) \cdots X(L))$ and $X(l) \in N^3$ ($l = 1, 2 \cdots L$) is the coordinate of the lth amino acid. In this paper, we chose the space coordinates. For example, Fig. 1 showed a conformation with 7 H-H interactions on 3D square lattice. Its conformation was denoted as $X = ((2, 3, 2), (3, 3, 2), (3, 4, 2), (3, 4, 3), (3, 3, 3), (2, 3, 3), (2, 2, 3), (3, 2, 3), (3, 2, 2), (3, 1, 2), (2, 1, 2), (2, 2, 2))$.

Based on the abstraction and minimum energy principle, we established the optimization model (OM) for protein structure prediction problem on 3D square lattice as following:

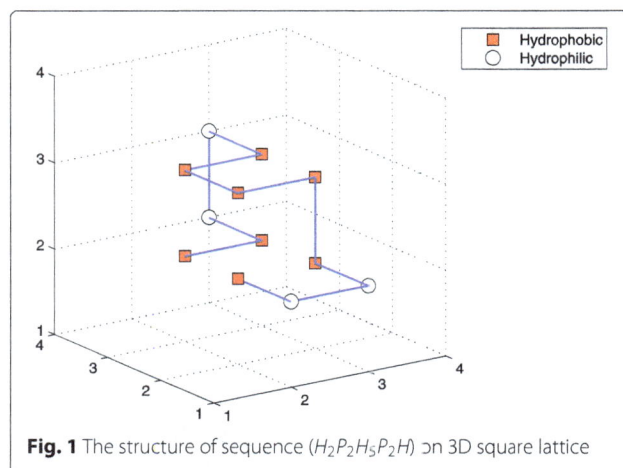

Fig. 1 The structure of sequence ($H_2P_2H_5P_2H$) on 3D square lattice

$$min \quad E(X) \qquad (1)$$

$$s.t. \quad \sum_{i=1}^{I}\sum_{j=1}^{J}\sum_{k=1}^{K} x_{i,j,k}(l) = 1 \qquad l = 1, 2 \cdots L \qquad (2)$$

$$0 \le \sum_{l=1}^{L} x_{i,j,k}(l) \le 1 \qquad l = 1, 2 \cdots L \qquad (3)$$

$$\sum_{d=1}^{3} |X(l+1)_d - X(l)_d| \cdot \|X(l+1) - X(l)\| = 1$$

$$l = 1, 2 \cdots L - 1 \qquad (4)$$

Here,

$$E(X) = -M(X) \qquad (5)$$

$$M = \sum_{i=1}^{I}\sum_{j=1}^{J}\sum_{k=1}^{K}\sum_{l=1}^{L} x_{i,j,k}(l)f(l)\sum_{r=1}^{L} f(r)[x_{i,j,k+1}(r) + x_{i,j+1,k}(r)$$
$$+ x_{i+1,j,k}(r)] - h \qquad (6)$$

$$h = \sum_{l=1}^{L-1} f(l)f(l+1) \qquad (7)$$

$$x_{i,j,k}(l) = \begin{cases} 1 & \text{if the } X(l) = (i,j,k) \\ 0 & \text{else} \end{cases} \qquad (8)$$

$$f(l) = \begin{cases} 1 & \text{if the } l^{th} \text{ amino acid is H} \\ 0 & \text{if the } l^{th} \text{ amino acid is P} \end{cases} \qquad (9)$$

Where, $E(X)$ is the free energy of protein conformation X, $X(l)_d$ is the dth component of $X(l)$, $M(X)$ is the number of H-H interactions in conformation X, r expresses the number of adjacent hydrophobic pairs in amino acid sequence and $\| \cdot \|$ is Hamming distance. Equations (2), (3) and (4) constrain that every amino acid occupies only one lattice point, each lattice point cannot be used more than once and adjacent amino acids in the chain occupy the adjacent points on the lattice. Equation (8) presents whether the lth amino acid occupies point (i,j,k). In Eq. (9), $f(l)$ translates the lth H (or P) of the amino acid sequence into 1 (or 0).

Solving the simplified HP model is NP-complete even on two dimensional lattice. Then we have to seek help from heuristic algorithms. Particle swarm optimization, one of the stochastic algorithm, serves as a powerful approximation method.

Hybrid algorithm
The basic PSO algorithm
Particle swarm optimization (PSO) is a heuristic framework that optimizes an objective function by iteratively improve a candidate solution. The motivation is to have a population of candidate particles, and move these particles around in the search-space according to simple mathematical formulae over the particle's position and velocity. Each particle's movement is influenced by its local best known position, but is also guided toward the best known positions in the search-space, which are updated as better positions are found by other particles. Finally it is expected to move the swarm toward the best solution. The advantage of PSO is that it makes no assumptions about the problem and can search very large spaces of candidate solutions.

In basic PSO algorithm (See Table 1), m particles search the optimal position simultaneously with dynamic velocity. Particle velocity is affected by iteration, own cognition, and social cognition of particle. Particularly, each particle can remember not only its own flight experience, but also the trajectories of all particles. In n dimensional search space, the position and velocity of the ith particle are represented as $X_i \in R^n$ and $V_i \in R^n$, respectively. They are updated by the following two equations:

$$V_i^{t+1} = \omega V_i^t + c_1 r_1 \left(P_{ib}^t - X_i^t \right) + c_2 r_2 \left(P_{gb}^t - X_i^t \right) \quad (10)$$

$$X_i^{t+1} = X_i^t + V_i^{t+1} \qquad (11)$$

Where P_{ib}^t and P_{gb}^t are the best position of the ith particle and the best position of all particles in the tth iteration, respectively. Inertia weight (ω), self confidence (c_1) and swarm confidence (c_2) are input parameters, r_1, r_2 are two separately generated uniformly distributed random numbers in the range [0,1].

The modified PSO algorithm
Definitions To solve the optimization model, we redefined position and velocity of PSO on 3D lattice. Particle position was orderly expressed by protein conformation (X). Velocity of particle was defined as a series of shift (j_1, j_2), which means that the j_1^{th} component of particle position becomes the j_2^{th} component, then the j_1^{th} component and the j_2^{th} component (including the j_2^{th} component) were changed subsequently. In addition, position X_1 was obtained by the sum of position X_2 and a series of shift, namely

Table 1 The process of basic PSO algorithm

Step 1	To **initialize** $\{X_i^0 | i = 1, 2 \cdots m\}$ and $\{V_i^0 | i = 1, 2 \cdots m\}$;
Step 2	To **calculate** $E(X_i^t)$, find P_{ib}^t and P_{gb}^t;
Step 3	To **update** X_i^t and V_i^t;
Step 4	To **output** P_{gb}.

$X_1 = X_2 + \{(j_p, j_q)\}$. For example, $V = \{(2,4), (3,1)\}$ and $X = (X(1), X(2), X(3), X(4))$, then

$$
\begin{aligned}
X + V &= (X(1), X(2), X(3), X(4)) + \{(2,4), (3,1)\} \\
&= (X(1), X(3), X(4), X(2)) + \{(3,1)\} \\
&= (X(4), X(1), X(3), X(2)).
\end{aligned}
$$

Clearly, $X + V$ is a new position. Nevertheless, the new position may not satisfy the constraints in the OM model. An adjustment strategy is needed to ensure the new position was valid.

Modified Tabu search strategy Premature convergence is one of the major difficulty to solve OM model by PSO algorithm. To further improve the modified PSO, we adopted the idea of Tabu search which was proposed by Glover [12]. This method was briefly described as follows.

Tabu search is a meta-heuristic method that maintains only one solution in the iteratively searching process. Given an initial solution X, the idea is to calculate and compare its neighboring solutions $N(X)$. The best solution is chosen as candidate solution X_c. If X_c is satisfied with the aspiration rule, it will replace the current solution X and be added to tabu list T_{list}; Otherwise, the current solution X will be replaced by the best one X' ($E(X') = min\{E(X)|X \in N(X), X \notin T_{list}\}$) and X' will be added to T_{list}. Generally, T_{list} is a first-in first-out (fifo) memory with limited length. So particles would not search the solutions which have been found for a while, simultaneously, the better solutions would not always be taboo.

Neighbourhood of solution and aspiration rule are the key components of Tabu search. In our 3D HP problem, feasible solution is a 3D self-avoiding path. It was not easy to figure out its neighboring solutions from a given solution. According to Eqs. (10) and (11), we got similar solutions by changing r_1, r_2 at the same iteration for the same particle in PSO, then these solutions constituted

a neighbourhood. When candidate solution was better than the current solution, we would ignore whether the candidate solution was taboo or not.

Pulling strategy The convergence rate of modified PSO with Tabu search strategy is not fast enough and the conformations obtained by this modified PSO may be too loose. The following strategy was designed in order to improve the algorithm.

In native protein structure, hydrophobic amino acids concentrate inside of conformation and they were surrounded by hydrophilic amino acids. If hydrophilic amino acids were pulled out of the central of protein structure, the structure will be more compact and more stable. Without changing structure's legitimacy, this strategy was defined as pulling strategy. In order to make pulled structure to satisfy the self-avoiding constraints, only one amino acid could be pulled to its vacant diagonal position once. Figure 2 showed the move and result of one pulling.

Table 2 The algorithm outline of TPPSO

Step 1	To **initialize** $\{X_i^0 | i = 1, 2 \cdots m\}$, $\{V_i^0 | i = 1, 2 \cdots m\}$ and $T_{list} = \emptyset$;
Step 2	To **calculate** $E(X_i^t)$, find P_{ik}^t and P_{gb}^t;
Step 3	To **update** $\{V_{ij}^t | j = 1, 2 \cdots s\}$ and $\{X_{ij}^t | j = 1, 2 \cdots s\}$;
Step 4	To **adjust** and pull $\{X_{ij}^t | j = 1, 2 \cdots s\}$;
Step 5	To **calculate** $E(X_{ic}^t) = min\{E(X_{ij}^t) | i = 1, 2 \cdots s\}$;
Step 6	**If** $E(X_{ic}^t) \le E(X_i^t)$ then $X_i^t = X_{ic}^t$;
Step 7	To **calculate** $E(P_{gbc}^t) = min\{E(X_i^t) | i = 1, 2 \cdots m\}$;
Step 8	**If** $E(P_{gbc}^t) < E(X_{gb}^t)$ then $X_i^t = X_{ic}, T_{list} = \emptyset$;
Step 9	**If** $E(P_{gbc}^t) = E(X_{gb}^t)$ and $P_{gbc}^t \notin T_{list}$ then $T_{list} = T_{list} + X_{gb}^t, X_{gb}^t = X_{gbc}^t$;
Step 10	To **output** P_{gb}.

Table 3 Sequences with 27 amino acids used in our study

Sequence ID	Amino acids sequence
A_1	$PHP_3HP_3P_2HPHP_{11}H_2P$
A_2	$PH_2P_{10}H_2P_2H_2P_2HP_2HPH$
A_3	$H_4P_4HP_4H_3P_9H$
A_4	$H_3P_2H_4P_3HPHP_2H_2P_2HP_3H_2$
A_5	$H_4P_4HP H_2P_3H_2P_{10}$
A_6	$HP_6HPH_3P_2H_2P_3HP_4HPH$
A_7	$HP_2HPH_2P_3HP_5HPH_2PHPHPH_2$
A_8	$HP_{11}HPHP_8HPH_2$
A_9	$P_7H_3P_3HPH_2P_3HP_2HP_3$
A_{10}	$P_5H_2PHPHPHPHP_2H_2PH_2PHP_3$
A_{11}	$HP_4H_4P_2HPHPH_3PHP_2H_2P_2H$

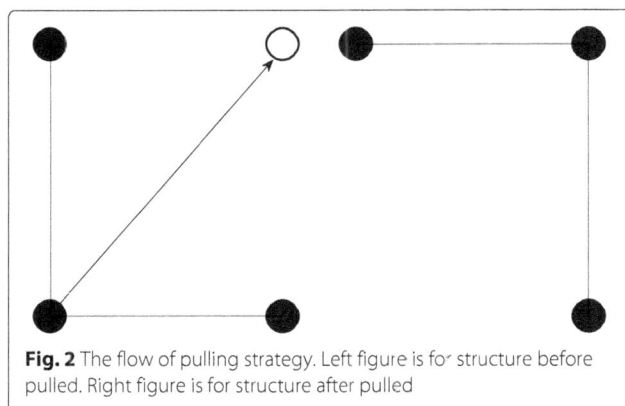

Fig. 2 The flow of pulling strategy. Left figure is for structure before pulled. Right figure is for structure after pulled

Table 4 Comparing four algorithms in eleven sequences with 27 amino acids

Sequence ID	EN	hELP	TPPSO[1]	TPPSO[2]
A_1	-9	-9(18009)	-9(1983)	-9(177)
A_2	-10	-10(9447)	-10(1304)	-10(439)
A_3	-8	-8(1420)	-8(1249)	-8(44)
A_4	-15	-15(2125)	-15(795)	-15(19)
A_5	-8	-8(2877)	-8(104)	-8(61)
A_6	-11	-12(2610)	-11(940)	**-12**(812)
A_7	-13	-13(3967)	-12(721)	**-13**(805)
A_8	-4	-4(1070)	-4(6)	-4(3)
A_9	-7	-7(363)	-7(389)	-7(14)
A_{10}	-11	-11(416)	-11(2784)	-11(83)
A_{11}	-14	-16(285)	-14(957)	**-16**(2672)

The number in parentheses is the iteration number before the lowest free energy values are found. TPPSO[2] can find the optimal results of all sequences. TPPSO[1] can't obtain the minimal free energies for sequence A_6, A_7, and A_{11} (highlighted in bold)

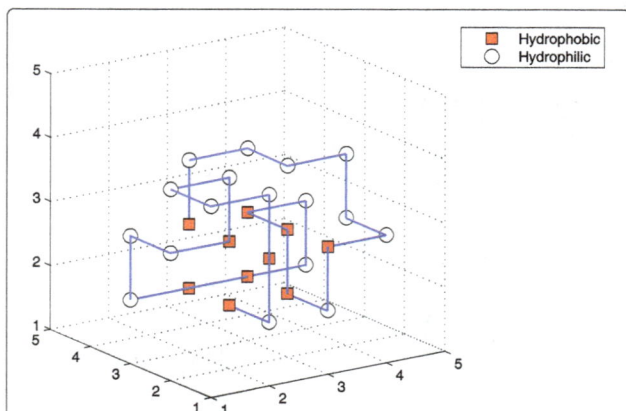

Fig. 4 This is one of structures for sequence A_7. This optimal conformation was simulated by TPPSO[2] with 13 H-H interactions. *Squares* are for hydrophobic amino acids, and *circles* are for hydrophilic amino acids. In this structure almost all hydrophobic amino acids are surrounded in center. It is stable with minimal free energy

Hybrid method A novel hybrid method was proposed by combining modified PSO with modified Tabu search strategy algorithm, denoted as TPPSO[1]. Another hybrid method was taken as TPPSO[2], which combined TPPSO[1] with pulling strategy. Both methods employed Tabu search strategy and were applied to solve protein structure prediction problem. In TPPSO[1] and TPPSO[2], when P_{ib} and P_{gb} were found, s alternative particles would be produced by Eqs. (10) and (11) for each particle.

We selected different r_1 and r_2 for finding alternative particles. These alternative particles might not satisfy the constraints, therefore they should be adjusted. Then the best alternative particle would replace the

previous particle and P_{gb} would be taboo in a period of time. Differently, pulling strategy has been used in TPPSO[2], so each particle could be closer to optimal position. Table 2 showed the detailed procedures of TPPSO[2].

Results

Numerical simulations

In order to test the feasibility of the hybrid algorithms (TPPSO[1] and TPPSO[2]) and explore the properties of algorithms, we calculated two groups of amino acids sequences, respectively.

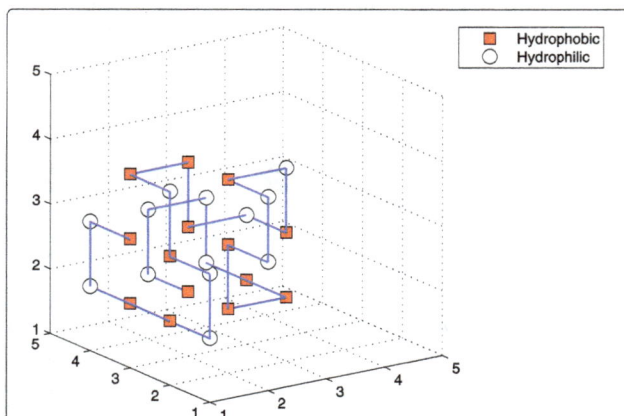

Fig. 3 This is one of structures for sequence A_6. This optimal conformation was simulated by TPPSO[2] with 12 H-H interactions. *Squares* are for hydrophobic amino acids, and *circles* are for hydrophilic amino acids. In this structure all hydrophobic amino acids are surrounded in center. It is stable with minimal free energy

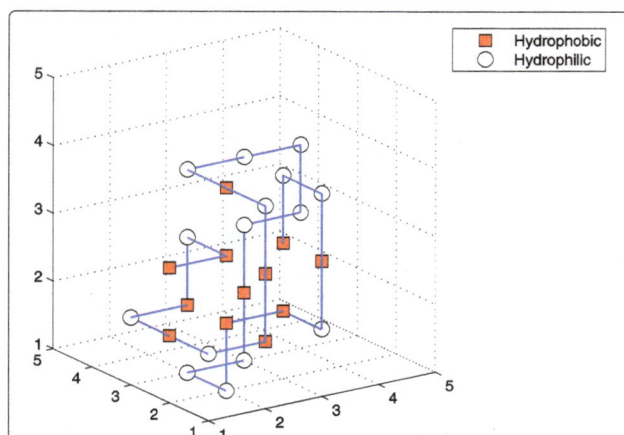

Fig. 5 This is one of structures for sequence A_{11}. This optimal conformation was simulated by TPPSO[2] with 16 H-H interactions. *Squares* are for hydrophobic amino acids, and *circles* are for hydrophilic amino acids. It is stable and compact with minimal free energy

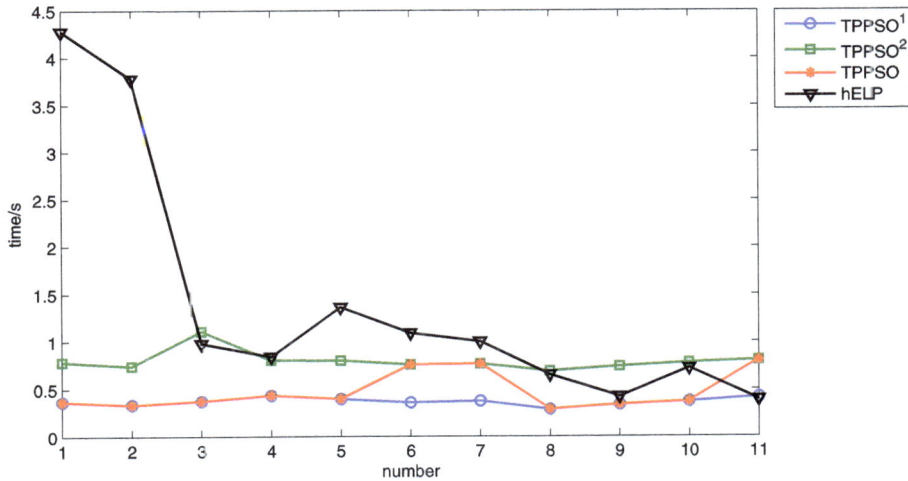

Fig. 6 The average CPU time of our methods and hELP. The abscissa is the number of sequence, and the ordinate is CPU time. Because TPPSO[1] can't obtain the minimal free energy of sequence A_6, A_7 and A_{11}, we chose smaller CUP time with minimal free energy for all sequences, denoted as TPPSO. In the figure, the CPU time of TPPSO[1] and TPPSO[2] are stable with respective optimal structure. CPU time of TPPSO also is stabler. But CPU time of hELP is fluctuant

Simulation of sequences with 27 amino acids

We selected 11 sequences with 27 amino acids (See Table 3) which were also computed by EN [13] and hELP [14]. These sequences were used to test the performances of TPPSO[1] (without pulling strategy) and TPPSO[2] (with pulling strategy), respectively. In TPPSO[1] and TPPSO[2], the inertia weight ω was updated by the following formula:

$$\omega = 0.1 - 0.05 \frac{Time}{Maxtime} \qquad (12)$$

The *Time* is the circular times and *Maxtime* is the maximum number of iterations which is 3000 in our implementation. For each particle, we chose $c_1 = c_2 = 1$, $r_{11} = rand(0.9, 1)$, $r_{12} = rand(0.82, 0.92)$, $r_{13} = rand(0.74, 0.84)$, $r_{21} = rand(0.9, 1)$, $r_{22} = rand(0.85, 0.95)$, $r_{23} = rand(0.8, 0.9)$ to produce three similar but not identical alternative particles. In this test, T_{list} only contained ten particles.

According to Table 4, we knows that all the sequences in Table 3 were simulated by EN, hELP and our method TPPSO[1] and TPPSO[2]. hELP and TPPSO[2] can obtain the minimal free energy of every sequence, but EN and TPPSO[1] can't find the minimal free energies of sequence A_6, A_7, and A_{11} which are bigger. It illustrated our method can successfully predict the protein

structure on 3D square lattice. The number in parentheses is the iteration number when the lowest free energy values are found. By comparing the results of hELP with TPPSO[1] and TPPSO[2], TPPSO[1] is superior to hELP, and TPPSO[2] can fold stable structures earlier than TPPSO[1].

Especially, TPPSO[2] found the lowest free energies of sequences A_6, A_7, and A_{11}, while TPPSO[1] or EN did not. So we enumerated the structures of these sequences (See Figs. 3, 4, and 5), which were computed by TPPSO[2]. It can be seen that these conformations were more native, furthermore, the hydrophobic amino acids were concentrated and surrounded by hydrophilic amino acids. Because pulling strategy of TPPSO[2] has not been employed in TPPSO[1], it is understandable that pulling strategy was able to accelerate the convergence of algorithm and optimize the protein structure.

The average CPU time of hELP was summarized in reference [14]. We also computed the average CPU time of TPPSO[1] and TPPSO[2]. The average CPU time of all methods were shown in Fig. 6. It is obvious that the average CPU time of every sequence of TPPSO[1] is the shortest, and that of TPPSO[2] is longer, because TPPSO[2] added the pulling strategy. For every sequence, the average CPU time

Table 5 IBE number of TPPSO[2]

Sequence ID	A_8	A_9	A_3	A_5	A_1	A_2	A_{10}	A_6	A_7	A_4	A_{11}
H-H[a]	4	7	8	8	9	10	11	12	13	15	16
IBE number[b]	3	14	44	61	177	439	83	812	805	19	2672

[a] H-H means the number of hydrophobic-hydrophobic amino acid interactions for optimal structure
[b] iteration numbers before the lowest free energy values are found

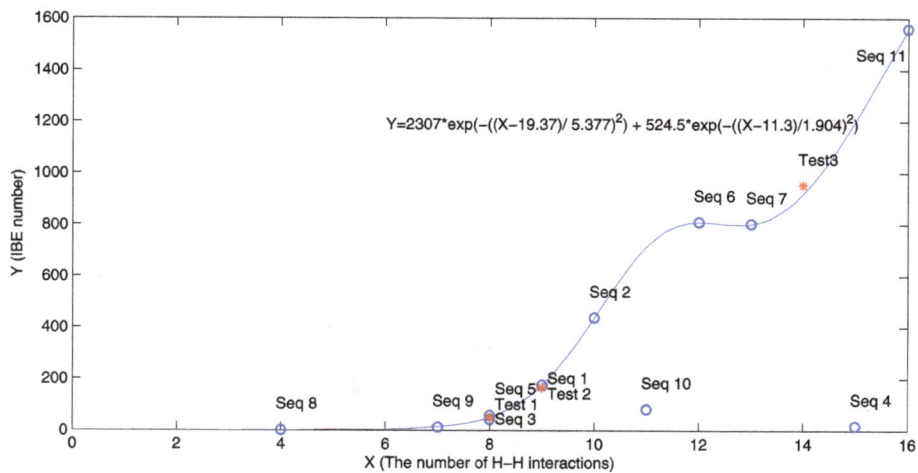

Fig. 7 The fitting figure of IBE number and H-H interaction for sequences in table 1 by TPPSO[2]. The abscissa is H-H interaction of every sequence, and the ordinate is IBM number. Almost all sequences are satisfied with this fitting figure. IBE number will increase with H-H interaction for sequences with the same length. The three stars is the IBE number of test sequence with the same length 27. Their fitting function values are almost matched to computed IBE numbers

of TPPSO[1] and TPPSO[2] are stable and vary around 0.4 and 0.8 s respectively. However, the average CPU time of hELP is not stable. Since TPPSO[1] can't obtain the lowest free energy of sequence A_6, A_7, and A_{11}, we made TPPSO as the method which can fold the optimal structures of all sequences by PSO. The average CPU time of TPPSO was taken as less average CPU time of TPPSO[1] and TPPSO[2], which was also showed in Fig. 6. We know that the average CPU times of all sequences of TPPSO and hELP are 0.475 and 1.41 s respectively. It indicated that our method TPPSO is faster.

Table 5 summarized the number of H-H interactions and iteration number before the lowest free energy values are found by TPPSO[2] (denoted as IBE number) for eleven sequences with 27 amino acids in the Table 3. It is obvious that the more H-H interactions, the more IBE numbers. The fitting function of the number of H-H interactions and IBE number was given as follows.

$$y = 2307 * e^{-\frac{(x-19.37)^2}{5.377^2}} + 524.5 * e^{-\frac{(x-3)^2}{1.904}} \quad (13)$$

where x is the number of H-H pairs, and y is the IBE number.

The figure of fitting function was exhibited in Fig. 7. Except for sequences A_4 and A_{10}, the IBE number of others are all close to the fitting function. The IBE number of sequence A_4 and A_{10} are not satisfied with the fitting function, because in these sequences H amino acids and P amino acids are very dispersive, but in other sequences H segments or P segments are longer. We believed that IBE number of TPPSO[2] is mainly affected by the number of H-H interactions for sequences with the same length. It tends to be larger with more H-H interactions. Moreover, the length of H or P segments will affect the IBE number.

In order to further verify the above conclusion, we simulated three test sequences with the same length (See Table 6). It is obvious that H segments and P segments of these test sequences are longer. IBE numbers for test sequences are close to the fitting curve (See Fig. 7) and all relative errors were showed in Table 6. It means that our inference about IBE number of TPPSO[2] is reasonable.

Table 6 Test sequences

Sequence ID	Amino acids sequence	H-H	IBE number	Relative error
Test 1	$H_4P_5HP_5H_3P_8H$	8	51 (52.3829)	0.0271
Test 2	$H_4P_5HP_5H_3P_4HP_3H$	9	167 (177.8417)	0.0649
Test 3	$(HP_2HP)_5HP$	14	956 (921.1219)	0.0365

IBE number is the iteration number of every sequence by TPPSO[2] before the minimal free energy was found. The number in parentheses is IBE number calculated by fitting function

Table 7 Sequences with different lengths

Sequence ID	Amino acids sequence	Length	H-H	IBE number
B	$H_4P_2H_7P_3H$	17	9	2
C	$HPHP_2H_2PHP_2HPH_2P_2HPH$	20	11	11
D	$P_2HP_2H_2P_4H_2P_4H_2P_4H_2$	25	9	139
E	$P_3H^2P_2H_2P_5H_7P_2H_2P_4H(HP_2)_2$	36	17	432
F	$P_2H(P_2H_2)_2P_5H_{10}P_6(H_2P_2)_2HP_2H_5$	48	29	976

H-H interactions were the same by TPPSO[1] and TPPSO[2]. IBE numbers were computed by TPPSO[2]. These sequences were simulated by different methods, and every method only folds a part of sequences. TPPSO[2] found the minimal free energy of all sequences

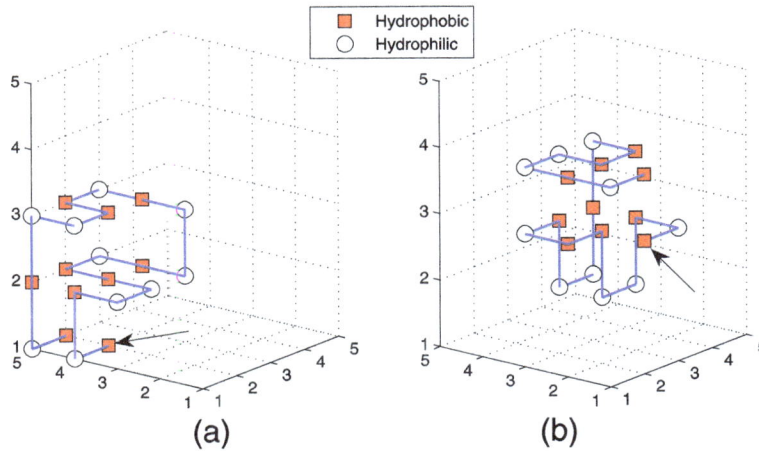

Fig. 8 These figures are the structures of sequence B with 20 amino acids. **a** is one of structures by TPPSO[1] with 11 H-H interactions. **b** is one of structures by TPPSO[2] with 11 H-H interactions. By comparing, *left* structure is more compact, *right* structure is looser

Simulation of sequences with different length

We also computed several sequences with different length which have not been solved by EN and hELP. Moreover, Table 7 recorded the H-H interactions and IBE number of TPPSO[2]. These sequences were simulated by TPPSO[1] and TPPSO[2] respectively. The results of two methods are the same (See Table 7). They have the same H-H interactions. But we knows that the CPU time of TPPSO[2] is shorter than one of TPPSO[1], because TPPSO[2] includes pulling strategy. For this reason, the structure obtained by TPPSO[2] is more compact. It is illustrated in Fig. 8.

These results shows that: a) TPPSO[2] is able to solve sequences with different length and the obtained characteristic of protein structure is significant. b) pulling strategy improved the performance. c) Tabu search strategy

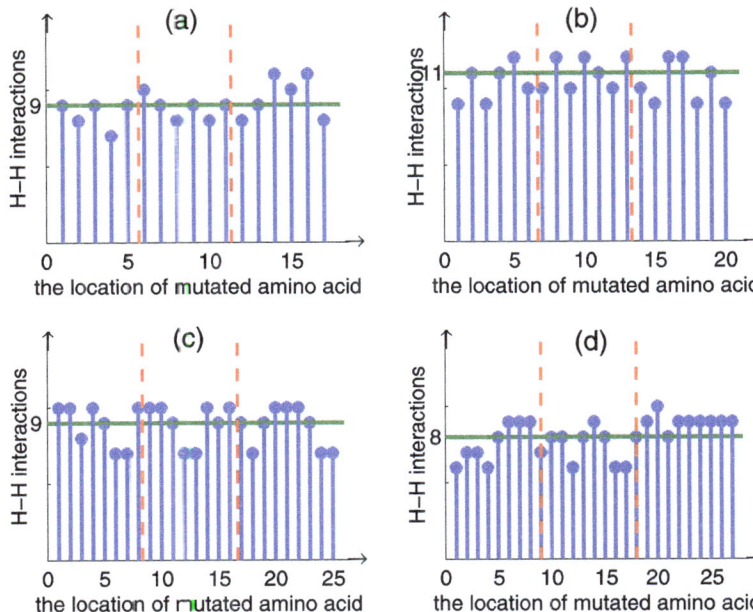

Fig. 9 These figures are H-H interactions of sequences after single amino acid mutation. The abscissa is the location of mutated amino acid, and the ordinate is the number of H-H interaction for mutated sequence. The *horizonal line* is the number of H-H interaction of original sequence. Two vertical lines spit sequence into three equal parts. **a** is mutational results of sequence B. In this figure, 100% pivotal amino acids locate at beginning or ending of sequence B. **b** is mutational results of sequence C. In this figure, 100% pivotal amino acids locate at beginning or ending of sequence C. **c** is mutational results of sequence D. In this figure, 71.4% pivotal amino acids locate at beginning or ending of sequence D. **d** is mutational results of sequence A_8. In this figure, 50% pivotal amino acids locate at beginning or ending of sequence A_8

Table 8 The single amino acid mutation results for sequence B

D-value	-3	-2	-1	0	1	2	3
Q-value	0	1	5	7	2	2	0
R-value	0%	**6%**	2%	41%	12%	**12%**	0%

There are 9 H-H interactions by TPPSO^2 for original sequence B. Every amino acid would be mutational, namely H (P) was changed into P (H). D-value is the deviation of H-H interactions between new sequence and original sequence when single amino acid was mutated. Q-value is the number of amino acids caused the deviation. R-value is the ratio of amino acids. The ratio of amino acids which caused the maximal deviation is 18% (summarization of the numbers highlighted in bold)

avoided prematurity effectively. d) For TPPSO^2, the longer the sequence, the more the IBE number.

Probing protein stability upon amino acid mutation

Protein stability determines whether a protein will be in its native folded conformation or a denatured state. The folded, biologically active conformation of a protein is believed more stable than the unfolded, inactive conformations [15]. Thus, making proteins more stable is important in medicine and basic research. Amino acid mutations are widely used in protein design and analysis techniques to increase or decrease stability. These mutations are carried out experimentally using site-directed mutagenesis and similar techniques. This is time-consuming and often requires the use of computational prediction methods to select the best possible combinations [16–19]. With the efficient hybrid method at hand, we aim to probe the protein stability on 3D lattice. Particularly, we will simulate how single-site or double amino acid mutation affects protein stability. i.e., predicting the protein stability changes upon amino acid mutations with TPPSO^2.

Single amino acid mutation

The hybrid method TPPSO^2 has been tested to solve protein structure prediction problem. Now, we focused on single amino acid mutation, whether and which amino acid affects the stability of protein structure. The experiments is designed as follows. We firstly calculate the optimal H-H interactions of original sequence by TPPSO^2. Then we choose one amino acid to mutate, i.e., we change it from H (P) into P (H). Then we calculate the optimal H-H interactions of mutated sequence by TPPSO^2. Finally the deviation of H-H interactions between mutated sequence and original sequence was recorded.

Table 10 The single amino acid mutation results for sequence D

D-value	-3	-2	-1	0	1	2	3
Q-value	0	7	1	6	11	0	0
R-value	0%	**28%**	4%	24%	44%	0%	0%

There are 9 H-H interactions by TPPSO^2 for original sequence D. Every amino acid would be mutational, namely H (P) was changed into P (H). The ratio of amino acids which caused the maximal deviation is 28% (summarization of the numbers highlighted in bold)

Sequences with different length In order to probe the stability of amino acid mutation, we chose four sequences with different lengths. These sequences were mentioned in the above section. They are sequence B, C, D and A_8.

Figure 9 recorded the H-H interactions of every single amino acid mutational sequence. From the results, we found that some mutational amino acids will result in a bigger deviation. We call those pivotal amino acids. The ratios of pivotal amino acids are 100%, 100%, 71.4% and 50% respectively for the four sequences. Also we deduced that those pivotal amino acids tend to locate at the beginning or end of sequence.

Tables 8, 9, 10 and 11 recorded the characters of mutated sequences including the deviation between mutated sequence and original sequence, the quantity of amino acid which mutated to cause the deviation and the ratio of every deviation. According to the results, we found that single amino acid mutation has the maximal and minimal deviation 2 and -2. The results also indicated that the ratio of maximal deviation is around 22%.

Table 12 summarized the effect of hydrophobic (hydrophilic) amino acid mutation on H-H interactions. We know that hydrophobic amino acid mutation would not make the H-H interactions increase and hydrophilic amino acid mutation would not lead the H-H interactions decrease. It means that hydrophilic amino acid mutation will result in more compact structure, while hydrophobic amino acid mutation will result in the looser structure. By comparing the results of H^2 and P^2 in the Table 12, we suppose that hydrophobic amino acid is more impressible than hydrophilic amino acid to reflect stability of protein structure for sequence with different lengths.

The structures of sequence B before and after mutation are showed in Fig. 10. Since the 14th amino acid was changed from P to H, the number of H-H interaction increases and the deviation is 2, which is the maximal

Table 9 The single amino acid mutation results for sequence C

D-value	-3	-2	-1	0	1	2	3
Q-value	0	5	5	4	6	0	0
R-value	0%	**25%**	25%	20%	30%	**0%**	0%

There are 11 H-H interactions by TPPSO^2 for original sequence C. Every amino acid would be mutational, namely H (P) was changed into P (H). The ratio of amino acids which caused the maximal deviation is 25% (summarization of the numbers highlighted in bold)

Table 11 The single amino acid mutation results for sequence A_8

D-value	-3	-2	-1	0	1	2	3
Q-value	0	5	3	7	11	1	0
R-value	0%	**19%**	11%	26%	41%	**3%**	0%

There are 8 H-H interactions by TPPSO^2 for original sequence A_8. Every amino acid would be mutational, namely H (P) was changed into P (H). The ratio of amino acids which caused the maximal deviation is 22% (summarization of the numbers highlighted in bold)

Table 12 Summary of the single mutation results

Sequence ID	H^0	P^0	H^1	P^1	H^2	P^2
B	12	5	12	5	1	2
C	10	10	10	10	5	0
D	9	9	16	16	7	0
A_8	6	21	6	21	5	1

H^0 and P^0 are the number of hydrophobic and hydrophilic in the original sequence. H^1 is the number of mutational hydrophobic amino acid whose H-H interactions is not more than the original one. P^1 is the number of mutational hydrophilic amino acid whose H-H interactions is not less than the original one. H^2 is the number of hydrophobic amino acid which caused the maximal deviation with original H-H pairs. P^2 is the number of hydrophilic amino acid which caused the maximal deviation with original H-H pairs

deviation. It is obvious that optimal structures of 14th amino acid mutation is more compact.

The structures of sequence D before and after mutation are showed in Fig. 11. Since the 6th amino acid was changed from H to P, the number of H-H interaction decreases and the deviation is 2 which is the maximal deviation. It is obvious that optimal structure of original sequence is more compact.

Sequences with the same length We selected five sequences from Table 3 to test what kind of protein structures are more stable upon single amino acid mutation by TPPSO[2]. We changed every amino acid of these sequences, then recalculated and recorded the H-H interactions of every mutated sequence.

Table 13 showed that: 1) For the sequences with the same length, UC number is likely larger with more mass number. The exception might be caused by the fact that UC number reflected the stability of protein structure. Namely, sequence with larger UC number was susceptible

to single amino acid mutation. 2) According to H-H interactions and the number of hydrophobic amino acids, the sequence with more hydrophobic amino acid usually had more H-H interactions. 3) From P→H and H→P, most of mutational hydrophobic amino acids made the H-H interactions changed. Relatively, only a part of hydrophilic amino acids affect the number of H-H interactions. We conclude that hydrophobic amino acid is more impressible than hydrophilic amino acid to reflect stability of protein structure for sequence with the same length.

All these results illustrated that: a) the more hydrophobic amino acids, the more H-H interactions; b) sequence with more H-H interactions tends to be more stable when single amino acid is mutated; c) hydrophobic amino acid mutation tends to alter the protein structure largely.

According to the above observations, we summarize that the sequence with more hydrophobic amino acids will be less susceptible to single amino acid mutation.

Double neighbouring amino acids mutation

Amino acid does not work alone and multiple amino acids coordinate to maintain stability and perform function. Our in-silicon simulation allows us to go beyond single amino acid mutation and explore the combinatorial effect of amino acid mutation. In this section, we explore the effect of double neighbouring amino acids mutation (two adjacent amino acids are mutated) in protein folding. Double neighbouring amino acids mutations were classified as HH → PP, PP → HH, HP → PH, PH → HP.

We simulated three sequences (B, C, D) with different length when adjacent amino acids mutated. The maximal deviation and locations of pivotal amino acids were conserved (See Tables 14, 15, 16).

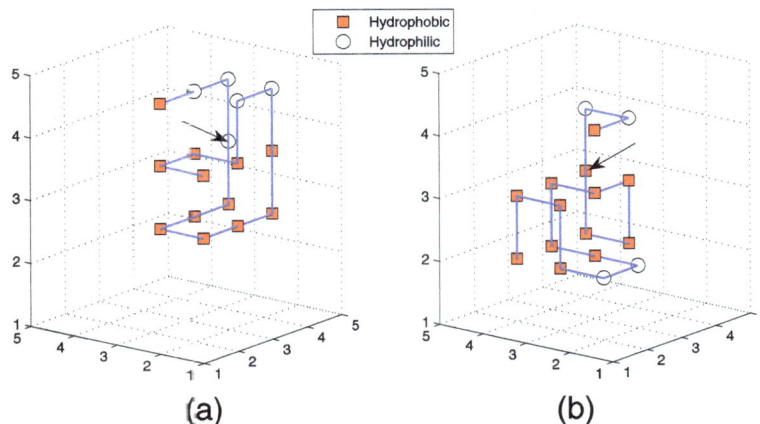

Fig. 10 These figures are the structures of sequence B before mutating and after mutating respectively. **a** is the optimal structure with 9 H-H interactions predicted for original sequence B by TPPSO[2]. **b** is one of optimal structures predicted for mutated sequence B by TPPSO[2], in which the 14th amino acid was mutated. The mutated amino acid is denoted by *arrow* in figures. Since the 14th amino acid was changed from P to H, the number of H-H interaction increases and the deviation is 2. It is obvious that right structure is more compact

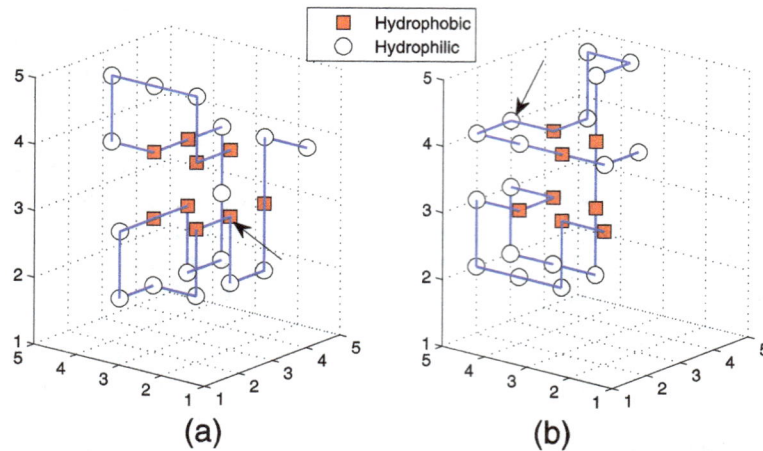

Fig. 11 These figures are the structures of sequence D before mutating and after mutating respectively. **a** is the optimal structure with 9 H-H interactions predicted for original sequence B by TPPSO². **b** is one of optimal structures predicted for mutated sequence C by TPPSO², in which the 6th amino acid was mutated. The mutated amino acid is denoted by *arrow* in figures. Since the 6th amino acid was changed from H to P, the number of H-H interaction decreases and the deviation is 2. It is obvious that left structure is more compact

Tables 14, 15 and 16 recorded the variation of H-H interactions and the position of pivotal double amino acids . According to these tables, we concluded that: a) If double amino acids mutation was HH → PP or PP → HH, the H-H interactions must be changed. But PH→HP and HP→PH maybe have variation. b)HH → PP and PP → HH must make the H-H interactions decrease and increase, respectively. c) The effect of double adjacent amino acids mutation which belongs to HP → PH or PH → HP was finite. d) The position of pivotal double adjacent amino acids mutation tend to locate be at the head or tail of sequence.

Double arbitrary amino acids mutation

We continued to explore the combinatorial effect of amino acid mutation. In this section, we check the effect of double amino acids mutations with arbitrary distance in protein folding. The amino acid mutations were classified ed as HH → PP, PP → HH, HP

→ PH, PH → HP. We simulated the sequence B in Table 7 with 20 amino acids. There are 10 hydrophobic amino acids and 10 hydrophilic amino acids in sequence B. We folded the conformations of all of mutation sequences.

Form Table 17, we knew that combinations of H+H and P+P are more sensitive and easier to affect the stability of protein structure. We simulated every mutational sequence. The results showed that a) all HH→PP mutations will decrease HH interactions, namely HH interactions won't increase; b) 43 PP→HH mutations will increase HH interactions, 2 PP→HH mutations won't change HH interactions, in other words, HH interactions won't decrease; c) HP→PH or PH→HP mutations will not influence HH interactions.

The simulation results in Table 18 indicated that **a)** closer H and H (or P and P) can result in D-value; **b)** amino acid H_{18} respectively matches amino acid H_{20} and amino acid P_{10} to obtain maximal deviation of structure, so amino acid H_{18} is the most sensitive amino acid, which

Table 13 Single amino acid mutation of sequences with 27 amino acids

Sequence ID	H-H	mass number	UC number	H	P	P→H	H→P
A_8	4	9	4	6	21	19	4
A_3	8	7	2	9	18	17	8
A_5	8	8	7	9	18	12	8
A_{10}	11	17	7	11	16	9	11
A_4	15	13	10	14	13	3	14

H-H is the number of H-H pairs of original sequence. The same continuous amino acids were taken as a mass. Mass number is for original sequence. UC number is the number of mutational amino acid which does not change the H-H interactions. H (P) is the number of hydrophobic (hydrophilic) amino acids in the original sequence. P→H (H→P) is the number of mutational hydrophilic (hydrophobic) amino acid which affected the H-H interactions

Table 14 Double amino acids mutation results for sequence B

D-vale	HH → PP (9)[a]	PP → HH (3)[a]	HP → PH (2)[a]	PH → HP (2)[a]
-2	4(H,M,T)[b]	0	0	0
-1	5	0	1	0
0	0	0	0	0
+1	0	1	1	1
+2	0	2(T)	0	1(T)

[a]The number in parentheses is the number of adjacent amino acids in sequences. D-value is the deviation of minimal free energy caused by neighboring mutational amino acid

[b]The position of mutational double amino acids. H (M,T) means that the mutational double amino acids is at head (middle, tail) of the sequence

Table 15 Double amino acids mutation results for sequence C

D-value	HH → PP (2)	PP → HH (3)	HP → PH (7)	PH → HP (7)
-3	1(H)	0	0	0
-2	1(T)	0	0	0
-1	0	0	3	3
0	0	0	4	4
+1	0	2	0	0
+2	0	1(H)	0	0

is at the tail of sequence; **c)** amino acid H_3 will cause D-value with hydrophilic (P), so amino acid H_3 is very sensitive to polar, it is at the head of sequence; **d)** matching amino acid H_7 with arbitrary amino acid P, HH interactions are invariable, so H_7 is obtuse, but by combining H_7 with arbitrary H, it is sensitive, this amino acid is in the middle of sequence; **e)** H_1 and H_{20} are impressible for other H, but they are stable for arbitrary P, the mutations of H_1 and H_{20} with all of P lead to decrease one more HH interaction, H_1 and H_{20} are at the head and tail of sequence, respectively.

According to the above observations, we summarized that a) double arbitrary amino acids mutation will be more sensitive to affect protein stability; b) double amino acids mutation with the same hydrophilic or hydrophobic property is more unstable than double amino acids mutation with different property; c) most of sensitive combinations are at the head or tail of sequence.

Discussion

As many research results indicate, HP model is very useful for modelling protein properties though it is simple and has many disadvantages. It captures the main difficulty of the real world problem. HP model has been applied in investigation of ligand binding to proteins [20]. The distinct influences of function, folding, and structure on the evolution of HP model are studied, by exhaustive enumeration of conformation and sequence space on a two dimensional lattice, which costs four week's computation [21]. These research all show that our effort to fold the HP chain by a hybrid method on 3D lattice is necessary and important.

Also we propose to use HP model to probe the protein stability. HP model serves as a very efficient tool here. The

Table 16 Double amino acids mutation results for sequence D

D-value	HH → PP (4)	PP → HH (11)	HP → PH (4)	PH → HP (5)
-2	4(H,M,T)	0	0	0
-1	0	0	0	3
0	0	0	2	1
+1	0	5	1	1
+2	0	6(M,T)	1(T)	0

Table 17 Double arbitrary amino acids mutation results for sequence B

Combination	O-num	V-num	V-rate
H+H	45	45(↓)	100%
P+P	45	43(↑)	96%
H+P	50	29(↑↓)	58%
P+H	50	18(↑↓)	38%

H+H(P+P) means that arbitrary double hydrophobic(hydrophilic) amino acids will be mutated. H+P(P+H) means that hydrophobic(hydrophilic) will match with hydrophilic(hydrophobic) behind of it to mutate. O-num is original combination number in sequence. V-num is the number of combinations with which minimal free energy were altered after mutating. V-rate is the rate of V-num. The arrows in parentheses indicate increase or decrease of the free energy

simplification of 20 amino acids to H, and P types dramatically reduce the possible mutation pattern. Especially we can easily perform the double mutation only considering four combinations. Those insights from the HP model can serve as novel hypothesis to guide experiments. We also need to point out that the protein stability results and conclusions are heavily depending on the optimal solution of 3D HP model. We demonstrate the results in some small scale problems. When we want to generalize the study, we need to further improve the hybrid algorithm.

In our study, the computational experiments show that the new hybrid algorithm is efficient for short sequences. When the input space is bigger, there will be some suboptimal solutions and more difficult to find the minimal energy configurations. It's really a challenge for large scale HP model. The conformation space grows rapidly as the chain length increases. A possible method is to introduce divide-and-conquer strategy. We can also consider to combine with other algorithms or start from a good initial point from biological view. It will be our future work in devising such an algorithm for large protein.

Conclusion

In this paper, we studied protein structure prediction problem on 3D square lattice. We summarize the findings of this work as follows. Firstly, we formulated the

Table 18 Combination D-value and pivotal amino acids results for sequence B

Combination D-value and pivotal amino acids		
H+H	-4	$H_1H_3, H_{18}H_{20}$
P+P	+2	$P_4P_5, P_4P_{13}, P_5P_8, P_8P_{17}, P_{1C}P_{13}, P_{11}P_{16}, P_{13}P_{16}, P_{16}P_{19}$
H+P	-2	$H_3P_{10}, H_3P_{16}, H_3P_{19}, H_6P_{19}$
P+H	-3	$P_{10}H_{18}$

H+H(P+P) means that arbitrary double hydrophobic(hydrophilic) amino acids will mutate. H+P(P+H) means that hydrophobic(hydrophilic) will match with hydrophilic(hydrophobic) behind of it to mutate. D-value is the maximal deviation of H-H interactions between new sequence and original sequence when double arbitrary amino acids mutation. H_iH_j means that the ith amino acid matches the jth amino acid to mutate, in which two mutational amino acids are hydrophobic

protein structure prediction problem on 3D lattice into a combinatorial optimization problem; secondly, basic PSO algorithm has been enhanced to deal with discrete optimization problem; thirdly, we proposed a novel hybrid method ($TPPSO^2$) and proved its feasibility by simulating; fourthly, we derived some interesting insights for protein stability via single and double amino acid mutation perturbation.

Acknowledgements

The open access fee was supported by the Strategic Priority Research Program of the Chinese Academy of Sciences (XDB13000000). This work is also supported by National Natural Fund under grant number 11601288, 11422108, 61621003, and 61304178.

About this supplement

This article has been published as part of *BMC Systems Biology* Volume 11 Supplement 4, 2017: Selected papers from the 10th International Conference on Systems Biology (ISB 2016). The full contents of the supplement are available online at https://bmcsystbiol.biomedcentral.com/articles/supplements/volume-11-supplement-4.

Authors' contributions

YG developed and implemented the methods. YW and FT participated in the development of the methods. YW conceived the protein stability experiment. All authors draft, read, and approved the final manuscript.

Competing interests

The authors declare that they have no competing interests.

Author details

[1]Department of Mathematics, Nanjing University of Aeronautics and Astronautics, 210000 Nanjing, People's Republic of China. [2]National Center for Mathematics and Interdisciplinary Sciences, Academy of Mathematics and Systems Science, Chinese Academy of Sciences, 100190 Beijing, People's Republic of China. [3]University of Chinese Academy of Sciences, 100049 Beijing, People's Republic of China. [4]University of Shanghai for Science and Technology, 200433 Shanghai, People's Republic of China. [5]Shanghai Key Laboratory of Intelligent Information Processing, Fudan University, 200433 Shanghai, People's Republic of China.

References

1. Dill KA, Bromberg S, Yue K, et al. Principles of protein folding a perspective from simple exact models. Protein Sci. 1995;4(4):561–602.
2. Guo YZ, Wu Z, Wang Y, et al. Extended particle swarm optimization method for folding protein on triangular lattice. IET Sys Bio. 2016;10(1):30–33.
3. Nardelli M, Tedesco L, Bechini A. Cross-lattice Behavior of General ACO Folding for Proteins in the HP Model. Proc of the 28th Annual ACM Symp on Appl Comput. 2013;18(22):1320–1327.
4. Zhang Y, Wu L. Artificial Bee Colony for Two Dimensional Protein Folding. Adv Electr Eng Syst. 2012;1(1):19–23.
5. Zhang XS, Wang Y, Zhan ZW, Wu LY, Chen LN. Exploring protein's optimal HP configurations by self-organizing mapping. J Bioinf Comput Biol. 2005;3(02):385–400.
6. García-Martínez JM, Garzón EM, Cecilia JM, et al. An efficient approach for solving the HP protein folding problem based on UEGO. J Math Chem. 2015;53(3):794–806.
7. Lin CJ, Su SC. Protein 3D HP model folding simulation using a hybrid of genetic algorithm and particle swarm optimization. Int J Fuzzy Syst. 2011;13:140–147.
8. Benítez CMV, Lopes HS. Protein structure prediction with the 3D-HP side-chain model using a master-slave parallel genetic algorithm. J Braz Comput Soc. 2010;16:69–78.
9. Tsay JJ, Su SC. An effective evolutionary algorithm for protein folding on 3D FCC HP model by lattice rotation and generalized move sets. Proteome Sci. 2013;11(1):1.
10. Eberhart RC, Kennedy J. A new optimizer using particle swarm theory. Proc of the sixth Int Symp on micro Mach Hum Sci. 1995;1:39–43.
11. Anfinsen CB. Principles that govern the folding of protein chains. Sci. 1973;181(4096):223–230.
12. Glover F. Tabu search-part I. J Comput. 1989;1(3):190–206.
13. Guo YZ, Feng EM. The simulation of the three-dimensional lattice hydrophobic-polar protein folding. J Chem Phys. 2006;125(23):234703.
14. Liu J, Li G, Yu J, et al. Heuristic energy landscape paving for protein folding problem in the three-dimensional HP lattice model. Comput Biol Chem. 2012;28:17–26.
15. Pascal L, Stefan G, Abdullah K, Valentina C, Paul JB, Christian M, Mering C, Paola P. Cell-wide analysis of protein thermal unfolding reveals determinants of thermostability. Science. 2017;355(6327):812.
16. Parthiban V, Michael MG, Schomburg D. CUPSAT: prediction of protein stability upon point mutation. Nucleic Acids Res. 2006;34(suppl 2):W239–W242.
17. Cheng J, Randall A, Baldi P. Prediction of Protein Stability Changes for Single Site Mutations Using Support Vector Machines, Proteins: Str. Func Biol. 2006;62:1125–32.
18. Shortle D, Stites WE, Meeker AK. Contributions of the large hydrophobic amino acids to the stability of staphylococcal nuclease. Biochem. 1990;29:8033–41.
19. Perl D, Mueller U, Heinemann U, Schmid FX. Two exposed amino acid residues confer thermostability on a cold shock protein. Nat Struct Bio. 2000;7(5):380–3.
20. Miller DW, Dill KA. Ligand binding to proteins: The binding landscape model. Prot Sci. 1997;6(10):2166–79.
21. Blackburne BP, Hirst JD. Evolution of functional model proteins. J Chem Phys. 2001;115(4):1935–42.

13

Comprehensive benchmarking of Markov chain Monte Carlo methods for dynamical systems

Benjamin Ballnus[1,2], Sabine Hug[1], Kathrin Hatz[3], Linus Görlitz[3], Jan Hasenauer[1,2] and Fabian J. Theis[1,2*]

Abstract

Background: In quantitative biology, mathematical models are used to describe and analyze biological processes. The parameters of these models are usually unknown and need to be estimated from experimental data using statistical methods. In particular, Markov chain Monte Carlo (MCMC) methods have become increasingly popular as they allow for a rigorous analysis of parameter and prediction uncertainties without the need for assuming parameter identifiability or removing non-identifiable parameters. A broad spectrum of MCMC algorithms have been proposed, including single- and multi-chain approaches. However, selecting and tuning sampling algorithms suited for a given problem remains challenging and a comprehensive comparison of different methods is so far not available.

Results: We present the results of a thorough benchmarking of state-of-the-art single- and multi-chain sampling methods, including Adaptive Metropolis, Delayed Rejection Adaptive Metropolis, Metropolis adjusted Langevin algorithm, Parallel Tempering and Parallel Hierarchical Sampling. Different initialization and adaptation schemes are considered. To ensure a comprehensive and fair comparison, we consider problems with a range of features such as bifurcations, periodical orbits, multistability of steady-state solutions and chaotic regimes. These problem properties give rise to various posterior distributions including uni- and multi-modal distributions and non-normally distributed mode tails. For an objective comparison, we developed a pipeline for the semi-automatic comparison of sampling results.

Conclusion: The comparison of MCMC algorithms, initialization and adaptation schemes revealed that overall multi-chain algorithms perform better than single-chain algorithms. In some cases this performance can be further increased by using a preceding multi-start local optimization scheme. These results can inform the selection of sampling methods and the benchmark collection can serve for the evaluation of new algorithms. Furthermore, our results confirm the need to address exploration quality of MCMC chains before applying the commonly used quality measure of effective sample size to prevent false analysis conclusions.

Keywords: Parameter estimation, Markov chain Monte Carlo, Sampling analysis, Benchmark collection, Ordinary differential equation, Systems biology

*Correspondence: fabian.theis@helmholtz-muenchen.de
[1] Helmholtz Zentrum München - German Research Center for Environmental Health, Institute of Computational Biology, Ingolstädter Landstraße 1, 85764 Neuherberg, Germany
[2] Technische Universität München, Center for Mathematics, Chair of Mathematical Modeling of Biological Systems, Boltzmannstraße 15, 85748 Garching, Germany
Full list of author information is available at the end of the article

Background

In the field of computational systems biology, mechanistic models are developed to explain experimental data, to gain a quantitative understanding of processes and to predict the process dynamics under new experimental conditions [1–3]. The parameters of these mechanistic models are typically unknown and need to be estimated from available experimental data. The parameter estimation provides insights into the biological processes and its quantitative properties.

The parameters of biological processes are often estimated using frequentist and Bayesian approaches [4, 5]. Frequentist approaches usually exploit optimization methods to determine the maximum likelihood estimate and its uncertainty, e.g., using bootstrapping or profile likelihoods [6–8]. Bayesian approaches often rely on the sampling of the parameter posterior distribution using MCMC algorithms [9–11]. Both, optimization and sampling, are challenging for a wide range of applications encountered in computational systems biology [5, 12]. Likelihoods and posterior distributions are frequently multi-modal and possess pronounced tails (see, e.g., [4, 5]), and many applications problems possess structural and practical non-identifiabilities (see, e.g., [13–16] and references therein). This is, among others, due to scares, noise-corrupted experimental data and a features of the underlying dynamical systems, such as bistability [17, 18], oscillation [19–21] and chaos [22–24].

For optimization, a large collections of benchmark problems were established to facilitate a fair comparison of methods (see, e.g. [25]). Furthermore, optimization toolboxes are available and provide access to a large number of different optimization schemes [26, 27]. The availability of both, benchmark problems and toolboxes, is more problematic for sampling methods. To the best of our knowledge, there is no collection of benchmarking problems for sampling methods featuring dynamical systems. For MATLAB, which is frequently used for dynamical modeling in systems biology, a selection of single-chain methods is implemented in the DRAM toolbox [28]. Standard implementations for state-of-the-art multi-chain methods do however not seem to be publicly available.

In this manuscript, we address the aforementioned needs by (i) providing generic MATLAB implementations for several MCMC algorithms and (ii) compiling a collection of benchmark problems. Our code provides implementations and interfaces to several single- and multi-chain methods, including *Adaptive - Metropolis* [29–32], *Delayed Rejection Adaptive Metropolis* [28], *Parallel Tempering* [32–35], *Parallel Hierarchical Sampling* [36] and *Metropolis - adjusted Langevin algorithm* [37] with or without a preceding *multi-start optimization* [12]. Furthermore, different initialization and adaptation schemes are provided. The sampling methods are

evaluated on a collection of benchmark problems – implementation provided in the Additional file 1 – featuring dynamical systems with different properties such as periodic attractors, bistability, saddle-node, Hopf and period-doubling bifurcations as well as chaotic parameter regimes and non-identifiabilities. The benchmark problems possess posterior distributions with different properties i.e., uni- and multi-modal, heavy tails and non-linear dependency structures of parameters. This collection of features which are commonly encountered in systems biology facilitates the evaluation of the sampling methods under realistic, challenging conditions. To ensure realism of the evaluations, knowledge about the posterior distribution, which is not available in practice, is not employed for selection, adaptation or tuning of methods.

To ensure a rigorous and efficient evaluation of sampling methods, we developed a semi-automatic analysis pipeline. This enabled us to evaluate $> 16,000$ MCMC runs covering a wide spectrum of sampling methods and benchmarks. This comprehensive assessment required roughly 300,000 CPU hours. The study among others revealed the importance of using multi-chain methods and appropriate adaptation schemes. In addition, our results for the benchmark problems indicated a strong dependence of the sampling performance on the properties of the underlying dynamical systems.

Methods

In this section, we introduce parameter estimation, sampling methods along with initialization and adaptation schemes. In addition, the analysis pipeline and the performance evaluation criteria are described.

Mechanistic modelling and parameter estimation

We focus on ordinary differential equation (ODE) models for the mechanistic description of biological processes. ODE models are used to study a variety of biological processes, including gene regulation, signal transduction and pharmacokinetics [11, 38]. Mathematically, ODE models can be defined as

$$\dot{x} = f(x, t, \eta), \qquad x(t_0) = x_0(\eta) \tag{1}$$

with time $t \in [t_0, t_{\max}]$, state vector $x(t) \in \mathbb{R}^{n_x}$ and a parameter vector $\eta \in \mathbb{R}^{n_\eta}$. The vector field $f(x, t, \eta)$ and the initial conditions $x_0(\eta)$ define the temporal evolution of the state variables as functions of η. For biological processes, experimental limitations usually prevent the direct measurement of the state vector $x(t)$. Instead, measurements provide information about the observable vector $y(t)$. The observables depend on the state of the process, $y = h(x, t, \eta)$, in which h denotes the output map $\mathbb{R}^{n_x} \times \mathbb{R} \times \mathbb{R}^{n_\eta} \to \mathbb{R}^{n_y}$. An exemplification of $f(x, t, \eta)$ can be found in "Benchmark problems" section for each of the benchmark problems.

The measurement of the observables y yields noise corrupted experimental data $\mathcal{D} = \{(t_k, \tilde{y}_k)\}_{k=1}^{n_t}$. In the following, we assume independent, additive normally distributed measurement noise

$$\tilde{y}_{ik} = y_i(t_k) + \epsilon_{ik}, \qquad \epsilon_{ik} \sim \mathcal{N}\left(0, \sigma_i^2\right) \tag{2}$$

in which σ denotes the standard deviation of the measurement noise and with $i = 1, \ldots, n_y$. An example for noisy measurement data is discussed and visualized in the "Results", "Application Of sampling methods to mRNA transcription model" subsection.

The standard deviations σ are usually unknown and part of the parameter vector, i.e., $\theta = (\eta, \sigma)$. The likelihood of observing the data $\mathcal{D}$ given the parameters θ is

$$p(\mathcal{D}|\theta) = \prod_{i=1}^{n_y} \prod_{k=1}^{n_t} \frac{1}{\sigma_i \sqrt{2\pi}} \exp\left(-\frac{(\tilde{y}_{ik} - y_i(t_k))^2}{2\sigma_i^2}\right), \tag{3}$$

in which $y(t_k)$ depends implicitly on η.

In Bayesian parameter estimation the posterior

$$p(\theta|\mathcal{D}) = \frac{p(\mathcal{D}|\theta)p(\theta)}{p(\mathcal{D})} \tag{4}$$

is considered, in which $p(\theta)$ denotes the prior and $p(\mathcal{D})$ denotes the marginal probability (being a normalization constant).

Sampling methods

The posterior $p(\theta|\mathcal{D})$ encodes the available information about the parameters θ given the experimental data $\mathcal{D}$ and the prior information $p(\theta)$ [39]. Accordingly, it also encodes information about parameter and prediction uncertainties. This information can be assessed by sampling from $p(\theta|\mathcal{D})$ using MCMC algorithms.

A well-known MCMC algorithm is the Metropolis-Hastings (MH) algorithm [40, 41]. The MH algorithm samples from the posterior via a weighted random walk. Parameter candidates are drawn from a proposal distribution and accepted or rejected based on the ratio of the posterior at the parameter candidate and the current parameter. The choice of the proposal distribution is a design parameter. In practice the distribution is frequently chosen to be symmetric, e.g., a normal distribution, and centered at the current point.

The MH algorithm has several shortcomings, including the need for manual tuning of the proposal covariance and high autocorrelation [39]. Accordingly, a large number of extensions have been developed. In the following, we introduce the three single-chain and the two multi-chain methods employed in this study. Figure 1 highlights the differences between the sampling methods employed in this study using a pseudo-code representation.

Adaptive Metropolis (AM): The AM algorithm is an extension of the standard MH algorithm. Instead of using a fixed proposal distribution which is tuned manually, the distribution is updated based on the already available samples. In particular, for posteriors with high correlation, this improves sampling efficiency by aligning the proposal with the posterior distribution [31]. In addition to the correlation structure, the scale of the proposal is also adapted. A commonly applied scaling scheme is based on the dimension of the problem [28, 29] while other possible schemes are based on the chain acceptance rate [34]. These scaling schemes are in the following indicated by 'dim' and 'acc', respectively.

Delayed Rejection Adaptive Metropolis (DRAM): To further decrease the in-chain auto-correlation, the AM algorithm has been combined with a delayed rejection method, yielding the DRAM algorithm [28]. When a candidate parameter is rejected, the algorithm tries to find a new point using the information about the rejected point. This is repeated multiple times until a certain number of tries is reached or a point is accepted. We employ the implementation provided in [28]. This implementation is exclusively based on the previously mentioned 'dim' adaption scheme.

Metropolis-adjusted Langevin Algorithm (MALA): Both AM and DRAM work best if the local and the global shape of the posterior are similar. Otherwise, the performance of the algorithm suffers, i.e. the in-chain auto-correlation increases. To circumvent this problem, the MALA makes use of the gradient, $\nabla_\theta p(\theta|\mathcal{D})$, and Fisher Information Matrix [37] of the estimation problem at the current point in parameter space. This information is used to construct a proposal which is adapted to the local posterior shape [37, 42]. Gradient and Fisher Information Matrix can be computed using forward sensitivity equations [43].

Parallel Tempering (PT): All of the algorithms, AM, DRAM and MALA, discussed so far are single-chain algorithms which exploit local posterior properties to tune their global movement. This can make transitions between different posterior modes unlikely if they are separated by areas of low probability density. To address the issue, PT algorithms have been introduced. These algorithm sample from multiple tempered versions of the posterior $p(\mathcal{D}|\theta)^{\frac{1}{\beta_l}} p(\theta)$, $\beta_l \geq 1$, $l = 1, \ldots, L$, at the same time [33–35]. The tempered posteriors are flattened out in comparison to the posterior, rendering transitions

---

Algorithm 1: MCMC algorithms used in this study

---

input : Initial point θ^0, lower bounds θ_{min}, upper bounds θ_{max} and number of samples N_{sample}
input : Initial covariance Σ^0 // This is required for all algorithms but MALA
input : Algorithm-specific options for AM, DRAM, PT, PHS and MALA
output: $\theta^1, ..., \theta^{N_{sample}}$

Initialize
for $i \leftarrow 1$ **to** N_{sample} **do**
   // AM, DRAM and MALA use a single chain $L = 1$ while PT and PHS use multiple chains $L > 1$. The chain index is denoted by $l = 1, ..., L$ and the chain position by $\theta^{l,i}$
   for $l \leftarrow 1$ **to** L **do**
     // AM, MALA, PHS and PT propose a single new candidate $\theta^{l,k}_{cand}$ in each iteration i per chain l ($N_{tries} = 1$). DRAM exploits multiple tries $N_{tries} > 1$ to decrease the auto-correlation.
     for $k \leftarrow 1$ **to** N_{tries} **do**
       Propose a candidate $\theta^{l,k}_{cand} \sim \mathcal{N}(\theta^{l,i-1}, \Sigma^{l,i-1})$. // All algorithms in this study use a normal distribution for proposing new candidates.
       if $\theta_{min} < \theta^{l,k}_{cand} < \theta_{max}$ **then**
         Evaluate the acceptance probability $p^{l,k}_{acc}$ *as a function of posterior values and transition probabilities* // For DRAM the acceptance probability accounts for the multiple tries. PT compares the tempered posterior values.
       else
         $p_{acc} \leftarrow -\infty$
       end if
       if $u \sim U(0,1) \leq p^{l,k}_{acc}$ **then**
         Accept candidate $\theta^{l,i} \leftarrow \theta^{l,k}_{cand}$
         break for // Necessary in case of DRAM.
       else if $k = N_{tries}$ **then**
         Reject candidate $\theta^{l,i} \leftarrow \theta^{l,i-1}$
       end if
     end for
     Calculate new proposal covariance matrix $\Sigma^{l,i}$. // For MH Σ is fixed. For AM, DRAM, PT and PHS, $\Sigma^{l,i}$ is calculated from $\Sigma^{l,i-1}$ and $\theta^{l,i}$. For MALA Σ is approximated using local gradient and Hessian information at the current point $\theta^{l,i}$.
     Adapt scaling factor η. // For AM, DRAM, PT and PHS, Σ is usually multiplied with a scalar factor η to ensure 23.4% acceptance of the chain.
   end for
   Swap chains // Only for PT and PHS. Swaps of PT chains are executed by chance, applying a swapping strategy as for example PTEE. PHS swaps its main chain ($l = 1$) with one of the auxiliary chains ($l \in \{2, ...L\}$) in each iteration. The auxiliary chain is chosen uniformly random.
   Adapt inverse temperatures $\beta^{1,...,L}$ // This is performed for some PT versions based on the acceptance rates of swaps between chains.
   Adapt the number of chains L // This is performed for some PT versions.
end for

Fig. 1 Pseudo-code for the MCMC methods used in this study. The pseudo-code highlights differences between MCMC methods using comments indicated by "//" and the *color-coded* name of the relevant algorithm either AM, DRAM, PT, PHS or MALA

between posterior modes more likely. Allowing the tempered chains to exchange their position by chance enables the untempered chain, which samples from the posterior, to 'jump'. For this study, we have implemented the PT algorithm as formulated by Lacki et al. [32] using AM with 'acc' adaption scheme or MALA for each tempered chain.

We considered different initial numbers L_0 of tempered chains, adaptive $L \leq L_0$ or fixed numbers $L = L_0$ and two different swapping strategies [32]:

- Swaps between all adjacent chains (aa)
- Swaps of chains with equal energy (ee)

are employed.

Parallel Hierarchical Sampling (PHS): An alternative to PT is PHS, which employs several chains sampling from the posterior [36]. Similar to PT, the idea is to start multiple auxiliary chains at different points in parameter space and to swap the main chain with a randomly picked one in each iteration. The main differences between PT and PHS are that all chains of PHS are sampling from the same distribution and that a swap between main and auxiliary chains is always accepted in PHS. The use of multiple chains can improve the mixing as different chains can employ different proposal distributions [5]. Here we apply AM('acc') for each of the auxiliary chains.

Initialization

The performance of sampling methods can depend on their initialization [39]. Here we consider two alternative initialization schemes: Initialization using samples from the prior distribution; and initialization using multi-start local optimization results. The methods are illustrated in Fig. 2.

Sampling From Prior Distribution (RND): In many applications, sampling is initialized with parameters drawn from the prior distribution. As the prior distributions are often available in closed-form, this is usually straightforward and computationally inexpensive.

Multi-start Local Optimization (MS): Sampling from the prior distribution frequently yields starting points with low posterior probability. Sampling methods started at these points can require a large number of iterations to reach a parameter regime with high posterior probabilities. To address this problem, initialization using multi-start local optimization has been proposed [5]. The results of multi-start local optimization provide a map of the local optima of the posterior distribution where the frequency of occurrence of a local optimum corresponds to the size of their basin of attraction. Single-chain methods are initialized at the local optima with the highest posterior probability. For multi-chain methods, we first filter the optimization results based on the difference to the best optimization result. From the remaining results initial conditions are sampled for each of the individual chains (please refer to the Additional file 1: Section 1 for further details of the initializations).

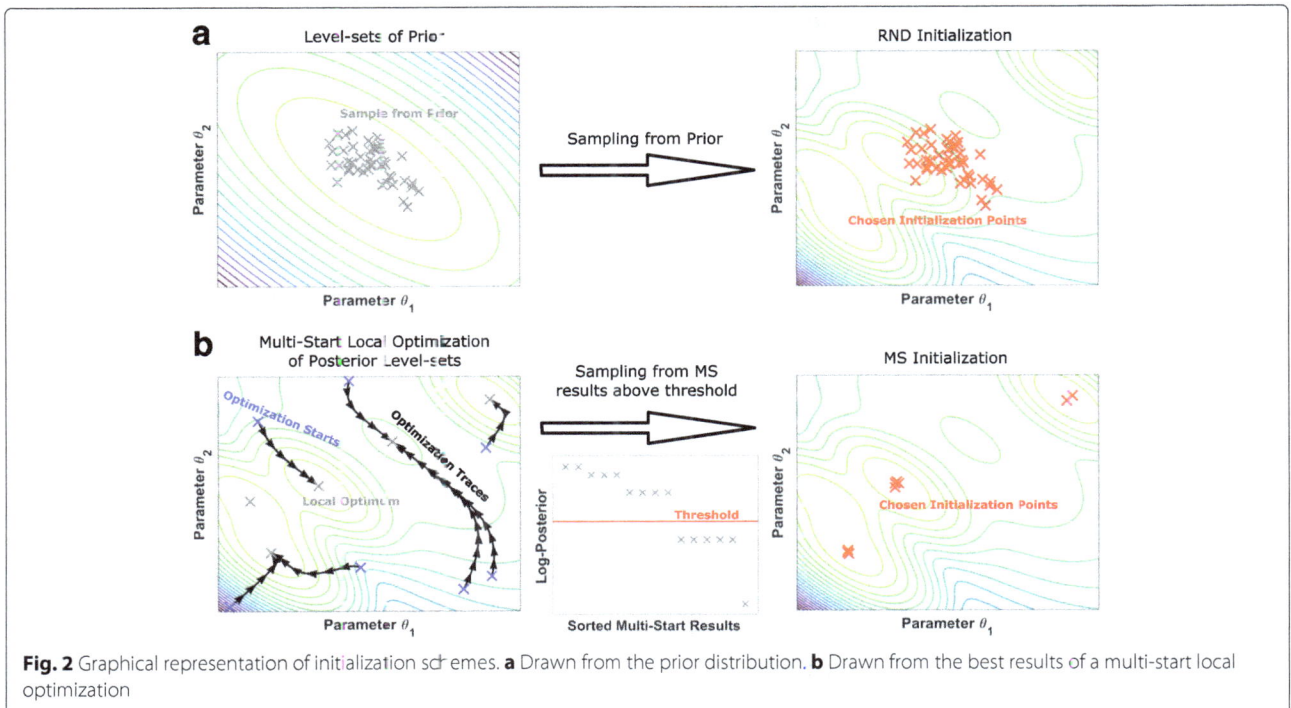

Fig. 2 Graphical representation of initialization schemes. **a** Drawn from the prior distribution. **b** Drawn from the best results of a multi-start local optimization

Run repetitions

We benchmark five state-of-the-art sampling approaches for multiple settings of tuning parameters in challenging, yet low dimensional benchmark problems. In the following, these combinations – of which we consider 23 – are denoted as *scenarios*. To obtain reliable evaluation results, we perform 100 runs for each scenario thus performing 2300 runs per benchmark problem (details about the benchmark problems can be found below). Each run comprises 10^6 iterations of a single- or multiple chains depending on the used algorithm.

Analysis pipeline

The sampling results for all benchmark problems and sampling strategies are analyzed using a combination of four measures: burn-in time, global exploration quality, effective sample size and computation time demand in seconds. The analysis pipeline is illustrated in Fig. 3. The pipeline exploits a combination of heuristics and statistical tests. General details are covered in the following while some further details regarding the statistical tests and heuristics can be found in Additional file 1: Sections 2 and 3.

Burn-In (BI): Often the first part of a Markov chain is strongly influenced by the starting point and, for adaptive methods, by the initial choice of the adaptation parameters [42]. While these effects will vanish asymptotically, for finite chain lengths there might be a large effect. To reduce these effects, the burn-in phase, in which the statistical sample mean changes substantially, is often discarded [44]. We denote the last of those iterations as n_{BI} and only the shortened chains with iteration numbers $n_{BI} + 1$ to 10^6 are considered for further analysis. The BI is typically estimated by a visual check and validated using the Geweke test [45], which is described below and illustrated in Fig. 4a. To circumvent a manual visual inspection, we developed an automatic approach for burn-in calculation using a sequence of Geweke tests taking Bonferroni-Holm adaptation [46] into account (see Additional file 1: Section 2 for further details).

Exploration Quality (EQ): An important quality measure for an MCMC algorithm is the fraction of runs which provide a representative sample from the posterior distribution for a given finite number of iterations. We denote this fraction as *EQ*.

While all MCMC algorithms considered in this manuscript converge asymptotically under mild conditions, for a finite number of samples, individual modes or tails of the posterior might be underrepresented in the chain. This problem is often adressed with statistical tests as Geweke [45] and the Gelman-Rubin-Brooks diagnostic [47]. While the Geweke test considers differences in the means of two signals (usually the beginning and the

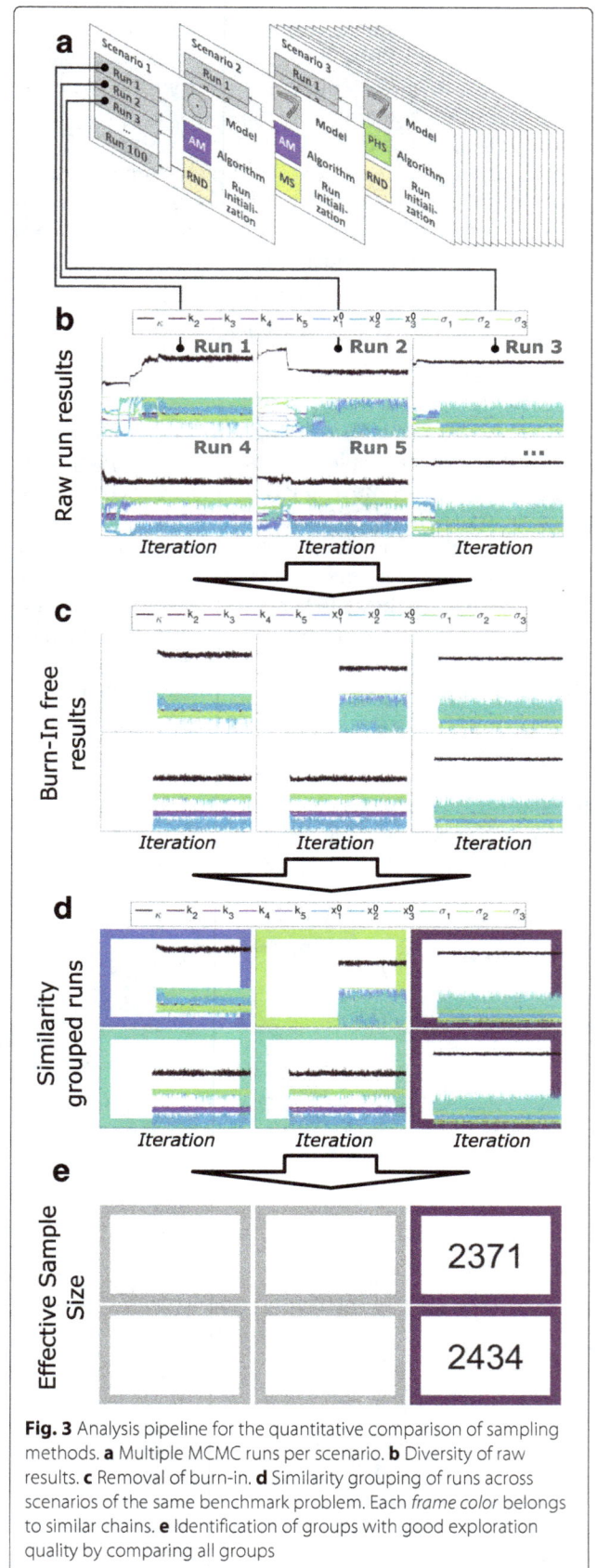

Fig. 3 Analysis pipeline for the quantitative comparison of sampling methods. **a** Multiple MCMC runs per scenario. **b** Diversity of raw results. **c** Removal of burn-in. **d** Similarity grouping of runs across scenarios of the same benchmark problem. Each *frame color* belongs to similar chains. **e** Identification of groups with good exploration quality by comparing all groups

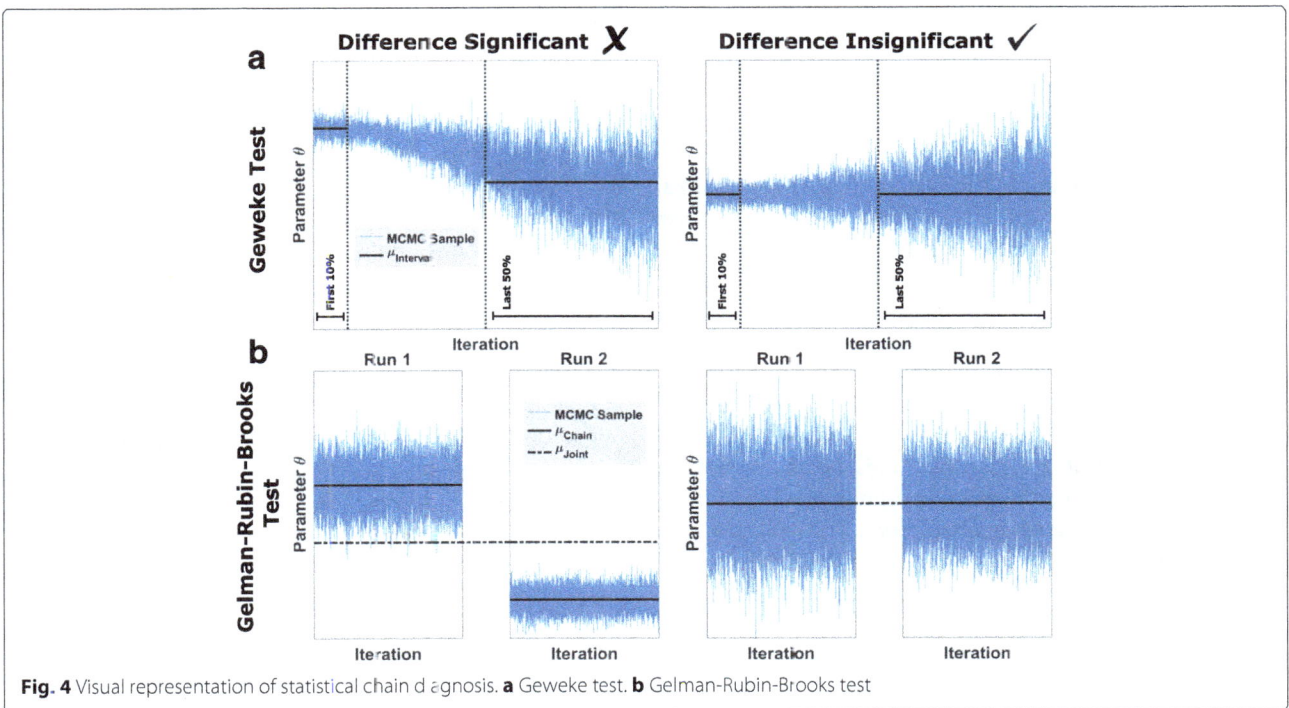

Fig. 4 Visual representation of statistical chain diagnosis. **a** Geweke test. **b** Gelman-Rubin-Brooks test

end of a MCMC chain), the Gelman-Rubin-Brooks diagnostic focuses on within-chain and between-chain variance comparison (see Fig. 4b for a visual representation). The convergence diagnostics consider selected summary statistics, mostly the sample means, and might miss differences which are easy to spot (see, e.g. the accepted cases in Fig. 4 (right panel) and the Additional file 1: Sections 2–3 for further details about the tests). Unfortunately, convergence diagnostics provide only necessary conditions for convergence and do not necessarily reveal problems. In particular for multi-modal posterior distributions, MCMC methods sampling only from one mode pass simple convergence tests [39]. For this reason, the assessment of chain convergence is still an active field of research.

In this manuscript, we determine the EQ by first grouping individual MCMC runs of the same benchmark problem and then identify groups with members which explored the relevant parameter space well. The inspection of groups replaces the inspection of individual chains, resulting in improved efficiency and decrease of subjective judgment regarding chain convergence. The grouping is based on a pairwise distance measure between chains using the afore-described multivariate Gelman-Rubin-Brooks and Geweke diagnostics [45, 47]. If both tests are passed, the corresponding runs are assumed to be similar. Each time two runs are similar they form a group. If one of the members of a group is classified as similar to a run not yet included in the group the latter run is assigned

to the entire group as well. For further details we refer to the Additional file 1: Section 3.

We compare 100 runs per scenario across algorithms (and tunings) thus evaluating 2300 runs per benchmark problem. Groups smaller than 115(5%) runs are neglected from further analysis. For each of the remaining groups we assess whether the posterior is explored by the group members by comparing the groups with each other. Therefore, we evaluate for each group if (i) all regions of high posterior probability and (ii) tails, found in the other groups, have been covered. In this way, we can tell if a group is not covering relevant parameter regimes found by others. This facilitates the selection of the group(s) with the best exploration properties (across algorithms). However, it can still not be ensured that the chains within the best exploring group have indeed explored the entire relevant parameter space properly.

Effective Sample Size (ESS): For the groups with well exploring members we compute the ESS [37, 42, 48]. The ESS accounts for the in-chain autocorrelation and is an important measure for the quality of the posterior approximation for individual chains. As the ESS is overestimated if chains sample only from individual modes of the posterior distribution, we only considered chains assigned to groups which explore the posterior well. For these chains, autocorrelation for individual parameters θ_i is determined using Sokal's adaptive truncated periodogram estimator [28, 49] which is implemented in the DRAM toolbox [28].

As this is a univariate measure, we take the maximum of the autocorrelation across all ϑ_i to determine the ESS and to thin the chain.

Computation Time: The different sampling methods demand different computational cost. MALA requires gradient information while multi-chain methods require multiple evaluations of the (tempered) posterior probability in each iteration. To account for these differences, we evaluate the ESS per central processing unit (CPU) second, which provides a comparable measure for computational efficiency. Furthermore, we consider the efficiency reduction caused by runs which lack proper exploration. Therefore, we multiply the ESS/s value of each run with the *EQ* of the scenario. This normalization is chosen because bad runs are sometimes much faster in execution than well behaving runs, e.g. a run only proposing parameter values outside the parameter bounds is extremely swift since neither cost function nor gradients are calculated.

Benchmark problems

For the evaluation of the sampling algorithms, we established six benchmark problems for ODE constrained parameter estimation. Each benchmark problem is related to a biologically motivated ODE model. The estimation problems considered are low dimensional, yet the ODE models possess properties such as structural non-identifiabilities, bifurcations, limit-cycle oscillations and chaotic behavior. This yields posterior distributions with pronounced tails, multi-modalities and rims which makes them difficult to sample. These are common scenarios for many application problems in systems biology [4, 5, 13–24] which are difficult to identify prior to the parameter estimation. A visual summary of the benchmark problems is depicted within Fig. 5 and described in the following.

(M1) mRNA Transfection: This model describes the transfection of cells with GFP mRNA, its translation and degradation [50]. The observable is the protein concentration. The posterior of the estimation problem is bimodal as the exchange of the degradation rates of mRNA and protein results in the same dynamics. This ODE model is studied for experimental data (M1a) and for artificial data (M1b).

(M2) Bistable Switch: This model describes a bistable switch [51], a frequent motif in gene regulation [52], neuronal science [53] and population dynamics [54]. Depending on the initial condition, for given parameters, the state orbit converges to one of two steady states. This leads to a steep rim in the posterior. In addition, (M2) possesses a saddle-node bifurcation resulting in the absence of the steep rim in certain parameter regimes.

(M3) Saturated Growth: This model describes the growth of a population in an environment with limited resources. It is widely used to model population dynamics, i.e. immigration-death processes [55], and a variety of extensions are available. Already for the simplest model, the parameters are strongly correlated and the posterior distribution possesses 'banana' shaped tails if the measurement is stopped before the steady state is reached [56]. This effect can be enhanced by decreasing the maximum measurement time t_{max} when creating data.

(M4) Biochemical System With Hopf Bifurcation: This model describes a simple biochemical reaction network [57] with a supercritical Hopf bifurcation [58–60] as found in many biological applications [54, 61, 62]. Depending on the parameter values, the orbit of the system approaches a stable limit cycle or a stable fixed point. The posterior distribution for this problem is multi-modal but most of the probability mass is contained in the main mode.

(M5) Driven Van Der Pol Oscillator: This model is an extension of the Van der Pol oscillator by an oscillating input [63–66]. The input causes deterministic chaos by creating a strange attractor. Chaotic behavior can be observed in biological applications e.g. in cardiovascular models with driving pacemaker compartment [61, 67]. The posterior distribution possesses a large number of modes of different sizes and masses. This effect can be increased by creating data with larger t_{max}. For chaotic systems sampling is known to be very challenging [68].

(M6) Lorenz Attractor: The Lorenz attractor provides an idealized description of a hydrodynamic process and can be interpreted as chemical reaction network [69]. Similar to (M5), this system is chaotic and thus possesses a multi-modal posterior distribution. However, its topology strongly differs from the one of (M5) and the chaotic behavior does not arise from a driving term.

Priors & data generation

We consider benchmark settings with measured data (M1a) or simulated data (M1b-M6). The simulated data is obtained by simulating the models for the parameters θ_{true} (Table 1) and adding normally distributed measurement noise. The prior distributions are uniform in the interval $\theta \in [\theta_{min}, \theta_{max}]$ and the data is created using an ODE solution at θ_{true}, absolute, normally distributed noise and equidistantly spaced points in time. Information about observables is provided in Fig. 5.

Implementation

We implemented the sampling algorithms and the benchmark problems in the Parameter EStimation TOolbox

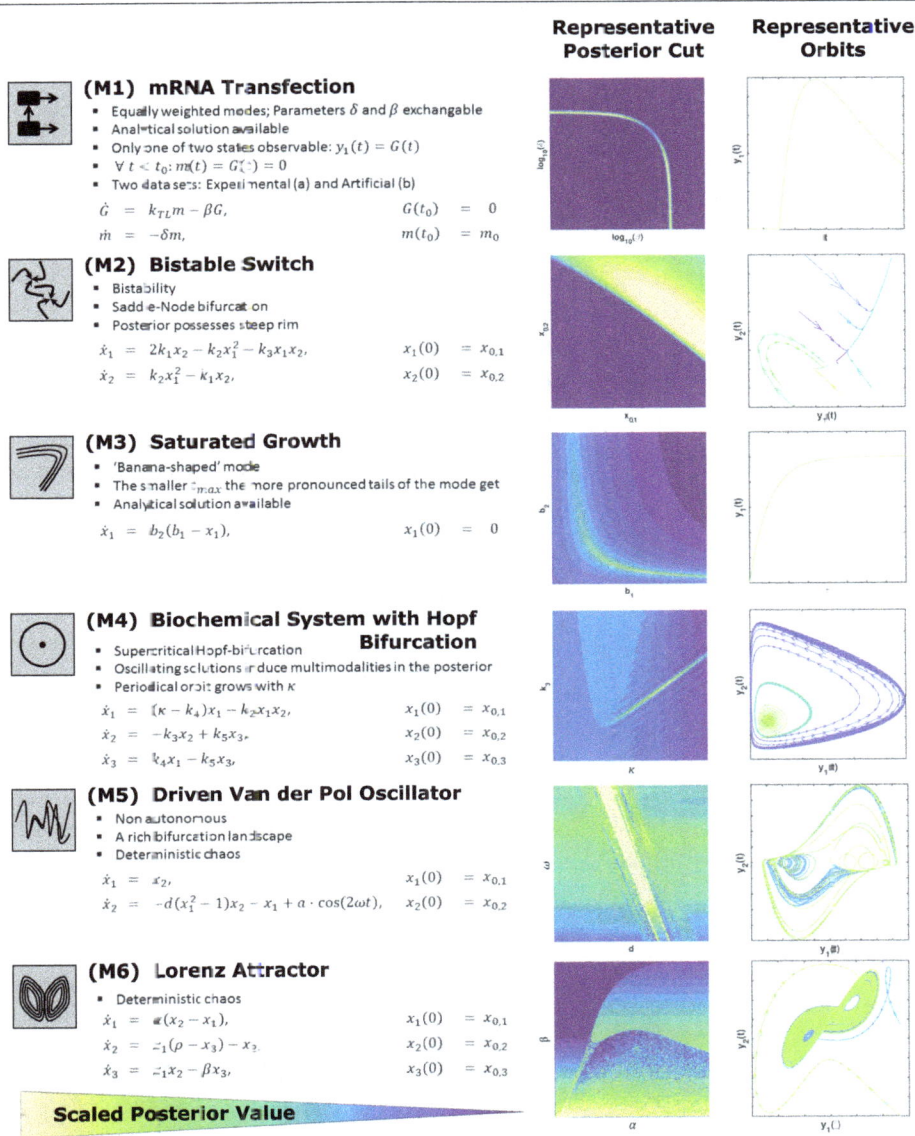

Fig. 5 Visual summary of benchmark problems. *Left* ODE model and its properties, e.g. bifurcations. *Right* Illustration of system dynamics using posterior cuts and orbits

(PESTO)(please refer to the "Availability of data and materials" section for a GitHub reference). This implementation in provided in Additional file 2. PESTO comes with a detailed documentation of all functionalities and the respective methods. For numerical simulation and sensitivity calculation we employed the Advanced MATLAB Interface for CVODES and IDAS (AMICI) [7, 70]. Both toolboxes are developed and available via GitHub and we provide the code used for this study in Additional file 2. The entire code basis could be transfered to other programming languages similar to MATLAB, such as Python, Octave or Julia, without major changes. A re-implementation of the tool in R would also be

conceptually possible and allow for the comparison with other packages, e.g. [71].

Results

In the following, we present the properties and the performance of sampling methods for an application problem as well as for the proposed benchmark problems.

Application of sampling methods to mRNA transcription model

To illustrate the behavior and the properties of the different sampling methods, we consider the process of mRNA transcription ((M1), Fig. 6a). This process has

Table 1 An overview on which priors were used and on how the data was created

	θ	θ_{min}	θ_{max}	θ_{true}
(M1a)	$\log_{10}(t_0)$	-2	1	-
	$\log_{10}(k_{TL}m_0)$	-5	5	-
	$\log_{10}(\beta)$	-5	5	-
$n_t = 150$	$\log_{10}(\delta)$	-5	5	-
$t \in [2, 27]$	$\log_{10}(\sigma)$	-2	2	-
(M1b)	$\log_{10}(t_0)$	-2	1	$\log_{10}(2)$
	$\log_{10}(k_{TL}m_0)$	-5	5	$\log_{10}(5)$
	$\log_{10}(\beta)$	-5	5	$\log_{10}(0.8)$
$n_t = 51$	$\log_{10}(\delta)$	-5	5	$\log_{10}(0.2)$
$t \in [0, 10]$	$\log_{10}(\sigma)$	-2	2	-1
(M2)	k_1	2	20	8
	k_2	0	5	1
	k_3	0	5	1
	k_4	0	5	1
	$x_{0,1}$	-3	3	2
	$x_{0,2}$	-3	3	0.25
$n_t = 101$	σ_1^0	10^{-3}	1	0.3
$t \in [0, 200]$	σ_2^0	10^{-3}	1	0.3
(M3)	b_1	0	5	1
$n_t = 101$	b_2	0	5	0.2
$t \in [0, 2.5]$	σ_1	10^{-3}	10^2	0.03
(M4)	κ	1	5	3.8
	k_2	0.8	1.2	1
	k_3	0.8	1.2	1
	k_4	0.8	1.2	1
	k_5	0.8	1.2	1
	$x_{0,1}$	0	2	1
	$x_{0,2}$	0	2	1
	$x_{0,3}$	0	2	1
	σ_1	10^{-2}	2	0.75
$n_t = 101$	σ_2	10^{-2}	2	0.32
$t \in [0, 200]$	σ_3	10^{-2}	2	0.46
(M5)	a	2	8	5
	d	2	8	5
	ω	2	8	2.464
	$x_{0,1}$	-1	3	0
	$x_{0,2}$	-1	3	0
	$x_{0,3}$	-1	3	1
	σ_1	10^{-2}	2	0.2

Table 1 An overview on which priors were used and on how the data was created (*Continued*)

	θ	θ_{min}	θ_{max}	θ_{true}
$n_t = 101$	σ_2	10^{-2}	2	0.8
$t \in [0, 200]$	σ_3	10^{-2}	2	0.2
(M6)	α	0	20	10
	β	0	10	$\frac{8}{3}$
	ρ	10	30	28
	$x_{0,1}$	0	35	26.61
	$x_{0,2}$	-10	10	-2.74
	$x_{0,3}$	-5	5	0.95
	σ_1	10^{-4}	10^2	1
$n_t = 101$	σ_2	10^{-4}	10^2	1
$t \in [0, 200]$	σ_3	10^{-4}	10^2	1

been modeled and experimentally assessed by Leonhardt et al. [50]. The ODE model possesses two state variables and five parameters. Structural analysis using the MATLAB toolbox GenSSI [15] indicated one structural non-identifiability but did not reveal its nature. Leonhardt et al. [50] derived the analytical solution of the ODE model and showed that the parameters β and δ can be interchanged without altering the output y. This implied that the parameters are locally but not globally structurally identifiable, giving rise to a bimodal posterior distribution (Fig. 6b, c). As the analytical solution is in general not available, we disregard the information about the interchangeability of β and δ for the initial assessment.

We sampled the posterior distribution using several single- and multi-chain methods as well as settings and initialization schemes. The analysis of the sampling results revealed that many methods fail to sample from both modes of the posterior within 10^6 iterations (see Fig. 6d, e). Accordingly, the exploration quality of many methods is low (Fig. 6f). We expected that the single-chain methods, AM, DRAM and MALA, always sample close to the starting point, which was indeed the case. Interestingly, we found that PHS often succeeded in moving its chain between both modes but failed to explore the modes tails properly. Merely PT, either MS or RND initialized, captured both modes in most runs (Fig. 6f). Thus, in (M1a) the conditional ESS – the ESS for the chains sampling both modes and the tails – was the highest for PT.

For most ODE constrained parameter estimation problems, information about the identifiability properties of parameters will not be available prior to the sampling. This is unfortunate as the sampling performance of all methods could be improved by exploiting such additional information. Models with parameter interchangeabilities

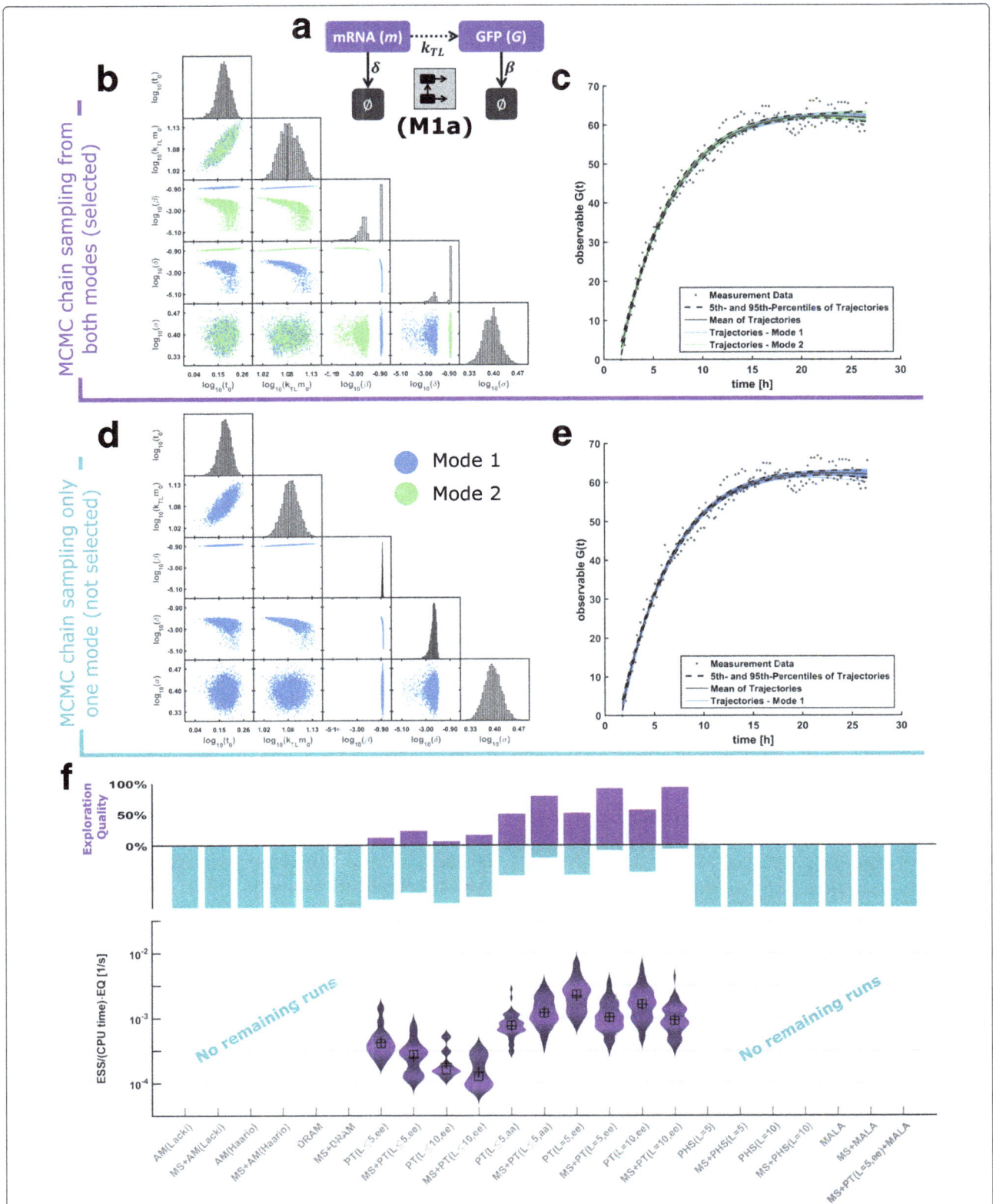

Fig. 6 Results from benchmark problem (M1a). **a** Sketch of the translation process. **b** A bivariate scatter plot of a chain which explored both modes. **c** The corresponding trajectories of the sampled parameter points of both modes. **d** A representative chain which was not able to cover both modes. **e** The corresponding trajectories of the sample of one mode. **f** Effective sample sizes of chains which explored both modes. For several methods, no chain explored both modes, implying an effective sample size of zero

such as (M1) are well studied in the context of mixture models. Tailored methods for such problems include post-processing methods or a random permutation sampler [72, 73]. For this simple ODE model, we evaluated the benefit of applying a post-processing strategy and found that having access to information about number and location of the posterior modes improved the sampling performance significantly for all sampling methods. (see Additional file 1: Section 4).

This application example highlights challenges arising from missing information about parameter identifiability and limitations of available sampling methods. Some of these limitations were not encountered in the manuscripts introducing the methods (e.g. [32] or [36]) as the study focused on different aspects or considered well-suited problems. The analysis of (M1a) demonstrates that even simple linear ODE models can give rise to posterior landscapes that are difficult to sample. This motivates the analysis of other (small-scale) benchmark problems.

Benchmarking of algorithms using simulated data

To facilitate a comprehensive evaluation of sampling methods, we considered the aforementioned benchmark problems (M1-6). These benchmark problems possess a wide range of different properties regarding the underlying dynamical system (e.g. mono- and bistable) as well as the posterior distribution (e.g. unimodal/multi-modal or with/without pronounced tails). This renders the collection presented suitable for the in-depth evaluation and will facilitate the derivation of guidelines for the a priori selection of the appropriate sampling scheme.

We sampled the posterior distributions of all benchmark problems using the algorithms introduced in the "Methods" section. Different tuning parameters and initialization schemes were employed to study their influence on the sampling efficiency. For each benchmark problem we performed 100 independent runs with 10^6 iterations. The large amount of sampling results was analyzed using the analysis pipeline illustrated in Fig. 3. The results for the individual problems (EQ and ESS) and some information about the memory usage of the different algorithms are provided in the Additional file 1: Figures S2, S4–S9.

Influence of posterior properties on sampling performance

Given the sampling results, we asked the question how EQ depends on the benchmark problem and its properties. We found that the size of the groups of runs identified by the analysis pipeline (Fig. 7a) and the EQ (Fig. 7b) varies strongly between the benchmark problems. For problems with uni-modal (M2-3) and weakly multi-modal (M4) posteriors, the average EQ of the sampling methods was higher than 50%. For the problems with bimodal posteriors (M1a,b), 79% of the runs sampled from one of the

modes and failed to explore the posterior, while 21% of the chains sampled from both modes and achieved a good exploration. For posteriors with strong multi-modalities (M5-6), all chains appear to be different and no large groups can be identified (Fig. 4a).

In terms of the dynamical properties of the underlying dynamical system, our results for the benchmark problems indicated that state-of-the-art sampling methods work well with multiple steady states and saddle-node bifurcations, as well as Hopf bifurcations and (limit cycle) oscillations resulting in weak multi-modality of the posterior, oscillating trajectories. However, these methods still fail in case of (aperiodic) oscillations/chaotic behavior and local non-identifiability resulting in strong multi-modality of the posterior.

The analysis on the level of sampling methods revealed that for (M2-4) most algorithms worked appropriately (Fig. 7b) while for (M5-6) all algorithms fail. For (M1), we observed a benefit for using PT and PHS. Since the EQ directly impacts the ESS, these observations hold true for the ESS per CPU second (Fig. 6f and Additional file 1: Figures S2–S8). Indeed, we found a strong correlation of exploration quality and sampling efficiency and identified it as the major limiting performance factor for (M1a,b) and (M5-6).

Comparison of single- and multi-chain methods

Following the analysis of the differences between benchmark problems, we compared single- and multi-chain methods. The average performance characteristics for single- and multi-chain methods were computed by averaging over sampling methods, initialization schemes and tuning parameter choices (Fig. 8). We found that for all considered benchmark problems, multi-chain methods achieved better EQs than single-chain methods (Fig. 8a). Indeed, for several problems, multi-chain methods provided representative samples from the posterior distributions while single-chain methods sampled only individual modes. Interestingly, the improved mixing of multi-chain methods outweighed the higher computational complexity even for benchmark problems with one mode. As a result, multi-chain methods produced higher effective samples sizes and were overall computationally more efficient (Fig. 8b).

Comparison of initialization schemes

In addition to characteristics of methods, we assessed the importance of initialization schemes. Therefore, the average performance characteristics for RND and MS initialization were computed by averaging over sampling methods and tuning parameter choices (Fig. 9). This revealed that multi-start local optimization substantially improved the EQ (Fig. 9a). The difference in the sampling efficiency (conditioned ESS per CPU second) was

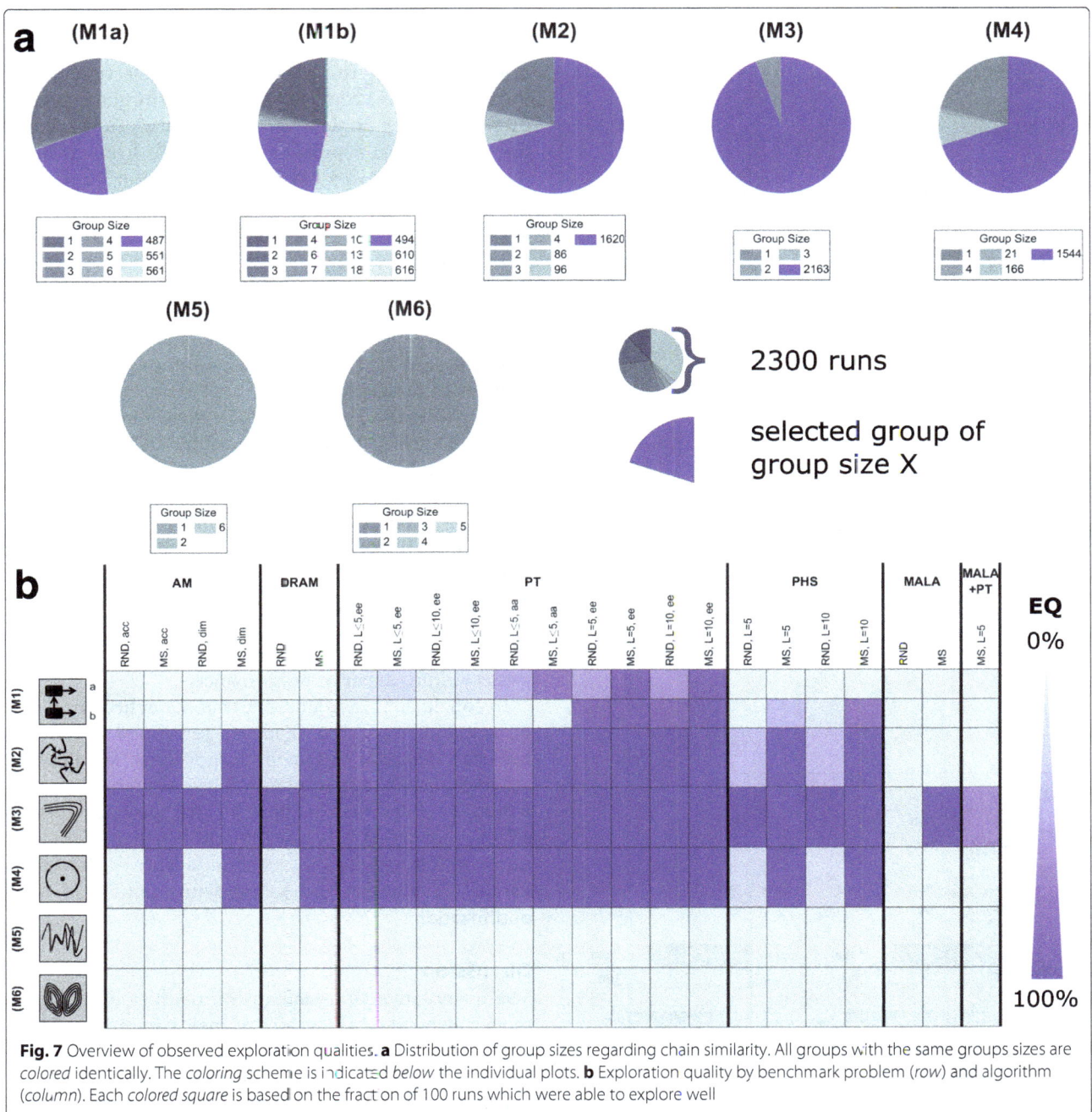

Fig. 7 Overview of observed exploration qualities. **a** Distribution of group sizes regarding chain similarity. All groups with the same groups sizes are *colored* identically. The *coloring* scheme is indicated *below* the individual plots. **b** Exploration quality by benchmark problem (*row*) and algorithm (*column*). Each *colored square* is based on the fraction of 100 runs which were able to explore well

less pronounced than for the EQ as multi-start local optimization required additional computation time (Fig. 9b).

A detailed analysis revealed that some methods were more sensitive to the initialization than others. The performance of PT appeared to be almost independent of the initialization scheme (Fig. 7b), making it a robust choice. PHS required initialization using multi-start optimization results to achieve good EQ (Fig. 7b). Indeed, PHS initialized using samples from the prior performed poorly while PHS initialized using multi-start

optimization outperformed the other methods in some cases.

Selection of tuning parameters and algorithm settings

To provide guidelines regarding tuning parameters and adaptation mechanisms, we carried out a fine-grained analysis of sampling method and subclasses of them. The assessment of single-chain samplers revealed that the adaptive Metropolis methods with acceptance rate dependent proposal scaling (AM(acc)) outperformed methods

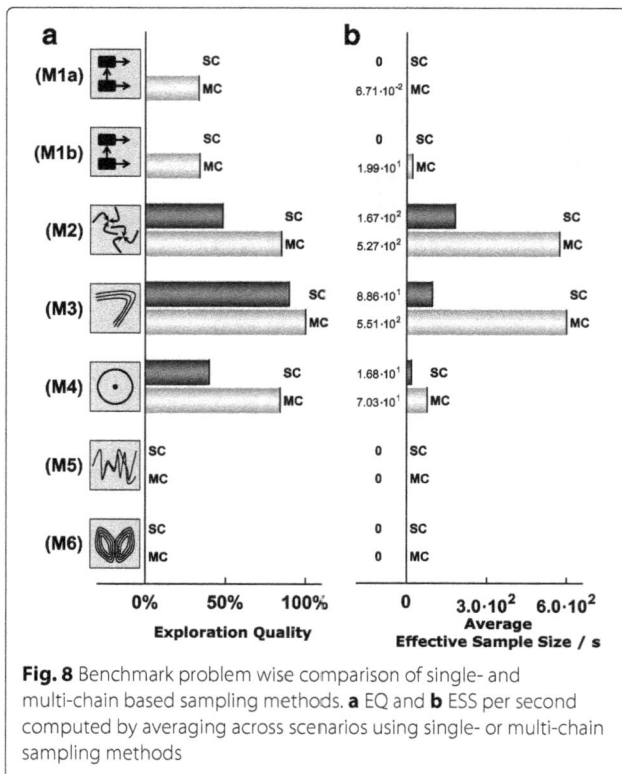

Fig. 8 Benchmark problem wise comparison of single- and multi-chain based sampling methods. **a** EQ and **b** ESS per second computed by averaging across scenarios using single- or multi-chain sampling methods

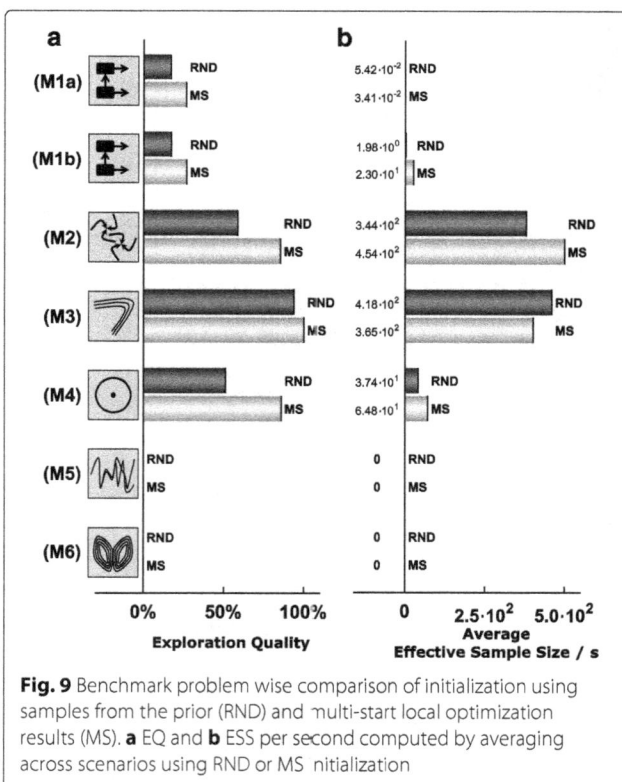

Fig. 9 Benchmark problem wise comparison of initialization using samples from the prior (RND) and multi-start local optimization results (MS). **a** EQ and **b** ESS per second computed by averaging across scenarios using RND or MS initialization

with dimension-dependent proposal scaling (AM(dim) and DRAM(dim)) as shown in Fig. 6f and the Additional file 1: Figures S2–S8. Delayed rejection implemented in DRAM could not compensate for the improved proposal scaling implemented in AM(acc). Furthermore, for the benchmark problems considered here, AM(acc) outperformed MALA. While AM(acc) worked for the benchmark problems with mono-modal posterior distributions, AM(dim), DRAM and MALA mostly failed to explore the posterior distribution (see Figs. 6f, 7b and Additional file1: Figures S2–S8).

The PT algorithms employed in this study used temperature and proposal density adaptation. We evaluated different swapping strategies and strategies to select the number of temperatures. The best performance characteristics were achieved with a large, fixed number of temperatures (see Fig. 6f and Additional file 1: Figures S2–S8). If few temperatures or an adaptive reduction of the number of temperatures are used, the methods are more likely to sample from a single mode. This indicates that the available methods for the reduction of the number of temperatures [32] — which worked for a series of simple examples — is not sufficiently robust. In contrast, the parallel tempering algorithms appeared to be robust with respect to the swapping strategy, with equi-energy (ee) swaps yielding superior performance.

To conclude, this section illustrated practical problems of sampling algorithms and we performed a comprehensive evaluation of sampling algorithms, initialization schemes and tuning parameters. The comprehensive evaluation provided information for the problem-specific selection of sampling strategies and beneficial combinations of settings, e.g. to combine adaptive Metropolis Parallel Hierarchical Sampling with multi-start local optimization.

Discussion

The quantitative and qualitative properties of biological models depend on the values of their parameters. These parameters values are usually inferred using optimization or sampling methods. For optimization schemes comprehensive benchmarking results are available [12, 25, 74, 75]. In this work we complemented these results and benchmarked a selection of sampling methods.

We studied a collection of small-sized benchmark problems for ODE constrained parameter estimation with oscillating, bifurcating and chaotic solutions as well as multi-stable steady states and non-identifiabilities. These model properties lead to pronounced tails, multiple modes and rims in the posterior distributions. Some of these challenges can be addressed by employing additional information about the model and tools like structural identifiability analysis (see "Application of sampling methods to mRNA transcription model" section).

However, in applications, it might not be possible to avoid non-identifiabilities, e.g., if the biological interpretation needs to be conserved or prediction uncertainties need to be quantified. By considering benchmark problems with a diverse set of features, this study provided an unbiased comparison for available sampling methods.

As a by-product of our presented benchmarking study we considered the effect of properties of the ODE model, such as Hopf-bifurcation and multi-stability, onto the performance of sampling algorithms. As most models of biological systems are nonlinear, high-dimensional and possess multiple positive and negative feedback loops [76], a single model can usually exhibit different properties in different parameter regimes. As the biologically relevant regimes in parameter spaces are usually unknown prior to the parameter estimation, knowledge about the dynamic properties cannot be employed and the use of robust sampling methods is beneficial. We previously expected bifurcations to strongly impact the sampling efficiency. This, however, was not the case. Instead, we observed that chaotic regimes have a strong influence on the sampling efficiency and might even render it intractable. This is consistent with previous finding and expected as "chaotic likelihood functions, while ultimately smooth, have such complicated small scale structure" [68].

To derive guidelines for sampling method selection, we assessed a range of single- and multi-chain samplers. This revealed that most state-of-the-art sampling methods require a large number of iterations to provide a representative sample from multi-modal posterior distributions even in low-dimensional parameter spaces. Multi-chain methods clearly outperformed single-chain methods, as reported earlier (see, e.g., [5, 21] and references therein), even for unimodal posterior distributions. The reliability and performance of all sampling methods except PT was substantially improved when initialized using optimization results instead of samples from the prior. Interestingly, for the benchmarks considered in this manuscript, PT performed better without novel adaptation schemes for the number of temperatures [32]. This is in contrast to results for posterior distributions in the original publication [32] – for which we achieved the same results using our implementation –, suggesting that additional research is required. Furthermore, this emphasizes the importance of realistic test problems. The comparison of dimension-dependent proposal scaling [28] and acceptance-rate-dependent proposal scaling [34], which was to the best of our knowledge not published before, revealed the superiority of the latter. From this insight a range of single- and multi-chain methods can benefit. Overall, PHS with optimization-based initialization performed best for uni-modal posterior landscapes while PT performed most robustly regarding all posteriors.

Beyond the evaluation of algorithms, the results demonstrate the importance of performing multiple independent runs of sampling methods starting from different points in parameter space [5]. Most algorithms provide merely a representative sample in a fraction of the runs. In addition to standard sampling diagnostics (e.g. convergence tests like Gelman-Rubin-Brooks [45]), our extended analysis pipeline takes into account the EQ while minimizing the need for subjective visual inspection. Our results confirm the need to evaluate sampling methods by not only taking into account the ESS of the generated runs but the overall EQ as important measure for algorithmic robustness.

The benchmark problems considered in this study are low-dimensional but resemble essential features of parameter estimation problems in systems biology. While the precise quantitative results might depend on the selection of the benchmarks, the qualitative findings should be transferable. To verify this, a range of application problems should be considered. Furthermore, while several classes of sampling methods have been considered, the study of additional methods would be beneficial. In particular the assessment of Hamiltonian Monte Carlo (HMC) based algorithms such as NUTS or Wormhole Monte Carlo [77, 78], region-based methods [79], Metropolis-in-Gibbs methods [80], Transitional MCMC [81], sequential Monte Carlo methods [82] or additional proposal adaptation strategies [71] would be valuable. For ODE models for which the full conditional distribution of the parameters can be derived, also Gibbs samplers might be used [83]. Furthermore, a comparison with non-sampling-based approximation methods, e.g. variational methods [84] or approximation methods [85] could be interesting.

Conclusion

In summary, our comprehensive evaluation revealed that even state-of-the-art MCMC algorithms have problems to sample efficiently from many posterior distributions arising in systems biology. Problems arose in particular in the presence of non-identifiabilities and chaotic regimes. The examples provided in manuscripts presenting new algorithms are often not representative and a more thorough assessment on benchmark collections should be required (as is common practice in other fields). The presented study provides a basis for future developments of such benchmark collections allowing for a rigorous assessment of novel sampling algorithms. In this study, we already used six benchmark problems with common challenges to provide practical guidelines for the selection of sampling algorithms, adaptation and initialization schemes. Furthermore, the presented results highlight the need to address chain exploration quality by taking into account multiple MCMC runs which can be compared with each other before calculating effective sample sizes. The availability of the code will simplify the extension of the methods and the extension of the benchmark collection.

Abbreviations

AM: Adaptive metropolis; aa: All adjacent; acc: Acceptance based adaption; BI: Burn-In; CPU: Central processing unit; dim: Dimension based adaption; DRAM: Delayed rejection adaptive metropolis; ee: Equal energy; ESS: Effective sample size; EQ: Exploration quality; MALA: Metropolis-adjusted Langevin algorithm; MCMC: Markov chain Monte Carlo; MH: Metropolis-hastings; MS: Multi-start local optimization; ODE: Ordinary differential equation; PT: Parallel tempering; PHS: Parallel hierarchical sampling; RND: Sampling from prior distribution

Acknowledgments

The authors acknowledge the technical support by Dennis Rickert.

Funding

The authors acknowledge financial support from the German Federal Ministry of Education and Research (BMBF) within the SYS-Stomach project (Grant No. 01ZX1310B) and the Postdoctoral Fellowship Program (PFP) of the Helmholtz Zentrum München.

Authors' contributions

BB and SH conceived the study. All authors contributed substantially to conception and design of the study. BB, SH, JH and FT coordinated the study. BB implemented the benchmark problems. BB, SH and JH implemented the methods. BB carried out the computations and analyzed the results. BB, SH and JH drafted the manuscript. All authors proofread and approved the final manuscript.

Competing interests

The authors declare that they have no competing interests.

Author details

[1]Helmholtz Zentrum München - German Research Center for Environmental Health, Institute of Computational Biology, Ingolstädter Landstraße 1, 85764 Neuherberg, Germany. [2]Technische Universität München, Center for Mathematics, Chair of Mathematical Modeling of Biological Systems, Boltzmannstraße 15, 85748 Garching, Germany. [3]Bayer AG, Engineering & Technologies, Applied Mathematics, Kaiser-Wilhelm-Allee, 51368 Leverkusen, Germany.

References

1. Gábor A, Banga JR. Robust and efficient parameter estimation in dynamic models of biological systems. BMC Syst Biol. 2015;9(1):74.
2. Klipp E, Nordlander B, Krüger R, Gennemark P, Hohmann S. Integrative model of the response of yeast to osmotic shock. Nat Biotechnol. 2005;23(8):975–82.
3. Kitano H. Computational systems biology. Nature. 2002;420(6912):206–10.
4. Raue A, Kreutz C, Theis FJ, Timmer J. Joining forces of Bayesian and Frequentist methodology: a study for inference in the presence of non-identifiability. Phil Trans R Soc A Math Phys Eng Sci. 2013;371(1984): 20110544.
5. Hug S, Raue A, Hasenauer J, Bachmann J, Klingmüller U, Timmer J, Theis FJ. High-dimensional bayesian parameter estimation: case study for a model of JAK2/STAT5 signaling. Math Biosci. 2013;246(2):293–304.
6. Joshi M, Seidel-Morgenstern A, Kremling A. Exploiting the bootstrap method for quantifying parameter confidence intervals in dynamical systems. Metab Engeneering. 2006;8:447–55.
7. Fröhlich F, Theis FJ, Hasenauer J. Uncertainty analysis for non-identifiable dynamical systems: Profile likelihoods, bootstrapping and more. In: Mendes P, Dada JO, Smallbore KO, editors. Proceedings of the 12th International Conference on Computational Methods in Systems Biology (CMSB 2014), Lecture Notes in Bioinformatics. Manchester: Springer; 2014. p. 61–72.
8. Raue A, Kreutz C, Maiwald T, Bachmann J, Schilling M, Klingmüller U, Timmer J. Structural and practical identifiability analysis of partially observed dynamical models by exploiting the profile likelihood. Bioinformatics. 2009;25(15):1923–9.
9. Wilkinson DJ. Bayesian methods in bioinformatics and computational systems biology. Brief Bioinform. 2007;8(2):109–16.
10. Xu TR, Vyshemirsky V, Gormand A, et al. Inferring signaling pathway topologies from multiple perturbation measurements of specific biochemical species. Sci Signal. 2010;3(113):20.
11. Krauss M, Burghaus R, Lippert J, Niemi M, Neuvonen P, Schuppert A, Willmann S, Kuepfer L, Görlitz L. Using Bayesian-PBPK modeling for assessment of inter-individual variability and subgroup stratification. In Silico Pharmacol. 2013;1(6):1–11.
12. Raue A, Schilling M, Bachmann J, Matteson A, Schelker M, Schelke M, Kaschek D, Hug S, Kreutz C, Harms BD, Theis FJ, Klingmüller U, Timmer J. Lessons learned from quantitative dynamical modeling in systems biology. PloS ONE. 2013;8(9):74335.
13. Raue A, Kreutz C, Maiwald T, Bachmann J, Schilling M, Klingmüller U, Timmer J. Structural and practical identifiability analysis of partially observed dynamical models by exploiting the profile likelihood. Bioinf. 2009;25(25):1923–9.
14. Balsa-Canto E, Alonso AA, Banga JR. An iterative identification procedure for dynamic modeling of biochemical networks. BMC Syst Biol. 2010;4:11. http://bmcsystbiol.biomedcentral.com/articles/10.1186/1752-0509-4-11.
15. Chiş O, Banga JR, Balsa-Canto E. GenSSI: a software toolbox for structural identifiability analysis of biological models. Bioinformatics. 2011;27(18): 2610–11.
16. Weber P, Hasenauer J, Allgöwer F, Radde N. Parameter estimation and identifiability of biological networks using relative data. In: Bittanti S, Cenedese A, Zampieri S, editors. Proc. of the 18th IFAC World Congress, vol. 18. Milano: Elsevier; 2011. p. 11648–53.
17. Gardner T, Cantor C, Collins J. Construction of a genetic toggle switch in escherichia coli. Nature. 2000;403(6767):242–339.
18. Ozbudak EM, Thattai M, Lim HN, Shraiman BI, van Oudenaarden A. Multistability in the lactose utilization network of Escherichia coli. Nature. 2004;427(6976):737–40.
19. Tyson JJ. Modeling the cell division cycle: cdc2 and cyclin interactions. Proc Nati Acad Sci USA. 1991;88:7328–32.
20. Kholodenko BN. Negative feedback and ultrasensitivity can bring about oscillations in the mitogen-activated protein kinase cascades. Eur J Biochem. 2000;267(6):1583–8.
21. Calderhead B. A study of population MCMC for estimating Bayes factors over nonlinear ODE models. Master thesis, University of Glasgow. 2007.
22. Kosuta S, Hazledine S, Sun J, Miwa H, Morris RJ, Downie JA, Oldroyd GE. Differential and chaotic calcium signatures in the symbiosis signaling pathway of legumes. Proc Natl Acad Sci. 2008;105(28):9823–28.
23. Ngonghala CN, Teboh-Ewungkem MI, Ngwa GA. Observance of period-doubling bifurcation and chaos in an autonomous ODE model for malaria with vector demography. Theor Ecol. 2016;9(3):337–51.
24. Braxenthaler M, Unger R, Auerbach D, Given JA, Moult J. Chaos in protein dynamics. Protein Struct Funct Genet. 1997;29(4):417–25.
25. Villaverde AF, Henriques D, Smallbone K, Bongard S, Schmid J, Cicin-Sain D, Crombach A, Saez-Rodriguez J, Mauch K, Balsa-Canto E, et al. BioPreDyn-bench: a suite of benchmark problems for dynamic modelling in systems biology. BMC Syst Biol. 2015;9:8.
26. Kronfeld M, Planatscher H, Zell A. The EvA2 Optimization Framework. Berlin: Springer; 2010.
27. Egea JA, Henriques D, Cokelaer T, Villaverde AF, MacNamara A, Danciu DP, Banga JR, Saez-Rodriguez J. MEIGO: an open-source software suite based on metaheuristics for global optimization in systems biology and bioinformatics. BMC Bioinforma. 2014;15:136.
28. Haario H, Laine M, Mira A, Saksman E. DRAM: efficient adaptive MCMC. Statistics and Computing. 2006;16(4):339–54.
29. Haario H, Saksman E, Tamminen J. An adaptive Metropolis algorithm. Bernoulli. 2001;7(2):223–42.
30. Roberts GO, Rosenthal JS. Examples of adaptive MCMC. J Comput Graph Stat. 2009;18(2):349–67.
31. Andrieu C, Thoms J. A tutorial on adaptive MCMC. Stat Comput. 2008;18(4):343–73.
32. Lacki MK, Miasojedow B. State-dependent swap strategies and automatic reduction of number of temperatures in adaptive parallel tempering algorithm. Stat Comput. 2015;26:1–14.

33. Sambridge M. A parallel tempering algorithm for probabilistic sampling and multimodal optimization. Geophys J Int. 2013;196:342.

34. Miasojedow B, Moulines E, Vihola M. An adaptive parallel tempering algorithm. J Comput Graph Stat. 2013;22(3):649–64.

35. Vousden W, Farr WM, Mandel I. Dynamic temperature selection for parallel tempering in Markov chain Monte Carlo simulations. Mon Not R Astron Soc. 2016;455(2):1919–37.

36. Rigat F, Mira A. Parallel hierarchical sampling: A general-purpose interacting Markov chains Monte Carlo algorithm. Comput Stat Data Anal. 2012;56(6):1450–67.

37. Girolami M, Calderhead B. Riemann manifold Langevin and Hamiltonian Monte Carlo methods. J R Stat Soc Ser B (Stat Methodol). 2011;73(2):123–214.

38. Klipp E, Herwig R, Kowald A, Wierling C, Lehrach H. Systems Biology in Practice. Weinheim: Wiley-VCH; 2005.

39. Andrieu C, De Freitas N, Doucet A, Jordan MI. An introduction to MCMC for machine learning. Mach Learn. 2003;50(1-2):5–43.

40. Metropolis N, Rosenbluth AW, Rosenbluth MN, Teller AH, Teller E. Equation of state calculations by fast computing machines. J Chem Phys. 1953;21(6):1087–92.

41. Hastings WK. Monte Carlo sampling methods using Markov chains and their applications. Biometrika. 1970;57(1):97–109

42. Calderhead B. Differential geometric MCMC methods and applications. PhD thesis, University of Glasgow. 2011.

43. Faue A, Karlsson J, Saccomani MP, Lirstrand M, Timmer J. Comparison of approaches for parameter identifiability analysis of biological systems. Bioinformatics. 2014;30(10):1440–48.

44. Brooks SP, Roberts GO. Assessing convergence of Markov chain Monte Carlo algorithms. Stat Comput. 1998;8(4):319–35.

45. Geweke J. Evaluating the accuracy of sampling-based approaches to the calculation of posterior moments. Bayesian Stat. 1992;4:169–88.

46. Holm S. A simple sequentially rejective multiple test procedure. Scand J Stat. 1979;6(2):65–70.

47. Brooks SP, Gelman A. General methods for monitoring convergence of iterative simulations. J Comput Graph Stat. 1998;7(4):434–55.

48. Schmidl D, Czado C, Hug S, Theis F, et al. A vine-copula based adaptive MCMC sampler for efficient inference of dynamical systems. Bayesian Anal. 2013;8(1):1–22.

49. Sacchi MD, Ulrych TJ, Walker CJ. Interpolation and extrapolation using a high-resolution discrete Fourier transform. IEEE Trans Signal Process. 1998;46(1):31–8.

50. Leonhardt C, Schwake G, Stögbauer TR, Rappl S, Kuhr JT, Ligon TS, Rädler JO. Single-cell mRNA transfection studies: delivery, kinetics and statistics by numbers. Nanomedicine Nanotechnol Biol Med. 2014;10(4):679–88.

51. Wilhelm T. The smallest chemical reaction system with bistability. BMC Syst Biol. 2009;3(1):90.

52. Chaves M, Eissing T, Allgöwer F. Bistable biological systems: A characterization through local compact input-to-state stability. IEEE Trans Autom Control. 2008;53:87–100.

53. Guevara MR. Bifurcations Involving Fixed Points and Limit Cycles in Biological Systems. New York: Springer; 2003.

54. Sgro AE, Schwab DJ, Noorbakhsh J, Mestler T, Mehta P, Gregor T. From intracellular signaling to population oscillations: bridging size- and time-scales in collective behavior. Mol Syst Biol. 2015;11(1):779.

55. Zimmer C, Sahle S, Pahle J. Exploiting intrinsic fluctuations to identify model parameters. IET Syst Biol. 2015;9(2):64–73.

56. Solonen A, Ollinaho P, Laine M, Haario H, Tamminen J, Järvinen H, et al. Efficient MCMC for climate model parameter estimation: Parallel adaptive chains and early rejection. Bayesian Anal. 2012;7(3):715–36.

57. Kirk PD, Toni T, Stumpf MP. Parameter inference for biochemical systems that undergo a Hopf bifurcation. Biophys J. 2008;95(2):540–9.

58. Crawford JD. Introduction to bifurcation theory. Rev Mod Phys. 1991;63(4):991.

59. Kuznetsov YA. Elements of Applied Bifurcation Theory. New York: Springer; 2013.

60. Dercole F, Rinaldi S. Dynamical systems and their bifurcations. Hoboken: Wiley; 2011. pp. 291–325. doi:10.1002/9781118007747.ch12. http://dx.doi.org/10.1002/9781118007747.ch12.

61. Heldt T, Shim EB, Kamm RD, Mark RG. Computational modeling of cardiovascular response to orthostatic stress. J Appl Physiol. 2002;92(3):1239–54.

62. Feinberg M, Horn FJ. Chemical mechanism structure and the coincidence of the stoichiometric and kinetic subspaces. Arch Ration Mech Anal 1977;66(1):83–97.

63. Tsatsos M. Theoretical and Numerical study of the Van der Pol equation. PhD thesis, Aristotle University of Thessaloniki. 2006.

64. Mettin R, Farlitz U, Lauterborn W. Bifurcation structure of the driven Van der Pol oscillator. Int J Bifurcation Chaos. 1993;3(06):1529–55.

65. Parlitz U, Lauterborn W. Period-doubling cascades and devil's staircases of the driven van der Pol oscillator. Phys Rev A. 1987;36(3):1428.

66. Leonov G, Kuznetsov N, Vagaitsev V. Localization of hidden Chua's attractors. Phys Lett A. 2011;375(23):2230–3.

67. Glass L, Guevara MR, Shrier A, Perez R. Bifurcation and chaos in a periodically stimulated cardiac oscillator. Phys D Nonlinear Phenom. 1983;7(1):89–101.

68. Du H, Smith LA. Rising Above Chaotic Likelihoods. SIAM/ASA J Uncertain Quantif. 2017;5(1):246–58.

69. Poland D. Cooperative catalysis and chemical chaos: a chemical model for the Lorenz equations. Phys D Nonlinear Phenom. 1993;65(1):86–99.

70. Fröhlich F, Kaltenbacher B, Theis FJ, Hasenauer J. Scalable parameter estimation for genome-scale biochemical reaction networks. PLoS Comput Biol. 2017;13(1):1–18.

71. Vihola M. Robust adaptive metropolis algorithm with coerced acceptance rate. Stat Comput. 2012;22(5):997–1008.

72. Jasra A, Holmes CC, Stephens DA. Markov chain Monte Carlo methods and the label switching problem in Bayesian mixture modeling. Stat Sci. 2005;20:50–67.

73. Papastamoulis P, Iliopoulos G. On the convergence rate of random permutation sampler and ECR algorithm in missing data models. Methodol Comput Appl Probab. 2013;15(2):293–304.

74. Moles CG, Mendes P, Banga JR. Parameter estimation in biochemical pathways: A comparison of global optimization methods. Genome Res. 2003;13:2467–74.

75. Hross S, Hasenauer J. Analysis of CFSE time-series data using division-, age- and label-structured population models. Bioinformatics. 2016;32(15):2321–29.

76. Alon U. An Introduction to Systems Biology: Design Principles of Biological Circuits. Boca Raton: CRC press; 2006.

77. Hoffman MD, Gelman A. The No-U-turn sampler: adaptively setting path lengths in Hamiltonian Monte Carlo. J Mach Learn Res. 2014;15(1):1593–623.

78. Lan S, Streets J, Shahbaba B. Wormhole Hamiltonian Monte Carlo. In: Proceedings of the AAAI Conference on Artificial Intelligence, vol. 2014. Rockville Pike: National Center for Biotechnology Information (NCBI); 2014. p. 1953.

79. Bai Y, Craiu RV, Di Narzo AF. Divide and conquer: a mixture-based approach to regional adaptation for MCMC. J Comput Graph Stat. 2011;20(1):63–79.

80. Bédard M. Hierarchical models: Local proposal variances for RWM-within-Gibbs and MALA-within-Gibbs. Comput Stat Data Anal. 2017;109:231–46.

81. Betz W, Papaioannou I, Straub D. Transitional Markov Chain Monte Carlo: Observations and Improvements. J Eng Mech. 2016;142(5):04016016.

82. Yanagita T, Iba Y. Exploration of order in chaos using the replica exchange Monte Carlo method. J Stat Mech Theory Exp. 2009;2009(02):02043.

83. Casella G, George EI. Explaining the gibbs sampler. Am Stat. 1992;46(3):167–74.

84. MacKay DJC. Information Theory, Inference, and Learning Algorithms, 7.2 ed. Cambridge: Cambridge University Press; 2005.

85. Fröhlich F, Hross S, Theis FJ, Hasenauer J. In: Mendes P, Dada JO, Smallbone KO, editors. Proceedings of the 12th International Conference on Computational Methods in Systems Biology (CMSB 2014). Manchester: Springer; 2014. pp. 73–85.

14

Parameter identifiability analysis and visualization in large-scale kinetic models of biosystems

Attila Gábor[1,2†], Alejandro F. Villaverde[1†] and Julio R. Banga[1*]

Abstract

Background: Kinetic models of biochemical systems usually consist of ordinary differential equations that have many unknown parameters. Some of these parameters are often practically unidentifiable, that is, their values cannot be uniquely determined from the available data. Possible causes are lack of influence on the measured outputs, interdependence among parameters, and poor data quality. Uncorrelated parameters can be seen as the key tuning knobs of a predictive model. Therefore, before attempting to perform parameter estimation (model calibration) it is important to characterize the subset(s) of identifiable parameters and their interplay. Once this is achieved, it is still necessary to perform parameter estimation, which poses additional challenges.

Methods: We present a methodology that (i) detects high-order relationships among parameters, and (ii) visualizes the results to facilitate further analysis. We use a collinearity index to quantify the correlation between parameters in a group in a computationally efficient way. Then we apply integer optimization to find the largest groups of uncorrelated parameters. We also use the collinearity index to identify small groups of highly correlated parameters. The results files can be visualized using Cytoscape, showing the identifiable and non-identifiable groups of parameters together with the model structure in the same graph.

Results: Our contributions alleviate the difficulties that appear at different stages of the identifiability analysis and parameter estimation process. We show how to combine global optimization and regularization techniques for calibrating medium and large scale biological models with moderate computation times. Then we evaluate the practical identifiability of the estimated parameters using the proposed methodology. The identifiability analysis techniques are implemented as a MATLAB toolbox called VisId, which is freely available as open source from GitHub (https://github.com/gabora/visid).

Conclusions: Our approach is geared towards scalability. It enables the practical identifiability analysis of dynamic models of large size, and accelerates their calibration. The visualization tool allows modellers to detect parts that are problematic and need refinement or reformulation, and provides experimentalists with information that can be helpful in the design of new experiments.

Keywords: Parameter estimation, Dynamic models, Identifiability, Global optimization, Regularization, Overfitting

*Correspondence: julio@iim.csic.es
†Equal contributors
[1]BioProcess Engineering Group, IIM-CSIC, Eduardo Cabello 6, 36208 Vigo, Spain
Full list of author information is available at the end of the article

Background

The development of mechanistic (kinetic) models in order to quantitatively describe the dynamics of biological phenomena is one of the core research themes in systems biology. During the last decade, fostered by the greater availability of the necessary experimental data, the development of large (up to genome-scale) kinetic models has become one of the main objectives in the field, as well as in related areas such as synthetic biology, metabolic engineering, or industrial biotechnology [1–10]. More recently, the first steps towards comprehensive whole-cell models have been taken [11], which has great potential for applications e.g. in personalized medicine [12]. However, the development of these large-scale integrated dynamic models poses severe challenges [13, 14]. Those associated with model building are common to the more general problem of reverse engineering of biological systems [15]. In this context, parameter estimation (i.e. model calibration) is arguably one of the most studied [16–19], yet more challenging step in model building.

Parameter estimation in nonlinear dynamic models can be an extremely hard problem mostly due to the following issues [15]: lack of identifiability, ill-conditioning, multi-modality and over-fitting. The latter three can be handled via global optimization and regularization methods, as reviewed and illustrated recently [20]. The present paper begins by continuing the line of work in [20], addressing these three issues. To this end we introduce a combination of a global optimization metaheuristic, eSS [21], and an efficient local search method, the adaptive algorithm NL2SOL [22]. By using this optimization technique jointly with regularization it is possible to *reduce the calibration times* of large dynamic models and simultaneously avoid over-fitting. We show this for models from the recently presented BioPreDyn benchmark collection [23]. Then we focus on the remaining issue, that is, *identifiability analysis* of large dynamic models. Our aim is to develop a methodology which (i) is able to characterize high-order relationships among parameters, and (ii) scales up well with model size. Thus, our objective goes beyond finding the subset of identifiable parameters: we also aim to systematically characterize the space of non-identifiable parameters, and to facilitate the advanced analysis of the results with scalable visualization tools.

Identifiability analysis aims at establishing whether it is possible to determine the values of the unknown model parameters [24]. It is common to distinguish between structural and practical identifiability. *Structural* or a priori identifiability analysis decides whether the model parameters are uniquely determinable based on the model formulation, which includes the dynamic equations, observation functions and stimuli [25]. A parameter θ of the model is structurally identifiable if $y(\theta) = y(\theta') \Leftrightarrow \theta = \theta'$, where y denotes the model predictions, which are observable in the experiments. A parameter θ is structurally *locally* identifiable if for almost any value θ^* there is a neighbourhood $V(\theta^*)$ in which the above relationship holds. It is *globally* identifiable if the relationship holds in all the range of values of the parameter. If there is some region with non-zero measure where the relationship does not hold, θ is structurally unidentifiable. Structural identifiability analyses usually involve a high computational burden, which makes them difficult to apply to large models [26–28]. Furthermore, structural identifiability is only a necessary but not sufficient condition for identifiability. Very often a structurally identifiable parameter is practically unidentifiable, that is, its value cannot be determined with precision due to limitations in the available data. This can be quantified using *practical* or *a posteriori* identifiability analysis, which provides confidence intervals of the parameter values. The two main sources of practical non-identifiability are (1) lack of influence of a parameter on the observables, and (2) interdependence among the parameters. Obviously, if a parameter does not influence the observables (case 1) it is not possible to determine its value. The second situation, in which the effect on the observables of a change in one parameter can be compensated by changes in other parameters, can also prevent parameter identification. Both problems are related to the sensitivities of the observables to changes in model parameters. While (1) is related to the average sensitivity of the model outputs to a specific parameter, (2) can be investigated based on the collinearity of the parametric sensitivities [29].

In this paper we combine global optimization and regularization techniques to calibrate medium and large scale biological models (in this context, we will use the term "medium-scale" for models with 10 to 50 parameters and "large-scale" for models with more than 50 parameters). Then we evaluate the practical identifiability of estimated model parameters using sensitivity analysis and collinearity measures. We determine the largest identifiable subsets of parameters, characterize the interplay among non-identifiable groups of parameters, and visualize the results using Cytoscape. The visualization tool shows the identifiable and non-identifiable groups of parameters together with the model structure in the same graph. In this way, modellers can detect parts that are problematic and need refinement or reformulation, and experimentalists obtain information that can be helpful in the design of new experiments. The methods for identifiability analysis and visualization presented here have been implemented as a MATLAB toolbox called VisId, which is available from GitHub (https://github.com/gabora/visid) and as Additional file 1.

Methods

Parameter estimation with regularization and global optimization

Mathematical model

We consider deterministic models of biological systems that can be described by nonlinear ordinary differential equations (ODEs) in the following form:

$$\frac{dx(t,\theta)}{dt} = f(x(t,\theta), u(t), \theta), \tag{1}$$

$$y(x,\theta) = g(x(t,\theta), \theta), \tag{2}$$

$$x(t_0) = x_0(\theta), \quad t \in [t_0, t_f]. \tag{3}$$

Here $x \in \mathbb{R}^{N_x}$ denotes the state vector (often concentrations), f describes the interactions among the state variables (often constructed from the reaction rate functions), and $u(t)$ denotes the input variables (stimuli). The parameter vector $\theta \in \mathbb{R}^{N_\theta}$ contains the (positive) parameters, e.g. reaction rate coefficients or Hill exponents. Their values are often unknown and must be estimated from data.

The model variables x are mapped to the measurable output variables $y \in \mathbb{R}^{N_y}$, also known as observables or model predictions, by the observation function g. These y signals are the quantities that can be experimentally measured. We will denote by y_{ijk} the model prediction for the j-th observed quantity in the k-th experiment at time $t_i \in [t_0, t_f]$. The corresponding measured data is denoted by $\tilde{y}_{ijk}$.

Parameter estimation

The goal of parameter estimation is to determine the values of the unknown parameter vector θ. This is usually done by minimizing a distance between model prediction y_{ijk} and measured data $\tilde{y}_{ijk}$. One of the simplest, but yet general, choices of this distance is the weighted sum-of-squares

$$Q_{\text{LS}}(\theta) = \sum_{k=1}^{N_e} \sum_{j=1}^{N_{y,k}} \sum_{i=1}^{N_{t,k,j}} w_{ijk} \left(y_{ijk} \left(x(t_i, \theta), \theta \right) - \tilde{y}_{ijk} \right)^2, \tag{4}$$

where N_e is the number of experiments, $N_{y,k}$ is the number of observed compounds in the k-th experiment, and $N_{t,k,j}$ is the number of measurement time points of the j-th observed quantity in the k-th experiment, and the weights are denoted by w_{ijk}. The total number of data in all experiments is denoted by $N_D = \sum_{k=1}^{N_e} \sum_{j=1}^{N_{y,k}} \sum_{i=1}^{N_{t,k,j}} 1$. In order to simplify the index triplet, from now on we will use only one index, i.e. the weights and observables are denoted by w_i and y_i for $i = 1, 2 \ldots N_D$.

Then the parameter estimation problem is formulated as an optimization problem in the following form:

$$\underset{\theta}{\text{minimize}} \; Q_{\text{LS}}(\theta) + \alpha \Gamma(\theta) \tag{5}$$

$$\text{subject to } \theta_{\min} \le \theta \le \theta_{\max}, \tag{6}$$

$$\frac{dx(t,\theta)}{dt} = f(u(t), x(t,\theta), \theta), \tag{7}$$

$$y(x,\theta) = g(x(t,\theta), \theta), \tag{8}$$

$$x(t_0) = x_0(\theta), \quad t \in [t_0, t_f]. \tag{9}$$

Here $\Gamma(\theta)$ is a a regularization term, which is described in the following subsection, and $\theta_{\min}$ and $\theta_{\max}$ are lower and upper bounds of the parameter values. The parameter vector $\hat{\theta}$ that solves this minimization problem is called the *optimal parameter vector* or the parameter estimates.

Regularization

Large scale dynamic models are often over-parametrized, turning the estimation of their parameters into an ill-posed problem [30]. This means that the minimum of the least-squares cost function (4) is non-unique, or that even a very small perturbation of the data results in very different estimated parameters. Furthermore, due to the large number of degrees of freedom, these models tend to capture the artificial dynamics of measurement noise. This is known as overfitting [31, 32] and it usually results in poor predictive capability of the calibrated model.

Regularization techniques incorporate a priori knowledge about the parameter values to make the problem well-posed. The regularization parameter α in (5) balances the strength of this knowledge; its value can be found by regularization tuning methods [33]. Here we followed the guidelines presented in [20] and chose a small regularization parameter ($\alpha = 0.1$), since we assume that we do not have good a priori estimates of the parameters.

Regarding the regularization function, $\Gamma(\theta)$, we chose the Tikhonov regularization framework to match the form of the penalty to the least squares formalism of the objective function. In this case the penalty is a quadratic penalty function,

$$\Gamma(\theta) = \left(\theta - \theta^{\text{ref}} \right)^T W^T W \left(\theta - \theta^{\text{ref}} \right), \tag{10}$$

where $W \in \mathbb{R}^{N_\theta \times N_\theta}$ is a diagonal scaling matrix and $\theta^{\text{ref}} \in \mathbb{R}^{N_\theta}$ is a reference parameter vector, which is problem dependent and determined by the available information about the model parameters.

Global optimization

We solve the minimization problem defined by (5)–(10) using optimization. Since the cost function (5) is usually multi-modal (i.e. it usually has several local minima) [34–37], it is necessary to use an efficient global optimization method. Deterministic global optimization methods [38–42] can guarantee global optimality of the solution. However, their computational cost increases exponentially with the number of parameters, which makes them unsatisfactory for large scale models. Stochastic and metaheuristic methods [17, 18, 35, 36, 43, 44], on the other hand, do not provide such guarantees, but are often capable of finding adequate solutions in reasonable computation times.

For this reason we use a method called enhanced scatter search (eSS) [21], which is an advanced implementation of a population-based algorithm called scatter search. The scatter search metaheuristic works by evolving a number of solutions (population members), which constitute the reference set (RefSet). Members of this set are selected due to their quality and diversity. They are updated at every iteration by combining them with other RefSet members and, occasionally, by applying an improvement method. This improvement consists of a local search to speed-up the convergence to optimal solutions. In the present work we have chosen NL2SOL [22] as a local method. NL2SOL is a quasi-Newton algorithm with trust region strategy that exploits the structure of the nonlinear least squares problem. Note that the combination of a global method (scatter search) with a local one makes eSS a hybrid algorithm.

Practical identifiability analysis

The shape of the cost function (5) in the surroundings of its optima determines the local identifiability of the parameters. We assess parametric identifiability in two consecutive steps:

1. First we calculate the sensitivity of the model outputs (observables) with respect to changes in the parameters. Those parameters which have no effect (or very little) on the observed signals are classified as non-identifiable. Note that this label is assigned on an individual basis, that is, taking only into account the effect of each parameter individually.
2. Even if a parameter influences the model output, it may still be unidentifiable if its effect can be compensated by changes in other(s) parameter(s). Hence in the second step we consider the interplay among parameters, aiming at finding groups of parameters which are non-identifiable due to their collinearity.

Note that, while it would be possible at least in principle to perform both steps simultaneously, in practice the curse of dimensionality hampers the application of such a global sensitivity approach to large models [45, 46].

Sensitivity analysis

The analysis of parametric sensitivity of kinetic models has a long tradition in model analysis [47, 48]. For the dynamical system (1)–(2), the parametric sensitivities of the observables can be accurately calculated by solving the forward sensitivity equations:

$$\frac{dX_i(t)}{dt} = \frac{\partial f(x,u,\theta)}{\partial x}X_i(t) + \frac{\partial f(x,u,\theta)}{\partial \theta} \quad \text{for } i = 1, \ldots, N_\theta \tag{11}$$

$$s_i(t) = \frac{\partial g(x,\theta)}{\partial x}X_i(t) + \frac{\partial g(x,\theta)}{\partial \theta} \quad \text{for } i = 1, \ldots, N_\theta \tag{12}$$

$$s_i(t_0) = \begin{cases} 0 & \text{if } \theta_i \text{ is a model parameter} \\ 1 & \text{if } \theta_i \text{ is an initial condition} \end{cases} \quad \text{for } i = 1, \ldots, N_\theta. \tag{13}$$

Here $X_i = \frac{\partial x}{\partial \theta_i}$ denotes the *sensitivity of the state vector* with respect to the i-th parameter and the vector $s_i = \frac{\partial y}{\partial \theta_i}$ is the *sensitivity of the observables* with respect to this parameter. This calculation requires the solution of the $N_x \times N_\theta$ ordinary differential Eq. (11) with initial conditions (13) for each experiment. The numerical solution is determined for the time points for which there are experimental data available, and then the algebraic Eq. (12) are evaluated. If the partial derivatives of the dynamic equations are not available, an alternative is to calculate the sensitivities using finite differences or automatic differentiation.

The sensitivities of the observables are scaled using the same weights as in Eq. (4), resulting in scaled sensitivities for an output j and a parameter i:

$$[\tilde{s}_i]_j = \sqrt{w_j}\frac{\partial y_j}{\partial \theta_i}. \tag{14}$$

For each parameter we calculate an overall scoring called root mean squared sensitivity, $\tilde{s}_i^{\text{msqr}}$, to take into account changes in time or across experiments [29, 49]:

$$\tilde{s}_i^{\text{msqr}} = \sqrt{\frac{1}{N_D}\sum_{j=1}^{N_D}\tilde{s}_{ij}^2} \quad \text{for } i = 1, \ldots, N_\theta . \tag{15}$$

Below a certain threshold the parameters are considered non-influential to the outputs. We set the threshold to four orders of magnitude smaller than the maximum root mean square value (15). Parameters whose sensitivity falls below this cut-off value are considered practically non-identifiable and they are kept out of further analysis. The procedure is summarized in Algorithm 1.

We remark that the outcome of the sensitivity calculations depends not only on the parameters, but also on the choice of initial conditions and external stimuli, which can have a strong influence in the practical identifiability of a

---

Algorithm 1 Finding sensitive model parameters

Require: Obtain vector of calibrated parameters $\rightarrow \hat{\theta} = [\hat{\theta}_1, \ldots, \hat{\theta}_{N_\theta}]$ (solve Eqs. (5)–(10))

1: Parameter index set $\mathcal{I} \leftarrow \{1, 2, \ldots, N_\theta\}$
2: Compute the sensitivity matrix at the optimal parameter vector $\rightarrow s(\hat{\theta})$ (solve Eqs. (11)–(13))
3: Compute the weighted sensitivities $\rightarrow \tilde{s}$ (solve Eq. (14))
4: Find the sensitive parameters by ranking the mean-square values of the sensitivity columns as in Eq. (15) and setting a cut-off value. The corresponding index set $\rightarrow \mathcal{I}_{\text{sensitive}} \subset \mathcal{I}$

---

model. If insufficiently excitatory stimuli or initial conditions result in poor practical identifiability, a solution – if it is possible to carry out additional measurements – is to design and perform a new experiment to generate maximally informative data [17].

Collinearity of parameters

Interplay among influential parameters can result in an unidentifiable model, because a variation in the cost function value due to a change in a parameter can be compensated by changes in other parameters. Pairwise interplay can be detected by plotting contours of the cost function versus pairs of parameters. Largely eccentric contours or "valleys" show that the cost function is almost unchanged in one direction, and the two parameters are highly correlated. This approach has two drawbacks: it involves a large computational effort and is limited to interplay between pairs of parameters. To compute higher dimensional interactions we use a different measure: the *collinearity* of parametric sensitivities.

To calculate collinearity we first normalize the scaled sensitivities (14) as follows:

$$\bar{s}_i = \frac{\hat{s}_i}{\|\hat{s}_i\|} \quad \text{for } i = 1, \ldots, N_\theta. \tag{16}$$

This normalization avoids biases caused by differences in the absolute values of the individual sensitivity vectors.

Let us consider a set K of k parameters and their corresponding sensitivity vectors. The parameters are linearly dependent if there exist k constants $\alpha_i \neq 0$ such that

$$\alpha_1 \bar{s}_{K_1} + \alpha_2 \bar{s}_{K_2} + \ldots \alpha_k \bar{s}_{K_k} = 0 \tag{17}$$

If the above relation does not hold, the set is independent. When the equality (17) holds only approximately, the parameters are nearly dependent or nearly collinear. The degree of collinearity among a set of parameters can be measured by the collinearity index, CI_K, which is defined as [29]:

$$\text{CI}_K = \frac{1}{\min_{\|\alpha\|=1} \|\bar{S}_K \alpha\|} = \frac{1}{\sqrt{\lambda_{K,\min}}}. \tag{18}$$

where $\bar{S}_k$ is the sensitivity matrix built from the k sensitivity vectors, $\bar{S}_K = [\bar{s}_{K_1}, \bar{s}_{K_2} \ldots \bar{s}_{K_k}]$, and $\lambda_{K,\min}$ is the smallest eigenvalue of $\bar{S}_K^T \bar{S}_K$. The larger the collinearity index is, the more dependent the corresponding parameters are. Brun and co-authors [29] proposed to classify a subset of parameters as identifiable if their collinearity index is smaller than a threshold which they chose as $\text{CI}_K < 20$. Roughly speaking, a value of 20 means that 95% of the variation in the model output caused by changing one of the parameters in the subset can be compensated by changing the other parameters in the set.

Other approaches for finding parameter correlations using sensitivity-based measures have been previously proposed in the literature. Li and Vu presented two methods [50, 51] that search for relationships among parameters in the context of a priori identifiability analysis (i.e. with noise-free, continuous data). The method in [50] provides a necessary but not sufficient condition for identifiability of nonlinear systems, which need to be fully observed (i.e. they must satisfy $y = x$). The method in [51] removes the requirement of measuring all the system states, replacing it with the restriction that the model must be linear. We remark that the method proposed in the present manuscript does not have these limitations: it can be applied to partially observed, nonlinear systems with noisy, discrete-time measurements.

Largest identifiable subset

As explained in the previous subsections, a subset of parameters is considered identifiable if its elements are influential and their sensitivity vectors are not collinear. We are interested in finding the largest set of parameters for which the collinearity of the corresponding sensitivity vectors is below the chosen threshold, $\text{CI}_K < 20$. Such a set of parameters represents all the degrees of freedom in the model. This means that perturbing a parameter *not* included in this set has an effect in the model predictions that can be compensated (at least by 95%) by changing other parameters in the set. However, a perturbation in a parameter belonging to the set cannot be compensated by changes in the remaining parameters.

Several methods have been developed for finding the group of identifiable parameters [30, 52, 53]. *Iterative selection methods* apply a step-wise procedure to select one parameter at a time, until no more parameters can be added to the identifiable set. In each step the parameter to be included is selected based on an optimality criteria. For example, the modified Gram-Schmidt orthogonalization method [54] projects all the remaining sensitivity vectors to the subspace spanned by the already selected sensitivity vectors, and includes the parameter corresponding to the one with the largest projection value. This step is repeated until the largest projection value falls below a threshold, which means that the next parameter would significantly interplay with the parameters already included. The computational cost of this method scales up well with the number of sensitivity vectors. However, the drawback of iterative procedures such as this one is that the solution might not be the global optimum, that is, it might fail to find the largest identifiable subset.

Alternatively, we propose to solve the problem of finding the largest identifiable subset of all the estimated parameters using *combinatorial optimization*. To this end we formulate it as a (nonlinear) integer optimization problem, where the goal is to maximize the number of sensitivity vectors included in the set, with the constraint that the corresponding collinearity index is below a threshold CI^*.

This algorithm can be stated as

$$\underset{i \in \{0,1\}^{N_\theta}}{\text{maximize}} \sum_{k=1}^{N_\theta} i_k \tag{19}$$

$$\text{subject to } S_i = \text{cat}(\{s_k \,|\, i_k = 1, \text{ for } k = 1, \ldots, N_\theta\}) \tag{20}$$

$$CI(S_i) < CI^* \tag{21}$$

$$i_k \text{ is a binary variable for } k = 1, \ldots, N_\theta \tag{22}$$

where the binary variable i_k indicates if the k-th parameter is included ($i_k = 1$) or not included ($i_k = 0$) in the identifiable group of parameters. The sensitivity matrix corresponding to the selected parameters is S_i, and 'cat' stands for the concatenation of the column vectors in the constraint (20). The collinearity index of this matrix is $CI(S_i)$ and it is determined by computing the minimum eigenvalue as in (17).

This combinatorial optimization problem has an exponentially scaling computational cost, and thus its solution requires an efficient algorithm. We chose the Variable Neighbourhood Search (VNS) technique [55], which is a heuristic global optimization method for integer optimization problems. We used the version of VNS included in the MEIGO Toolbox [56], which is implemented in MATLAB.

We modified this initial formulation of the problem described in Eqs. (19)–(21) after finding that its solution is often not unique: even after maximizing the number of parameters in the subset, there may be multiple subsets that yield a collinearity index below the threshold CI^*. Indeed, we found large variability in the solutions if no initial guess was specified. Therefore, we reformulated the optimization problem in two ways, as described in the following paragraphs.

As a first modification, we transformed the collinearity requirement (21) from a 'hard' to a 'soft' constraint (or penalty). The modified optimization problem reads as

$$\underset{i \in \{0,1\}^{N_\theta}}{\text{maximize}} \sum_{k=1}^{N_\theta} i_k - P_1(i) - P_2(i) \tag{23}$$

$$\text{subject to } S_i = \text{cat}(\{s_k \,|\, i_k = 1, \text{ for } k = 1, \ldots, N_\theta\}) \tag{24}$$

$$P_1(i) = \frac{1}{2}CI(S_i)/CI^* \tag{25}$$

$$P_2(i) = \begin{cases} 0 & \text{if } CI(S_i) < CI^* \\ \alpha \left(CI(S_i) - CI^*\right)^\beta & \text{otherwise} \end{cases} \tag{26}$$

$$i_k \text{ is a binary variable for } k = 1, \ldots, N_\theta \tag{27}$$

As above, the binary variable i_k indicates if the k-th parameter is included (1) or not included (0) in the selected group of parameters. The penalty P_1 is a monotone increasing (linear) function of the collinearity index $CI(S_i)$, such that P_1 is 0.5 when the collinearity equals to the threshold. Due to this small value, P_1 does not influence the size of the largest subset below the threshold. In this way, when multiple sets of the same size co-exist, the set with smaller collinearity index is always favoured. This results in an unique solution of the optimization problem if there are no sets with identical collinearity index. The second penalty function P_2 represents a soft constraint that is active when the collinearity exceeds the threshold. The steepness of this constraint is tuned by the values of α and β, which we set to $\alpha = 1$ and $\beta = 2$.

Our second improvement of the formulation of the optimization problem consists in providing a good initial guess of the solution using QR decomposition. The rank revealing QR decomposition algorithm, or rrqr [57], rewrites a matrix S as

$$\Pi S = QR, \tag{28}$$

where Q is an orthogonal matrix, R is an upper triangular matrix, and Π is a permutation matrix. Due to the properties of this decomposition, the permutation matrix defines a reordering of the columns of S. In this re-ordered matrix $S_{\text{ro}} = \Pi S$, the most orthogonal columns are located in the left. In other words, the first n columns of the reordered matrix define a linear subspace, and the $(n + 1)$-th column has the largest projection value on this subspace among the remaining $N_\theta - n$ columns located to the right of the n-th column. The outcome of the rrqr technique is similar to that of the aforementioned Gram-Schmidt orthogonalization method, but its implementation is more efficient.

We applied rank revealing QR decomposition to the sensitivity matrix, following the procedure described in Algorithm 2. Then, we used the resulted ordering of the sensitivity vectors to initialize the global optimizer. In this way we improved the performance of the global optimizer, which often found larger sets with collinearity index below the threshold value. The whole procedure for identifying the largest non-collinear subset of parameters is summarized in Algorithm 3.

Algorithm 2 Finding the largest identifiable subset of parameters by rank revealing QR decomposition (rrqr)

Require: Find sensitive parameters by Algorithm 1 $\rightarrow$ $\mathcal{I}_{\text{sensitive}}$

Require: Define collinearity threshold: CI^*

1: Number of sensitive parameters: $N_{\text{sp}} = $ cardinality($\mathcal{I}_{\text{sensitive}}$)
2: **for all** $i \in \mathcal{I}_{\text{sensitive}}$ **do**
3:    Normalize the sensitivity columns: $\bar{s}_i \leftarrow \frac{\hat{s}_i}{||\hat{s}_i||}$ (Eq. (16))
4: **end for**
5: Form $\bar{S} \leftarrow \text{cat}(\{\bar{s}_i \mid i \in \mathcal{I}_{\text{sensitive}}\})$, where 'cat' stands for concatenation of a set of column vectors.
6: $[Q, R, p, r] \leftarrow \textbf{rrqr}(\bar{S})$, where vector p contains the permutation vector
7: **for** $i = 2$ **to** N_{sp} **do**
8:    $S_{ss} \leftarrow \bar{S}(:, p(1:i))$
9:    $CI_{ss} = \text{collinearity}(S_{ss})$
10:    **if** $CI_{ss} > CI^*$ **then**
11:       indexLargestIdSetQR $= p(1:i-1)$
12:       **break**
13:    **end if**
14: **end for**
15: **return** indexLargestIdSetQR

Algorithm 3 Finding the largest identifiable subset of parameters by VNS

Require: Find sensitive parameters by Algorithm 1 $\rightarrow$ $\mathcal{I}_{\text{sensitive}}$

Require: Find largest set by Algorithm 2 $\rightarrow$ indexLargestIdSetQR

Require: Define collinearity threshold: CI^*

1: number of sensitive parameters: $N_s = $ cardinality($\mathcal{I}_{\text{sensitive}}$)
2: $x_{\text{init}} = \text{zeros}(1, N_s)$
3: $x_{\text{init}}(\text{indexLargestIdSetQR}) = 1$
4: solve optimization (23)–(27) using x_{init} as initial guess

---

The procedure presented in this subsection has similarities with the one proposed by Chu and Hahn [54]. One difference is that we maximize the subset size for a given collinearity threshold, whereas Chu and Hahn adopted the opposite approach, i.e., maximizing parametric identifiability for a pre-specified subset size. Additionally, both methods differ in the optimization technique: we use Variable Neighbourhood Search, which has better scalability than the genetic algorithm chosen in [54]. Recently, Nienałtowski et al. [58] have proposed a method for finding clusters of correlated parameters using so-called canonical correlation analysis (CCA). CCA is an extension of Pearson correlation for measuring multidimensional correlations between groups of parameters. Given two groups of parameters of sizes m and n, with $m < n$, calculation of the canonical correlations provides m measures, which are summarized in a single measure, called MI-CCA. This similarity measure represents the mutual information between the two groups, although it should be noted that average mutual information is equivalent to canonical correlation only if the random variables follow an elliptically symmetric probability model. Nienałtowski et al. use MI-CCA to cluster parameters until an identifiable subset is reached. This approach is sequential and yields a single parameter subset, which is possibly not maximal. In contrast, the methodology described here combines an initial sequential phase with a subsequent combinatorial optimization procedure. The second phase yields several identifiable parameter subsets and usually improves the initial solution.

Finding all largest subsets

As mentioned above, the largest non-collinear subset of parameters is not unique. To realize this, imagine that we have a non-collinear set of the parameters, and consider an additional pair of highly collinear parameters. Since we may add either of these two parameters to the set, but not both of them, we have two potential solutions. The optimization algorithm described above would choose the option with a lower collinearity index.

However, we may also be interested in enumerating *all* the possible sets, instead of only one. Finding all the largest subsets is a combinatorial problem too, which is computationally expensive. A naive approach for solving it could be to generate all possible sets of parameters and compute the corresponding collinearity index. However, note that if two parameters θ_1 and θ_2 are collinear, then any sets including the pair $\{\theta_1, \theta_2\}$ are highly collinear. Using this fact, we developed an incremental procedure for the systematic determination of the sets. We start by considering all possible pairs of parameters and determining their collinearity. Then we extend only those pairs which have a small collinearity index, by considering all possible combinations of a third parameter. This procedure is repeated until either all the sets are highly collinear, or there is only one set containing all the parameters. In this way, summarized in Algorithm 4, we can find all the largest subsets of non-collinear parameters.

Algorithm 4 Finding all the largest identifiable subset of parameters

---

Require: Sensitivity matrix S at the optimal parameters
Require: Define collinearity threshold: CI^*

1: Given the sensitivity matrix $S = [s_1 \ldots s_{N_\theta}]$ and a subset of column indexes $K \subseteq \mathcal{I} = \{1 \ldots N_\theta\}$ of S. Then let S_K the sub-matrix of S containing the columns specified by indices in K, i.e. $S_K := \mathrm{cat}(\{s_i \mid i \in K\})$

2: Generate all combinations of pairs of parameter indexes: $\mathcal{I}_2 = \{(i,j) \mid i, j \in \mathcal{I}, i < j)\}$

3: Compute the collinearity index for each element of the set: $CI_2 = \{CI(S_K) \mid K \in \mathcal{I}_2\}$

4: Find sets with small collinearity: $\mathcal{I}_2^* = \{K \mid K \in \mathcal{I}_2, CI(S_K) < CI^*\}$

5: **for** setSize = 3 **to** n_θ **do**

6:     generate all the extension sets $\mathcal{I}_{\mathrm{setSize}} = \{K \cup i \mid K \in \mathcal{I}_{\mathrm{setSize}-1}^*, i \in \mathcal{I}, i \notin K\}$

7:     Compute the collinearity index for each element: $CI_{\mathrm{setSize}} = \{CI(S_K) \mid K \in \mathcal{I}_{\mathrm{setSize}}\}$

8:     Find sets with small collinearity: $\mathcal{I}_{\mathrm{setSize}}^* = \{K \mid K \in \mathcal{I}_{\mathrm{setSize}}, CI(S_K) < CI^*\}$

9:     **if** cardinality($\mathcal{I}_{\mathrm{setSize}}^*$) = 0 **then**

10:         report $\mathcal{I}_{\mathrm{setSize}-1}^*$ and $CI_{\mathrm{setSize}-1}$

11:         **break;**

12:     **end if**

13: **end for**

Partitioning the non-identifiable parameters

The two procedures presented above can be used for finding (i) the largest, least collinear subset of parameters, and (ii) all the largest subsets; in both cases, restricted to those subsets whose collinearity falls below a threshold. However, it is often important to understand why certain parameters are *not* identifiable. For example, a parameter may be unidentifiable because the model output has very low sensitivity to changes in its value. But it could also be because it is highly correlated with another parameter, even when both parameters have high sensitivities. Finding small groups of highly collinear parameters can be helpful in determining the exact source of unidentifiability.

The collinearity of a subset always increases when a new parameter is added to the set. For example, considering three parameters, the collinearity of the triplet is always higher than the collinearity of any pairs. Therefore, if a larger set of parameters contains a collinear pair, then the collinearity index of the large set is also large.

If we are interested in *finding the smallest groups of highly collinear parameters*, we can proceed as follows. First we generate all possible pairs of parameters, and compute the collinearity of the corresponding sensitivity vectors. Then we evaluate all possible triplets. The procedure can be extended for the analysis of larger

sets. However, due to the combinatorial explosion of the computational cost, this method can be applied only to models of moderate size (with a maximum of roughly 20 parameters).

Visualization of identifiable subsets

It can be useful to represent the identifiability results graphically, because such visualization can provide modellers with insight about how to reformulate their models and/or design new experiments in order to avoid non-identifiable parameters.

With this aim, we display the model structure in the natural network visualization technique. An example is shown in Fig. 1c. The model structure is represented as a graph whose nodes are state variables, observables, stimuli, and model parameters. The edges – which can be directed (arrows) or undirected – have the following meaning: an arrow from node A to node B indicates that node B appears in the equation of A. For example, if the dynamic equation of a state x_1 is $\dot{x}_1 = p_1 \cdot x_2$, the corresponding graph would show two arrows $x_2 \to x_1$ and $p_1 \to x_1$.

More formally, we determine how the state, input variables, stimuli and parameters are connected and influence each other through symbolic manipulation of the model Eq. (1). For this purpose we compute: (i) the Jacobian matrix with respect to the states: $J_{i,j}^{ss} = \frac{\partial f_i}{\partial x_j}$, (ii) the Jacobian of the observation functions with respect to the states: $J_{i,j}^{so} = \frac{\partial g_i}{\partial x_j}$, (iii) the Jacobian of the systems dynamics with respect to the stimuli $J_{i,j}^{si} = \frac{\partial f_i}{\partial u_j}$, and (iv) the Jacobian with respect to the parameters $J_{i,j}^{sp} = \frac{\partial f_i}{\partial \theta_j}$. All these matrices are evaluated symbolically, and then the expressions are converted to a logical 1 (if the symbolic expression is non zero) or 0 when the symbolic result is zero.

Additionally, we can connect parameters by undirected edges if their collinearity is larger than the collinearity threshold.

Implementation: the VisId software tool

We implemented the techniques proposed in subsections "Practical identifiability analysis" and "Visualization of identifiable subsets" as a MATLAB software package called VisId, which is provided as Additional file 1 and can also be downloaded from GitHub (https://github.com/gabora/visid). It is free software, made available under the terms of the GNU General Public License version 3. The VisId toolbox relies on three other MATLAB toolboxes, which are also freely available: AMIGO2 [59] (https://sites.google.com/site/amigo2toolbox/download), which is used to to store, simulate and calibrate the models; MEIGO [56] (http://www.iim.csic.es/~gingproc/meigo.html), which implements the Variable Neighbouring

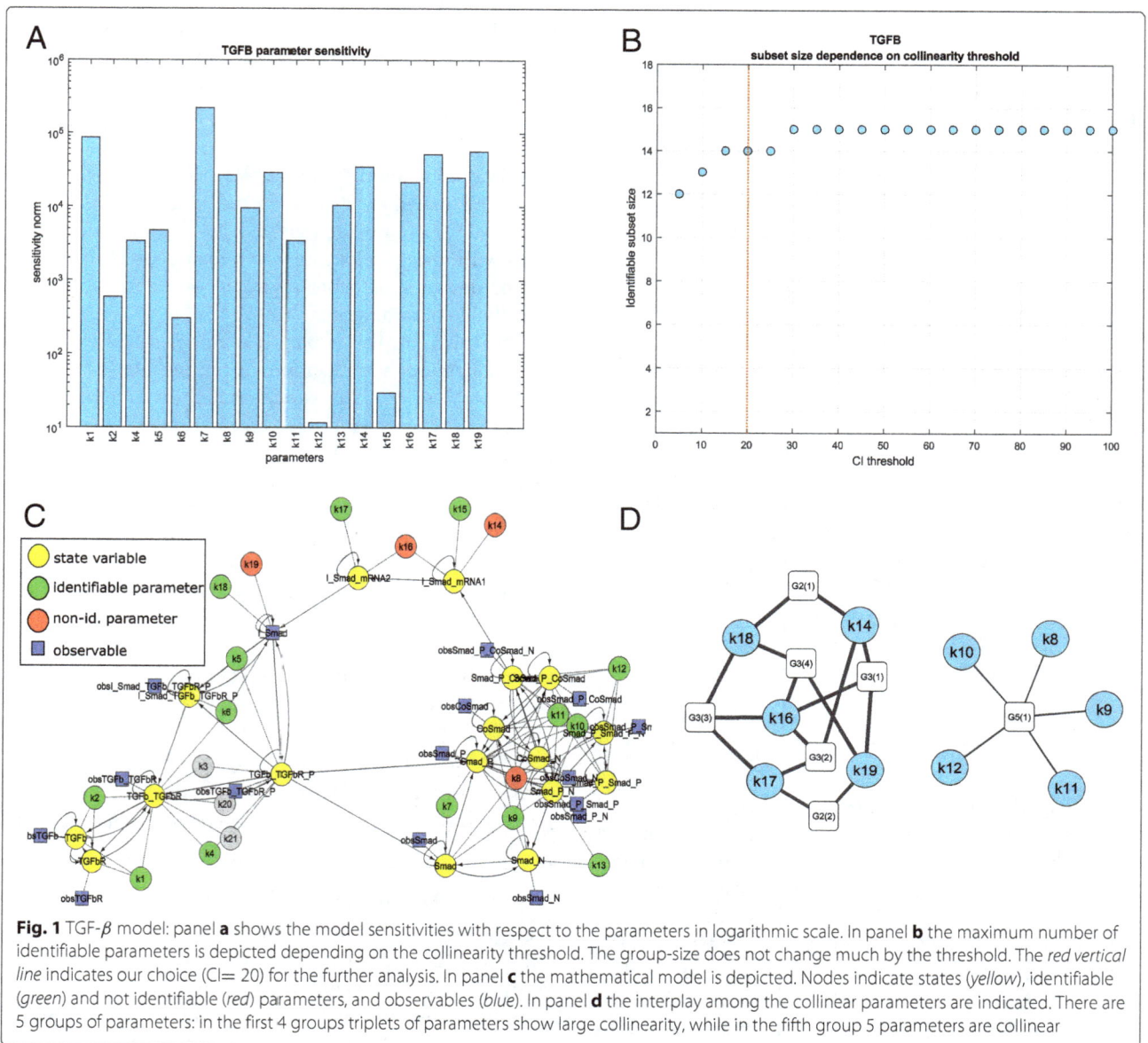

Fig. 1 TGF-β model: panel **a** shows the model sensitivities with respect to the parameters in logarithmic scale. In panel **b** the maximum number of identifiable parameters is depicted depending on the collinearity threshold. The group-size does not change much by the threshold. The *red vertical line* indicates our choice (CI= 20) for the further analysis. In panel **c** the mathematical model is depicted. Nodes indicate states (*yellow*), identifiable (*green*) and not identifiable (*red*) parameters, and observables (*blue*). In panel **d** the interplay among the collinear parameters are indicated. There are 5 groups of parameters: in the first 4 groups triplets of parameters show large collinearity, while in the fifth group 5 parameters are collinear

Search (VNS) optimization method; and (optionally) RRQR (https://www.mpi-magdeburg.mpg.de/1094756/rrqr), which performs the rank revealing QR decomposition used to initialize the global optimizer. Network visualization is performed with Cytoscape [60] (http://www.cytoscape.org/). Further details can be found in Section 4 of Additional file 2.

Results

In this section we demonstrate the application of the methodology presented in the previous section using several dynamic systems biology models of different type and complexity. Their main characteristics are given in Table 1. First we present detailed results of identifiability analysis and visualization for a model of the TGF-β

signalling pathway. We also provide similar results for the genetic network that controls the circadian clock in *Arabidopsis thaliana*. Due to their complexity and yet relatively moderate size, these models are well suited as case studies for illustrating the identifiability methodology in depth.

Then we study two large scale benchmark problems included in the BioPreDyn-bench collection [23]. Since the analysis of these latter models is more challenging due to their larger size, we start by demonstrating the performance improvements that can be achieved during parameter estimation using the model calibration procedure proposed in Section "Parameter estimation with regularization and global optimization". Then we perform identifiability analysis and report the corresponding

Table 1 List of models used as case studies and their characteristics

	TGF-β	Circadian	B2	B4
Description	TGF-β signaling pathway	Gene network, A. thaliana	Central Carbon Metabolism, E. coli	Metabolic model, Chinese Hamster Ovary
Reference	[61]	[62]	[23, 63]	[23, 64]
Parameters	18	27	116	117
States	21	7	18	34
Outputs	16	2	9	13

results, including the graphical representation of the identifiable subset using the natural network visualization.

TGF-β signalling pathway

The dynamic model of the TGF-β signaling pathway was presented in [61] as a tutorial example for model calibration. It has 18 dynamic states and 21 kinetic parameters (k_1–k_{21}), of which 18 need to be estimated. Following [61], we assumed that all the concentrations, except the Smad RNAs ($C_{I\_Smad\_mRNA1}$ and $C_{I\_Smad\_mRNA2}$), can be measured in the experiments. The algebraic Equations of the reaction kinetics and the dynamic equations are provided in the Additional file 2.

For the purpose of testing the methodology we generated a training dataset by simulating the model equations using the nominal values of the parameters k_1–k_{21} (numerical values are listed in Additional file 2: Table S1). Then we sampled the simulated trajectories at equidistant time points, and added normally distributed random numbers to the data to mimic measurement errors. Finally, we estimated the model parameters from the generated data set. This approach is widely used for testing calibration methods and assessing the extent to which they recover the nominal parameters. It should be noted that, as the amount of noise in the dataset increases, the information/signal ratio decreases, making the estimation problem more ill-conditioned. This makes it more difficult to recover the correct value of the parameters, but has a small effect in computation times. The numerical values of the estimated parameters are reported in Additional file 2: Table S2.

We started the identifiability analysis by computing the sensitivities of the observations with respect to the estimated model parameters, according to Algorithm 1. We found that all the parameters have a non-negligible influence on the model outputs, thus there are no individually non-identifiable parameters (see Fig. 1a).

Next, following Algorithm 2, we applied QR decomposition and ranked the parameters according to their orthogonality. We then solved the optimization problem (23)–(27) by initializing the variable neighboring search method with the results of the QR decomposition (Algorithm 3). Setting the threshold level for the

collinearity index to CI $=$ 20 yielded 14 identifiable parameters, which are shown as green nodes in the network in Fig. 1c. Parameters not present in the identifiable subset are shown as red nodes. Parameters are connected by arrows to state variables (represented by yellow nodes) if they appear in the equation of the corresponding dynamic equation. States which directly influence each other are also connected by directed edges in the same manner. Blue squares represent measurements; a state is connected to a blue square if it appears in the corresponding observation function.

To see how the size of the identifiable subset is influenced by the choice of the collinearity index threshold (CI), we solved the optimization problem for a range of threshold values. The results are depicted in Fig. 1b. As the collinearity index threshold decreases, less parameters are considered identifiable. We can see that the identifiability results are quite robust to the choice of threshold level: the number of identifiable parameters is always between 12 and 15, and it is constant ($=$ 14) for a very wide range of CI, $15 \leq CI \leq 25$.

The results presented so far tell us that the 14 parameters are not correlated. However, they do not inform of the relationships among identifiable and non-identifiable parameters. To investigate this point, we computed the smallest correlated subsets as described in Section "Partitioning the non-identifiable parameters", up to groups of 6 parameters. Figure 1d shows such groups; parameters are depicted as blue circles connected with group identifying nodes (white squares). These nodes are labeled as GX(Y), where X indicates the number of parameters in the group and Y is the group index for a given number of parameters (e.g. G3(2) stands for the second group of three correlated parameters). We found that the large pairwise collinearity between $k_{14} - k_{18}$ and $k_{17} - k_{19}$ explains the non-identifiability of the model parameters only partially. There are 4 groups of triplets and a group of 5 parameters which are highly correlated. The members of the groups and the corresponding collinearity index are reported in Table 2.

It is important to note that collinearity might arise among multiple parameters, even if they are pairwise independent. For example, despite the fact that none of

Table 2 TGF-β model: highly collinear parameter sets. A set ID indicates the number of parameters involved in the collinearity group. They are also depicted in Fig. 1d

Set ID.	CI	Parameters				
G2(1)	2.87e+07	k14	k18			
G2(2)	41.4	k17	k19			
G3(1)	110	k14	k16	k19		
G3(2)	1.37e+03	k14	k16	k17		
G3(3)	1.37e+03	k16	k17	k18		
G3(4)	110	k16	k18	k19		
G5(1)	22.9	k8	k9	k10	k11	k12

the pairs in the group of k_{14}, k_{16} and k_{17} has a high pairwise collinearity, the collinearity index of the triplet is extremely large.

Algorithm 4 found 40 different sets of identifiable parameters with collinearity index ranging between 12.4 and 16, less than the threshold (CI = 20). The sets are reported with the corresponding collinearity index in Additional file 2: Table S3. We can see that parameters $\{k_1 - k_7, k_{13}, k_{15}\}$ are members of all the groups, and they do not participate in any of the small correlated groups in Fig. 1d. From each correlated group of size K, only $K - 1$

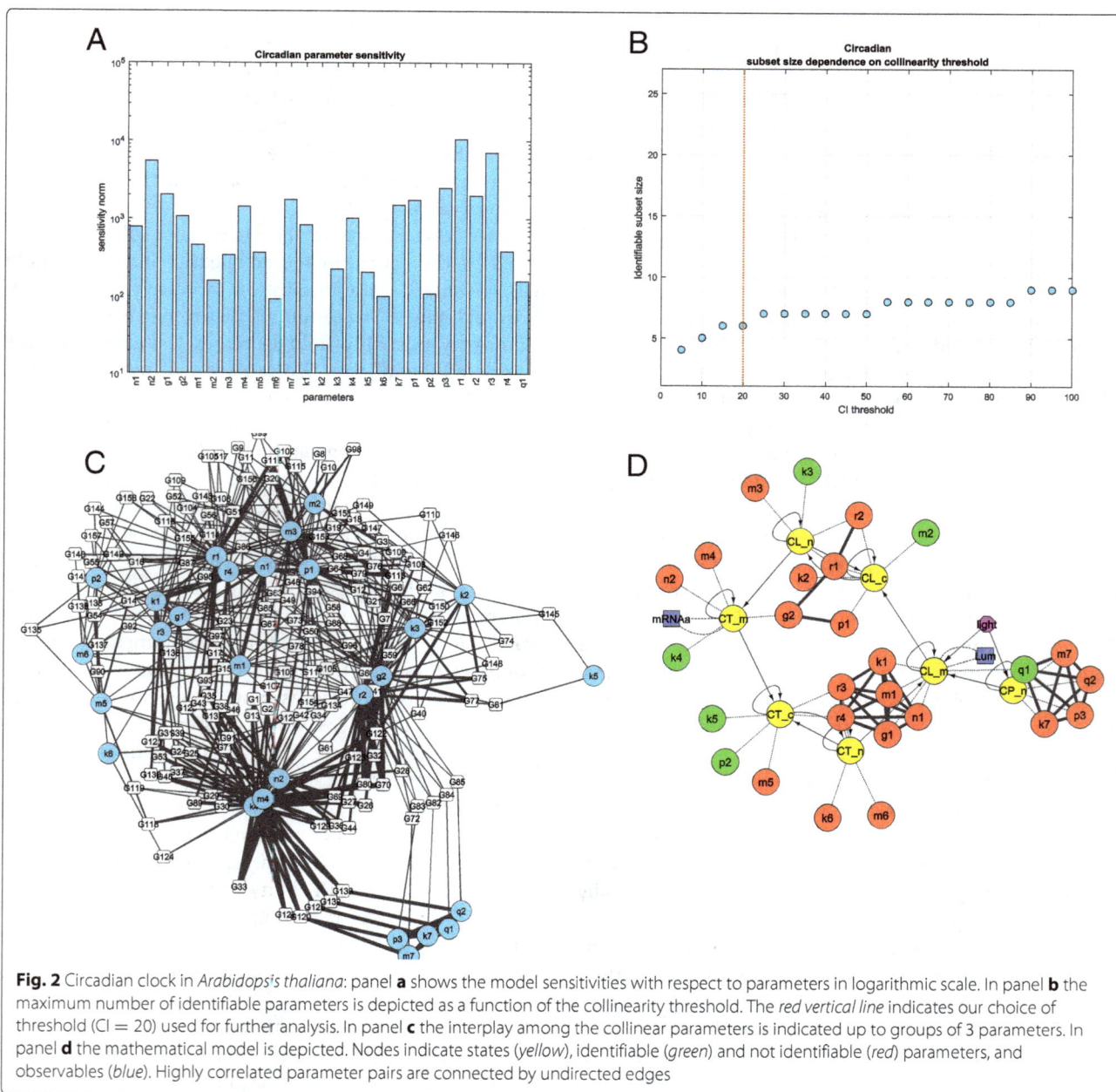

Fig. 2 Circadian clock in *Arabidopsis thaliana*: panel **a** shows the model sensitivities with respect to parameters in logarithmic scale. In panel **b** the maximum number of identifiable parameters is depicted as a function of the collinearity threshold. The *red vertical line* indicates our choice of threshold (CI = 20) used for further analysis. In panel **c** the interplay among the collinear parameters is indicated up to groups of 3 parameters. In panel **d** the mathematical model is depicted. Nodes indicate states (*yellow*), identifiable (*green*) and not identifiable (*red*) parameters, and observables (*blue*). Highly correlated parameter pairs are connected by undirected edges

parameters can participate in the largest set of identifiable parameters.

The aforementioned identifiability procedures can be carried out in a few seconds. Detailed computational costs are shown in Table S6 of the Additional file 2 for all the case studies considered in this paper.

Circadian clock in *Arabidopsis thaliana*

Locke and co-authors [62] described the genetic network controlling the circadian clock in *Arabidopsis thaliana*; the dynamic equations of this model are provided in the Additional file 2.

We generated training data by simulating the model equations with the nominal parameters (Additional file 2: Table S4) in two experimental conditions. In the first one, the model input was kept constant ($\theta_{light} = 1$), representing continuous light stimulation of the plant. In the second experiment the input was changed pulse-wise in 12-hour

cycles, repeated 5 times. As in the previous example, the trajectories were sampled at equidistant time-points and disturbed by pseudo-random noise. Only two states, CT_m and CL_m, were observed. The estimated model parameters are collected in Additional file 2: Table S4.

Although the model outputs showed sensitivity to all the parameters (Fig. 2a), i.e. there were no zero sensitivity vectors, we found that most of the model parameters are non-identifiable due to heavy collinearities. The largest identifiable subset contains only 6 of the 27 parameters, depicted in Fig. 2d by green nodes. The enumeration of the largest sets of identifiable parameters by Algorithm 4 identified 1331 parameter sets.

Benchmarks B2 and B4 from the BioPreDyn-bench collection

In this subsection we analyze two large scale benchmark problems taken from the BioPreDyn-bench collection

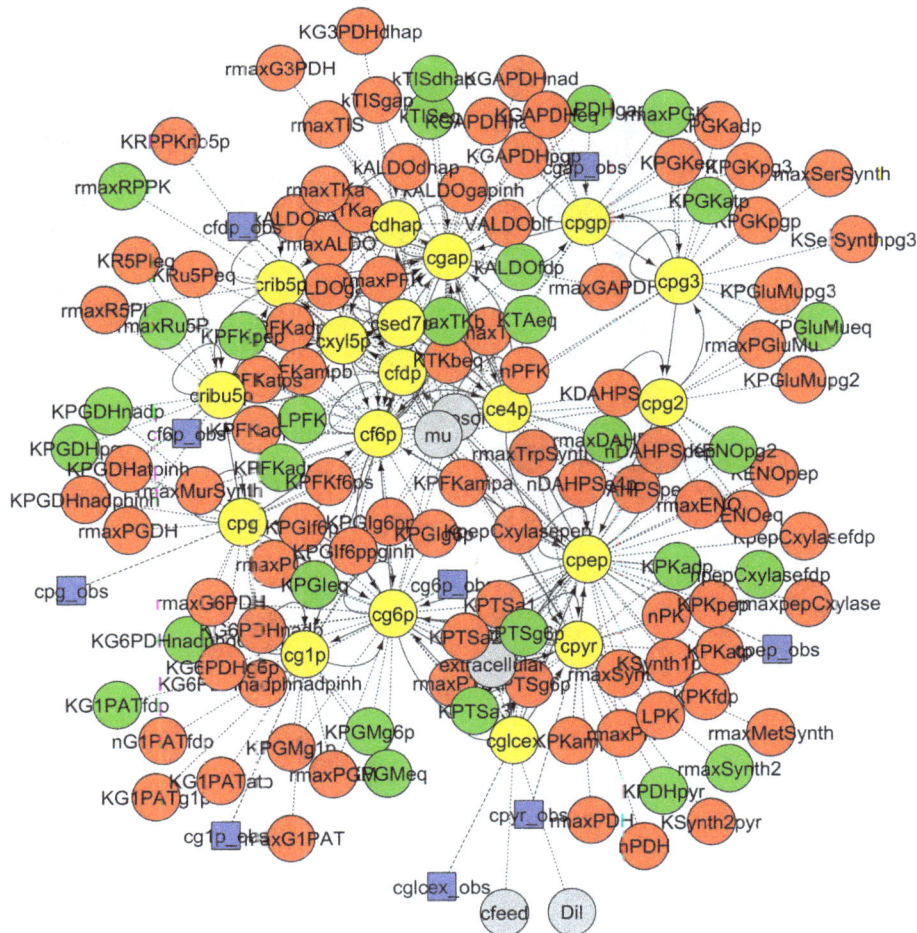

Fig. 3 Representation of the connections in the B2 model using the network diagram formalism. Nodes indicate states (*yellow*), identifiable (*green*) and not identifiable (*red*) parameters, observables (*blue*), and inputs (*grey*). The source file of this figure is provided with the VisId toolbox; using Cytoscape, the user can navigate through it and zoom on different areas to improve the visibility

[23]: the metabolic models of *Escherichia coli* (B2) and Chinese Hamster Ovary cells (B4). They are highly non-linear, partially observed systems with more than 100 unknown parameters, which pose serious challenges for parameter identification. In B4 the calibration data was generated by model simulation and disturbed by random noise, while in B2 it was experimentally measured. Further details about the models and the parameter estimation challenge can be found in [23].

First we use these benchmarks to illustrate the benefits of the parameter estimation strategy proposed in subsection "Parameter estimation with regularization and global optimization", comparing it with the one used in [23]. Both approaches use a hybrid method, eSS [21], which combines a global optimization algorithm (scatter search) with a local search. In [23] the local method of choice was FMINCON; here we compare that configuration with NL2SOL (with and without regularization). Global optimization algorithms use pseudo-random numbers. Hence

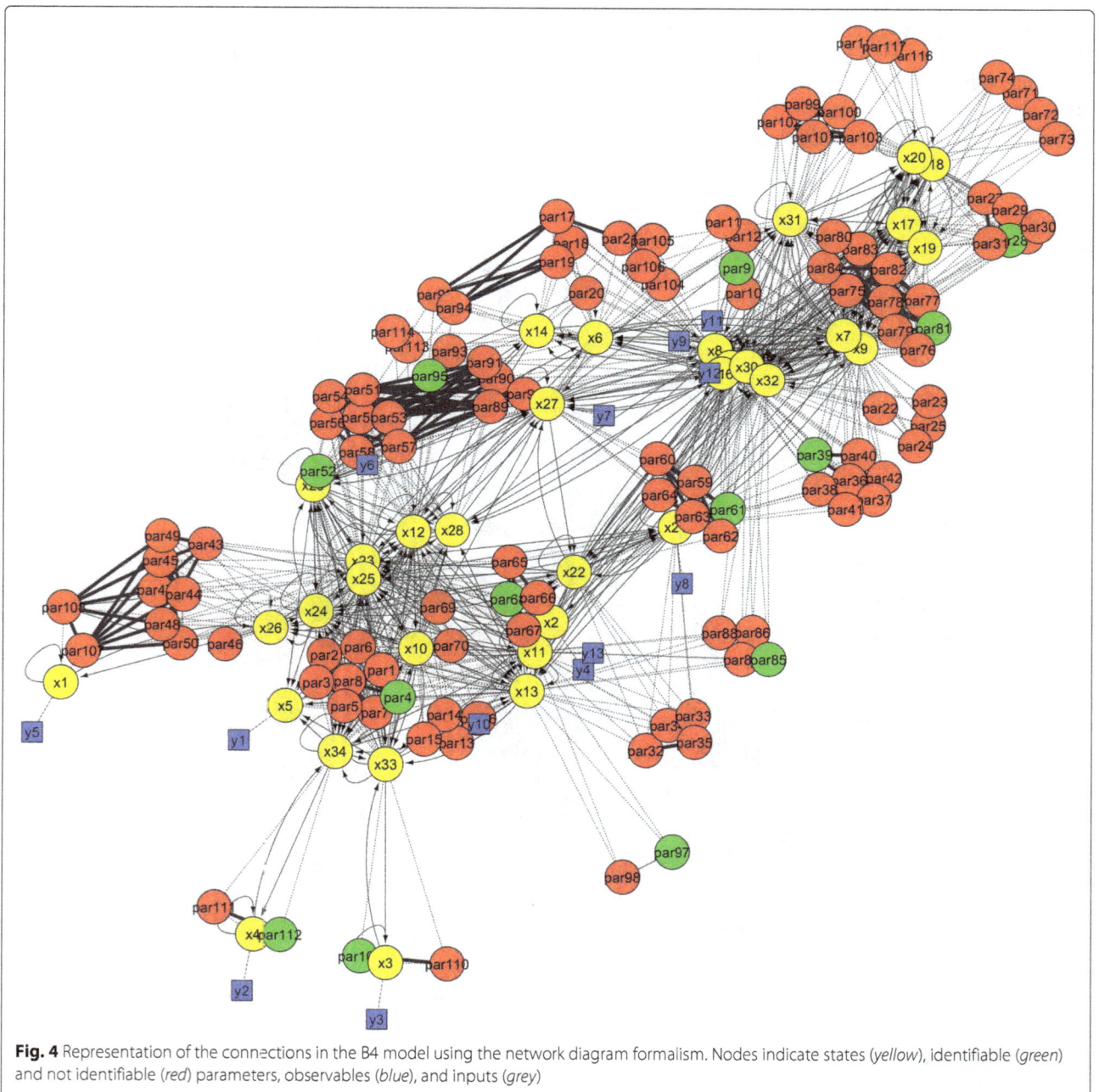

Fig. 4 Representation of the connections in the B4 model using the network diagram formalism. Nodes indicate states (*yellow*), identifiable (*green*) and not identifiable (*red*) parameters, observables (*blue*), and inputs (*grey*)

their performance changes at every run, and the calibration problem should be solved several times to obtain more robust results. Since each optimization takes several hours we limited the number of runs to five for each problem. We used the approximate computation time (CPU time) reported in [23] as the stopping criterion for the model calibration. Convergence curves depict the best objective function value found versus CPU time, and can be used to compare the performance of different algorithms. An optimization method is preferred if it achieves a lower objective function value at earlier CPU time. The best convergence curves (out of 5) corresponding to B2 and B4 are shown in the Additional file 2 for 3 algorithms: (1) eSS-FMINCON, as reported in [23]; (2) eSS-NL2SOL; and (3) eSS-NL2SOL using regularization as recommended in Section "Parameter estimation with regularization and global optimization". From those curves we see that the algorithm (3) proposed here converged earlier than the others to the optimal objective function value (note that log-log scale is used in these curves). We stress that the main purpose of regularization is to avoid overfitting: we do not wish to obtain an excessively good fit, which would indicate that we are reproducing noise instead of the true dynamics. Therefore, regularization should *not* achieve a smaller objective function value.

Next, we apply the identifiability analysis procedures presented in subsections "Practical identifiability analysis" and "Visualization of identifiable subsets" to these two models. For B2 they yield an identifiable subset of size 29, and for B4 of size 13 (recall that both models have a total of 116 parameters). The corresponding networks are shown in Figs. 3 and 4, respectively. It is also possible to find small groups of highly correlated parameters for models of this size; e.g. for B4 we obtained those depicted in Fig. 5.

The aforementioned results show that both models are poorly identifiable in practice for the considered datasets; more informative data would be needed in order to obtain accurate estimates of their parameters.

Discussion and conclusions

In this paper we have presented a workflow to efficiently estimate the parameters of dynamic models and analyze their practical identifiability. Our approach combines an advanced optimization technique, which reduces computation times in parameter estimation, and several identifiability analysis procedures, which can find subsets of identifiable and unidentifiable parameters. Results are visualized using network diagrams, which provide an intuitive representation of the findings and facilitate their analysis and understanding.

Many approaches have been applied to study identifiability of kinetic models, but they suffer from lack of scalability. An advantage of the integrated method presented here is its moderate computational cost, which enables its application to large-scale models; complete results can be obtained in a few hours for models of more than a hundred parameters. Another important aspect is the integration of identifiability analysis with visualization, which presents the results in a way that is easily interpretable for

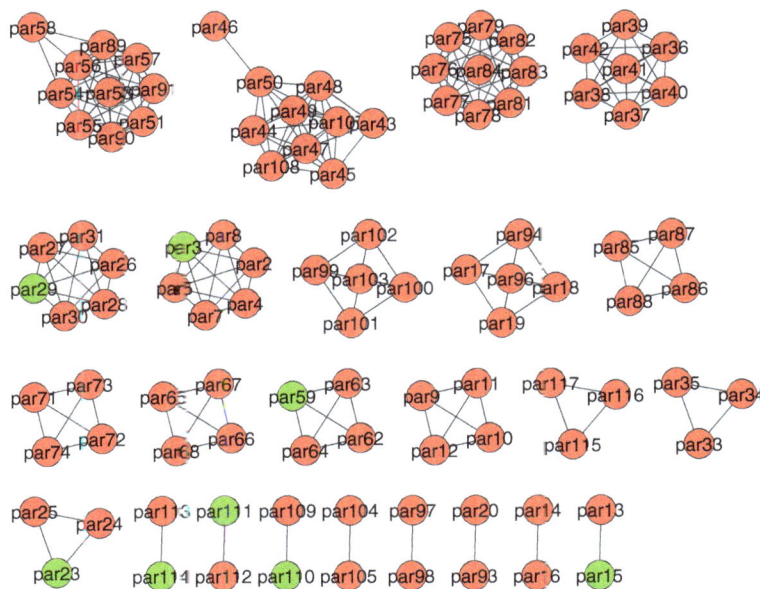

Fig. 5 Visualization of the relationships among highly collinear parameters in the B4 model. The figure shows small groups, whose sizes range between 2 and 10 parameters. Unidentifiable parameters are shown in *red*; identifiable parameters in *green*. Highly correlated pairs are connected by lines

modelers and experimentalists. Currently, its main limitation arises when trying to find *all* the different existing groups of highly correlated parameters: the combinatorial explosion of this particular task makes it feasible only for models of moderate size, i.e. of a few dozens of parameters. However, all the remaining steps of the workflow presented in this manuscript scale up well up to several hundred parameters.

The usefulness of the methodology and workflow presented here goes beyond basic parameter identifiability analysis. The procedure not only (i) determines the largest subset of identifiable parameters, but also (ii) informs about the characteristics of the space of non-identifiable parameters, reporting small groups of highly correlated parameters, and (iii) presents all these results in a coherent and scalable way using visualization techniques, facilitating the understanding of the underlying complex interactions. Uncovering these higher order relationships helps in determining the causes of unidentifiability and provides guidelines for remedying them, e.g. by reformulating the model or by collecting new data through a new experimental design. All this information can be readily used to improve the iterative model-building cycle.

A MATLAB implementation of the identifiability and visualization methodology, which we have called the VisId software package (Additional file 1), is available from GitHub (https://github.com/gabora/visid) as free, open source software. This distribution includes the case studies discussed above.

Abbreviations
CCA: Canonical correlation analysis; CI: Collinearity index; CPU: Central processing unit; eSS: Enhanced scatter search; MI: Mutual information; ODE: Ordinary differential equation; QR: Decomposition of a matrix into an orthogonal matrix (Q) and an upper triangular matrix (R); RNA: Ribonucleic acid; RRQR: Rank revealing QR decomposition; TGF-β: Transforming growth factor beta; VNS: Variable neighbourhood search

Acknowledgements
Not applicable.

Funding
This project has received funding from the European Union's Horizon 2020 research and innovation programme under grant agreement No 686282 ("CANPATHPRO"), from the EU FP7 project "NICHE" (ITN Grant number 289384), and from the Spanish MINECO project "SYNBIOFACTORY" (grant number DPI2014-55276-C5-2-R).

Authors' contributions
JRB and AG conceived of the study. JRB coordinated the study. AG implemented the methods and carried out all the computations. AFV assisted in the development of the methodology. All authors analysed the results, drafted the manuscript, and read and approved the final manuscript.

Competing interests
The authors declare that they have no competing interests.

Author details
[1] BioProcess Engineering Group, IIM-CSIC, Eduardo Cabello 6, 36208 Vigo, Spain. [2] JRC-COMBINE, RWTH Aachen University, Photonics Cluster, Level 4, Campus-Boulevard 79, 52074 Aachen, Germany.

References
1. Wiechert W, Noack S. Mechanistic pathway modeling for industrial biotechnology: challenging but worthwhile. Curr Opinion Biotechnol. 2011;22(5):604–10.
2. Menolascina F, Siciliano V, Di Bernardo D. Engineering and control of biological systems: a new way to tackle complex diseases. FEBS Lett. 2012;586(15):2122–8.
3. Smallbone K, Mendes P. Large-scale metabolic models: From reconstruction to differential equations. Ind Biotechnol. 2013;9(4):179–84.
4. Chakrabarti A, Miskovic L, Soh KC, Hatzimanikatis V. Towards kinetic modeling of genome-scale metabolic networks without sacrificing stoichiometric, thermodynamic and physiological constraints. Biotechnol J. 2013;8(9):1043–57.
5. Almquist J, Cvijovic M, Hatzimanikatis V, Nielsen J, Jirstrand M. Kinetic models in industrial biotechnology–improving cell factory performance. Metab Eng. 2014;24:38–60.
6. Link H, Christodoulou D, Sauer U. Advancing metabolic models with kinetic information. Curr Opin Biotechnol. 2014;29:8–14. doi:10.1016/j.copbio.2014.01.015.
7. Miskovic L, Tokic M, Fengos G, Hatzimanikatis V. Rites of passage: requirements and standards for building kinetic models of metabolic phenotypes. Curr Opinion Biotechnol. 2015;36:146–53.
8. Srinivasan S, Cluett WR, Mahadevan R. Constructing kinetic models of metabolism at genome-scales: A review. Biotechnol J. 2015;10(9):1345–59.
9. Evangelista PT. Novel approaches for dynamic modelling of e. coli and their application in metabolic engineering PhD thesis, Universidade do Minho. 2016.
10. Vasilakou E, Machado D, Theorell A, Rocha I, Nöh K, Oldiges M, Wahl SA. Current state and challenges for dynamic metabolic modeling. Curr Opin Microbiol. 2016;33:97–104.
11. Karr JR, Sanghvi JC, Macklin DN, Gutschow MV, Jacobs JM, Bolival B, Assad-Garcia N, Glass JI, Covert MW. A whole-cell computational model predicts phenotype from genotype. Cell. 2012;150(2):389–401.
12. Bordbar A, McCloskey D, Zielinski DC, Sonnenschein N, Jamshidi N, Palsson BO. Personalized whole-cell kinetic models of metabolism for discovery in genomics and pharmacodynamics. Cell Syst. 2015;1(4):283–92.
13. Karr JR, Takahashi K, Funahashi A. The principles of whole-cell modeling. Curr Opin Microbiol. 2015;27:18–24.
14. Macklin DN, Ruggero NA, Covert MW. The future of whole-cell modeling. Curr Opin Biotech. 2014;28:111–5.
15. Villaverde AF, Banga JR. Reverse engineering and identification in systems biology: strategies, perspectives and challenges. J R Soc Interface. 2014;11(91):20130505.
16. Jaqaman K, Danuser G. Linking data to models: data regression. Mol Cell Biol. 2006;7(11):813–9. doi:10.1038/nrm2030.
17. Banga JR, Balsa-Canto E. Parameter estimation and optimal experimental design. Essays Biochem. 2008;45:195–210.
18. Ashyraliyev M, Fomekong-Nanfack Y, Kaandorp JA, Blom JG. Systems biology: parameter estimation for biochemical models. FEBS J. 2009;276(4):886–902.
19. Chou IC, Voit EO. Recent developments in parameter estimation and structure identification of biochemical and genomic systems. Math Biosci. 2009;219(2):57–83. doi:10.1016/j.mbs.2009.03.002.
20. Gábor A, Banga JR. Robust and efficient parameter estimation in dynamic models of biological systems. BMC Syst Biol. 2015;9(1):74.
21. Egea JA, Martí R, Banga JR. An evolutionary method for complex-process optimization. Comput Oper Res. 2010;37(2):315–24.

22. Dennis JE, Gay DM, Welsch RE. An Adaptive Nonlinear Least-Squares Algorithm. ACM Trans Math Softw. 1981;7(3):348–68.

23. Villaverde AF, Henriques D, Smallbone K, Bongard S, Schmid J, Cicin-Sain D, Crombach A, Saez-Rodriguez J, Mauch K, Balsa-Canto E, Mendes P, Jaeger J, Banga JR. BioPreDyn-bench: a suite of benchmark problems for dynamic modelling in systems biology. BMC Syst Biol. 2015;9(1):8. doi:10.1186/s12918-015-0144-4.

24. Villaverde AF, Barreiro A. Identifiability of large nonlinear biochemical networks. MATCH Commun Math Comput Chem 2016;76(2):259–96.

25. Walter E, Pronzato L. Identification of Parametric Models from Experimental Data. Communications and Control Engineering Series. London, UK: Springer; 1997.

26. Miao H, Xia X, Perelson AS, Wu H. On identifiability of nonlinear ODE models and applications in viral dynamics. SIAM Rev. 2011;53(1):3–39.

27. Chiş O-T, Banga JR, Balsa-Canto E. Structural identifiability of systems biology models: a critical comparison of methods. PLoS One. 2011;6(11): 27755.

28. Villaverde AF, Barreiro A, Papachristodoulou A. Structural identifiability of dynamic systems biology models. PLOS Comput Biol. 2016;12(10): 1005153.

29. Brun R, Reichert P, Künsch HR. Practical identifiability analysis of large environmental simulation models. Water Resour Res. 2001;37(4):1015–30.

30. López D, Barz T, Körkel S, Wozny G. Nonlinear ill-posed problem analysis in model-based parameter estimation and experimental design. Comput Chem Eng. 2015;77:24–42.

31. Ljung L. System Identification: Theory for User. New Jersey: PTR Prentice Hall; 1987. doi:10.1016/0005-1098(89)90019-8.

32. Hawkins DM. The problem of overfitting. J Chem Inform Comput Sci. 2004;44(1):1–12. doi:10.1021/ci0342472.

33. Bauer F, Lukas MA. Comparingparameter choice methods for regularization of ill-posed problems. Math Comp t Simul. 2011;81(9): 1795–841. doi:10.1016/j.matcom.2011.01.016.

34. Schittkowski K, Vol. 77. Numerical Data Fitting in Dynamical Systems: a Practical Introduction with Applications and Software. Dordrecht: Springer; 2002, pp. 1–405.

35. Moles CG, Mendes P, Banga JR. Parameter Estimation in Biochemical Pathways: A Comparison of Global Optimization Methods. Genome Res. 2003;13:2467–74. doi:10.1101/gr.1262503.

36. Chen WW, Niepel M, Sorger PK. Classic and contemporary approaches to modeling biochemical reactions. Genes Dev. 2010;24(17):1861–75.

37. Ljung L, Chen T. Convexity issues in system identification. In: 10th IEEE International Conference on Control and Automation. IEEE; 2013. p. 1–9. doi:10.1109/ICCA.2013.6565206, http://ieeexplore.ieee.org/document/6565206/.

38. Esposito WR, Floudas CA. Global optimization for the parameter estimation of differential-algebraic systems. Ind Eng Chem Res. 2000;39: 1291–310.

39. Papamichail I, Adjiman CS. Global optimization of dynamic systems. Comput Chem Eng. 2004;28:403–15.

40. Singer AB, Taylor JW, Barton PI, Green Jr WH. Global dynamic optimization for parameter estimation in chemical kinetics. J Phys Chem. 2006;110(3):971–6.

41. Chachuat B, Singer AB, Barton PI. Global methods for dynamic optimization and mixed-integer dynamic optimization. Ind Eng Chem Res. 2006;45(25):8373–92.

42. Miró A, Pozo C, Guillén-Gosálbez G, Egea JA, Jiménez L. Deterministic global optimization algorithm based on outer approximation for the parameter estimation of nonlinear dynamic biological systems. BMC Bioinformatics. 2012;13(1):90.

43. Rodriguez-Fernandez M, Mendes P, Banga JR. A hybrid approach for efficient and robust parameter estimation in biochemical pathways. Bio Syst. 2006;83(2–3):248–65. doi:10.1016/j.biosystems.2005.06.016.

44. Sun J, Garibaldi JM, Hodgman C. Parameter estimation using metaheuristics in systems biology: a comprehensive review. Comput Biol Bioinformatics, IEEE/ACM Trans. 2012;9(1):185–202.

45. Saltelli A, Ratto M, Andres T, Campolongo F, Cariboni J, Gatelli D, Saisana M, Tarantola S. Global Sensitivity Analysis: the Primer. Chichester: John Wiley & Sons; 2008. http://eu.wiley.com/WileyCDA/WileyTitle/productCd-0470059974.html.

46. Rodriguez-Fernandez M, Banga JR. Senssb: a software toolbox for the development and sensitivity analysis of systems biology models. Bioinformatics. 2010;26(13):1675–6.

47. Turányi T. Sensitivity analysis of complex kinetic systems. Tools and applications. J Math Chem. 1990;5(3):203–48.

48. Saltelli A, Tarantola S, Campolongo F. Sensitivity analysis as an ingredient of modeling. Statist. Sci. 2000;15(4):377–95.

49. Weijers SR, Vanrolleghem PA. A procedure for selecting best identifiable parameters in calibrating Activated Sludge Model No. 1 to full-scale plant data. Water Sci. Technol. 1997;36(5):69–79. ISSN:0273-1223, http://dx.doi.org/10.1016/S0273-1223(97)00463-0.

50. Li P, Vu QD. Identification of parameter correlations for parameter estimation in dynamic biological models. BMC Syst Biol. 2013;7:91. doi:10.1186/1752-0509-7-91.

51. Li P, Vu QD. A simple method for identifying parameter correlations in partially observed linear dynamic models. BMC Syst Biol. 2015;9(1):92.

52. McLean KA, Wu S, McAuley KB. Mean-squared-error methods for selecting optimal parameter subsets for estimation. Ind Eng Chem Res. 2012;51(17):6105–15.

53. Kravaris C, Hahn J, Chu Y. Advances and selected recent developments in state and parameter estimation. Comput Chem Eng. 2013;51:111–23. doi:10.1016/j.compchemeng.2012.06.001.

54. Chu Y, Hahn J. Parameter Set Selection for Estimation of Nonlinear Dynamic Systems. AIChE J. 2007;53(11):2858–70. doi:10.1002/aic.

55. Mladenović N, Hansen P. Variable neighborhood search. Comput Oper Res. 1997;24(11):1097–100.

56. Egea J, Henriques D, Cokelaer T, Villaverde A, MacNamara A, Danciu DP, Banga J, Saez-Rodriguez J. Meigo: an open-source software suite based on metaheuristics for global optimization in systems biology and bioinformatics. BMC Bioinf. 2014;15:136.

57. Bischof CH, Quintana-Ortí G. Algorithm 782: Codes for Rank-Revealing QR Factorization of Dense Matrices. ACM Trans Math Softw. 1998;24(2):254–7.

58. Nienałtowski K, Włodarczyk M, Lipniacki T, Komorowski M. Clustering reveals limits of parameter identifiability in multi-parameter models of biochemical dynamics. BMC Syst Biol. 2015;9(1):65.

59. Balsa-Canto E, Henriques D, Gábor A, Banga JR. Amigo2, a toolbox for dynamic modeling, optimization and control in systems biology. Bioinformatics. 2016;32(21):3357–9. doi:10.1093/bioinformatics/btw411.

60. Shannon P, Markiel A, Ozier O, Baliga NS, Wang JT, Ramage D, Amin N, Schwikowski B, Ideker T. Cytoscape: a software environment for integrated models of biomolecular interaction networks. Genome Res. 2003;13(11):2498–504.

61. Geier F, Fengos G, Felizzi F, Iber D. Analyzing and Constraining Signaling Networks: Parameter Estimation for the User In: Liu X, Betterton MD, editors. Computational Modeling of Signaling Networks. Methods in Molecular Biology. Totowa, NJ: Humana Press; 2012. p.23–40. doi:10.1007/978-1-61779-833-7. http://www.springerlink.com/index/10.1007/978-1-61779-833-7.

62. Locke JCW, Millar aJ, Turner MS. Modeling genetic networks with noisy and varied experimental data: the circadian clock in Arabidopsis thaliana. J Theor Biol. 2005;234(3):383–93. doi:10.1016/j.jtbi.2004.11.038.

63. Chassagnole C, Noisommit-Rizzi N, Schmid JW, Mauch K, Reuss M. Dynamic modeling of the central carbon metabolism of Escherichia coli. Biotechnol Bioeng. 2002;79(1):53–73. doi:10.1002/bit.10288.

64. Villaverde AF, Bongard S, Mauch K, Müller D, Balsa-Canto E, Schmid J, Banga JR. A consensus approach for estimating the predictive accuracy of dynamic models in biology. Comput Methods Programs Biomed. 2015;119(1):17–28.

15

Reconstruction of the microalga Nannochloropsis salina genome-scale metabolic model with applications to lipid production

Nicolás Loira[1,2]* (iD), Sebastian Mendoza[1,2], María Paz Cortés[1,2,4], Natalia Rojas[2], Dante Travisany[1,2], Alex Di Genova[1,2], Natalia Gajardo[3], Nicole Ehrenfeld[3†] and Alejandro Maass[1,2†]

Abstract

Background: *Nannochloropsis salina* (= Eustigmatophyceae) is a marine microalga which has become a biotechnological target because of its high capacity to produce polyunsaturated fatty acids and triacylglycerols. It has been used as a source of biofuel, pigments and food supplements, like Omega 3. Only some *Nannochloropsis* species have been sequenced, but none of them benefit from a genome-scale metabolic model (GSMM), able to predict its metabolic capabilities.

Results: We present iNS934, the first GSMM for *N. salina*, including 2345 reactions, 934 genes and an exhaustive description of lipid and nitrogen metabolism. iNS934 has a 90% of accuracy when making simple growth/no-growth predictions and has a 15% error rate in predicting growth rates in different experimental conditions. Moreover, iNS934 allowed us to propose 82 different knockout strategies for strain optimization of triacylglycerols.

Conclusions: iNS934 provides a powerful tool for metabolic improvement, allowing predictions and simulations of *N. salina* metabolism under different media and genetic conditions. It also provides a systemic view of *N. salina* metabolism, potentially guiding research and providing context to *-omics* data.

Keywords: Genome-scale Metabolic model, Nannochloropsis salina, TAG, Microalgæ

Background

In the last few years, interest in microalgæ has risen because of their ability to produce a wide range of compounds, such as carotenoids [1–3], lipids [4–6], hydrogen [7, 8], proteins [9, 10] and starch [11]. These algal compounds have numerous relevant applications, from fine natural chemicals to biofuels and food additives. However, it is still a challenge to optimize algal biomass and specific lipid composition to reach an economically feasible bulk production of these compounds [12]. Understanding the complexity of algal metabolism is key to tackling this problem.

Metabolic networks provide an efficient framework to describe cellular metabolism and have become an important tool in metabolic engineering, facilitating strain optimization and reducing the need for expensive in vivo experiments [13]. In addition, metabolic models integrated with *omics* data, such as transcriptional profiling, allows development of a meaningful systemic representation of metabolism [14]. Genome-scale metabolic network models (GSMMs) have been successfully reconstructed for several model species and a few biotechnologically relevant organisms like *Escherichia coli* [15, 16], *Saccharomyces cerevisiae* [17] and *Arabidopsis thaliana* [18].

Several efforts have been made to model algæ metabolism [19]. Green algæ have received special attention,

*Correspondence: nloira@dim.uchile.cl
†Equal contributors
[1]Mathomics, Center for Mathematical Modeling, Universidad de Chile, Beauchef 851, 7th Floor, Santiago, Chile
[2]Center for Genome Regulation (Fondap 15090007), Universidad de Chile, Blanco Encalada 2085, Santiago, Chile
Full list of author information is available at the end of the article

with eight metabolic models for *Chlamydomonas rein-hardtii* [20–27], one for *Botryococcus braunii* [28], five for the genera Chlorella [29–33] and two for *Ostreococcus* [34]. Additionally, seven models for the diatom *Phaeodactylum tricornutum* [35–41], one for the multicellular brown algæ *Ectocarpus siliculosus* [42] and one for the coccolithophore *Emiliania huxleyi* [43]. One important alga that is absent in this list is the marine specie *Nannochloropsis salina*. Nannochloropsis species has emerged as a leading microorganism for biodiesel production, due to their high photoautotrophic biomass accumulation rates [44] and high lipid content [45], either in open ponds or enclosed systems. Additionally, successful cultivation of *Nannochloropsis* species on a large scale using natural sunlight has been achieved by several companies [46]. Moreover, *Nannochloropsis* has gained great interest because of its potential for bio-production of eicosapentaenoic acid (EPA) which is a relevant additive for human health [47–49] and nutrition [50–52]. EPA is one of the major fatty acids produced by *Nannochloropsis*. Indeed, it could represent over 30% of total fatty acid content under heterotrophic conditions [53] and over 28% under autotrophic conditions [54] in *Nannochloropsis*.

We present here iNS934, the first genome-scale functional metabolic model for *N. salina*, built with a strategy that integrates metabolic knowledge from several related species, genomic and transcriptomic data. In particular, we generated transcriptomic data for *N. salina* which allowed us to confirm coding sequences (CDS) in its genome and also discover new ones. iNS934 provides a detailed description of biosynthesis of lipids for the *Nannochloropsis* genus. Specifically, it describes reactions for the biosynthesis of polyunsaturated fatty acids such as EPA, arachidonic acid (ARA) and eicosatetraenoic acid (ETA). iNS934 was validated both qualitative and quantitatively, with an average error of 15% in the latter. Moreover, the model was used to propose knockouts that could improve the production of triacylglycerols (TAGs).

Methods

N. salina transcriptome

Setting up culture conditions for RNA extraction *N. salina* cells were obtained from Commonwealth Scientific and Industrial Research Organization (CSIRO) and identified as CS-190 *Nannochloropsis salina* CCAP 849/2. They were cultured in Artificial Sea Water (ASW), supplemented with f/2 medium [55] at 20 °C, with an illumination of white-blue leds (30 μE photons m-2s-1) on a 24 h light/day cycle, primary in batch cultures, as described by Chen et al. [56]. For mRNA extraction, *N. salina* cells were collected at the exponential growth phase ($\sim 5 \times 10^6$ cell/mL) at the following conditions: (1) Dark, Low CO_2 (DLC): 4 h dark with 1 L/min air influx (CO_2 0.03%), (2) High light, Low CO_2 (HLLC): 2 h high light

(1000μE photons m-2s-1) and 1 L/min air influx (0.03% CO_2) and (3) High light, High CO_2 (HLHC): 2 h high light (1000μE photons m-2s-1) and 1 L/min air influx (1.5% CO_2). Total RNA was extracted from frozen cells, which were ground using a mortar and pestle, using TRIzol RNA Isolation Reagents (Invitrogen) according to the manufacturer. Total RNA and mRNA integrity were analyzed by running them on agarose gel and in an Agilent Bioanalyzer to evaluate its quality before sending it to library construction.

EST collection/library construction and sequencing

To obtain a good coverage of the *N. salina* transcriptome, two different sequencing techniques were adopted: GS FLX+ System (Roche), sequencing a normalized cDNA library, and Illumina sequencing a cDNA library. Regular and normalized library construction and sequencing was performed by Eurofin MWG Operon, USA.

For the Roche GS FLX sequencing, we made a RNA pool, including all 3 conditions previously described (DLC; HLLC; HLHC). To build the normalized cDNA library construction, from a total RNA sample, poly(A)+ RNA was isolated and used for cDNA synthesis. The poly(A)+ was fragmented by ultrasound (1 pulse of 30 s at 4 °C). First-strand cDNA synthesis was primed with a N6 randomized primer. Then 454 adapters A and B were ligated to the 5' and 3' ends of the cDNA. The cDNA was finally amplified with 13 PCR cycles using a proof reading enzyme. Normalization was carried out by one cycle of denaturation and reassociation of the cDNA, resulting in N1-cDNA. Reassociated ds-cDNA was separated from the remaining ss-cDNA (normalized cDNA) by passing the mixture over a hydroxylapatite column. After hydroxylapatite chromatography, the ss-cDNA was amplified with 14 PCR cycles. For Titanium sequencing the cDNA in the size range of 500–700 bp was eluted from a preparative agarose gel. Half a plate of GS FLX+ System (Roche) was sequenced.

Library preparation of total RNA from conditions HLLC and HLHC was carried out using the Illumina TruSeq kit. Cluster formation and sequencing on HiSeq2000 were done according to the manufacturer's instructions. Two samples, one from each condition, were prepared in 250 bp paired-end sequenced in 2 different lanes, delivering around 40 million reads per lane.

***De novo* transcriptome assembly** We divided this process in 4 steps: (1) Illumina raw data was error corrected, and then sequences were assembled using the Trinity package. (2) Roche 454 GS FLX raw reads were cleaned and trimmed with Figaro [57]. (3) Using BLASTN, 454 reads were mapped to Illumina contigs in order to avoid redundancy between corrected Illumina contigs and 454 data. (4) We generated 3 sets of data: The first was constructed using all Illumina transcripts without mapped 454 reads; the second set of data was constructed using 454 reads that had a hit against Illumina contigs and the corresponding Illumina reads. This data was reassembled using the Phrap[1] software. The third dataset corresponds to the transcripts assembled by wgs-assembler, using 454 reads that did not have hits against Illumina transcripts. Transcripts from these three sets of data with less than 30 mapping reads or length under 300 bp were discarded. The remaining transcripts constitute our *de novo* transcriptome.

Mapping transcriptome to reference genome We mapped this *de novo* transcriptome assembly to the *N. salina* CCMP537 genome assembly [58] (NCBI BioProject ID: PRJNA62503) using GMAP. Transcripts aligned to reference gene models with a coverage of at least 70% and and identity of 95% or higher were assigned to those genes. We identified coding regions on the remaining transcripts using TransDecoder[2].

Functional annotation of coding regions from the reference genome and the supplementary transcripts from the *de novo* transcriptome was performed using BLAST searches (with an e-value threshold of 1e-10 and keeping the best ten hits) against Swissprot, KEGG, PRIAM and NR protein databases. Moreover, InterProScan (default parameters) was used to identify protein domains and GO numbers. In order to build a consensus annotation for each gene, the Gene name (Swissprot, KEGG), EC number (KEGG, PRIAM), KO number (KEGG), GO number (Interpro), InterPro number (Interpro) and protein product (Swissprot, KEGG, NR) attributes were obtained from BLAST and Interpro results; using an in-house PERL script; In parenthesis we show the databases from which attributes were obtained. Afterwards, a single value for each attribute was defined by picking the most frequent from the set of 10 best hits times the number of databases the attribute was parsed (i.e EC numbers can be obtained from KEGG and PRIAM results; then, we can count the most frequent from a list of 20 possible values). One exception to the previous rule was the protein product attribute; we chose it prioritizing the databases result in the following decreasing priority order: Swissprot, KEGG and NR.

The complete set of annotated CDS is included as Additional file 1.

Reconstruction of *N. salina* metabolic model

The reconstruction of *N. salina* metabolic network was generated following the five stages described in the protocol for generating high-quality GSMMs [59]. We used as a reference *Chlamydomonas reinhardtii* and its iRC1080 metabolic model [22] in addition to *N. salina* genome

annotation and literature. First, we searched for orthologs between our *N. salina* CDS and the *C. reinhardtii* protein sequences, using Inparanoid [60] and OrthoMCL [61]. Then we built a draft model for *N. salina* using *Pantograph* [62], taking as a template the iRC1080 model for *C. reinhardtii*. Then, we looked at the list of genes not present in the draft model, but which annotation included an Enzyme Commission (EC) number and we mapped those ECs to BiGG reaction identifiers. We imported those BiGG reactions into our model, using the BiGG web API[3]. Afterwards, we manually determined a list of reactions using different resources of information such as BIGG, MetaCyc and KEGG. These reactions were added to our reconstruction with their corresponding identifiers and *N. salina* gene associations. For reactions and metabolites without BiGG identifiers, BiGG-like identifiers were assigned. After adding reactions, we manually curated compartments, changed reversibility for some reactions, moved species among compartments, renamed and pruned unused elements, among other changes. In order to generate a functional model, we used `meneco` [63, 64] to look for BiGG reactions that could fill gaps in the model. `meneco` provided us with candidate reactions that were handed to the manual curators, who approved their inclusion into the model. We took metabolites available in the media as sources for gap-filling and the requirements for biomass as targets.

Formulas and charges for metabolites in our model were revised and all reactions in our model were subjected to mass and charge balances. Except light and exchange reactions, all reactions in our model are mass balanced and only 6% could not be charge balanced. The complete list of metabolites in the model with their formula, charge, BiGG ID and ID from the external database MetaNetX [65] is included in Additional file 2.

We produced a version of our model in Systems Biology Markup Language (SBML) format in order to analyze it with compatible existing tools, and share it with the community. The model can be obtained as Additional file 3 and a diagram summarizing the reconstruction process is depicted in Fig. 1.

Growth experiments for model validation In order to validate our metabolic network, we prepared a battery of simulation tests based on 32 previously published growth experiments (see Table 1). These experiments comprise 23 cases in mixotrophic condition, 2 cases in heterotrophic condition and 1 in autotrophic condition. Additionally, 6 knockout experiments from Killian et al. (2011) were also included in this battery. We complemented this evidence by conducting experiments that could help us to improve and validate our model. We analyzed growth of *N. salina* in the presence of different nitrogen sources, phosphate and glucose. For experimental cultures we considered that

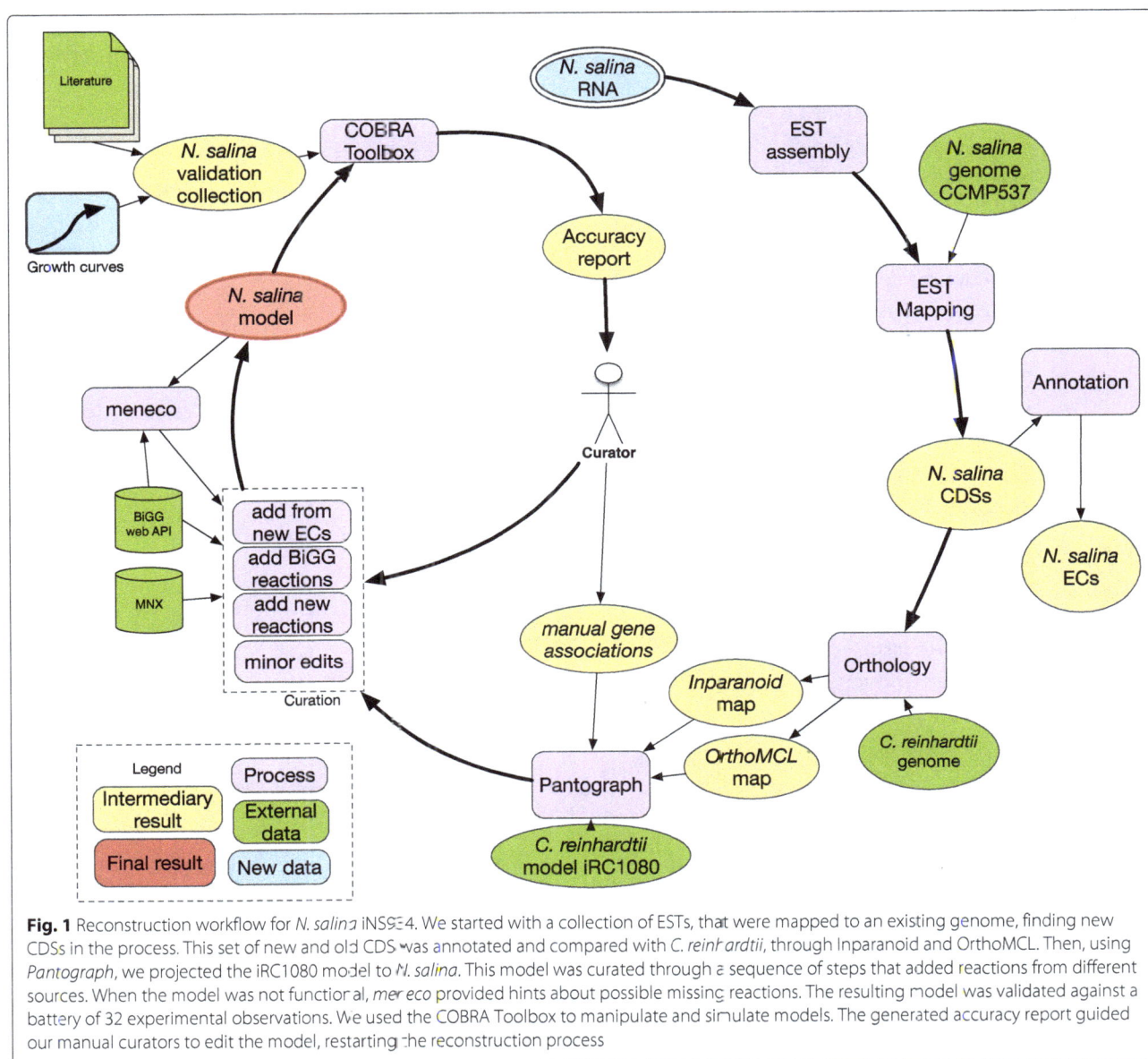

Fig. 1 Reconstruction workflow for *N. salina* iNS934. We started with a collection of ESTs, that were mapped to an existing genome, finding new CDSs in the process. This set of new and old CDS was annotated and compared with *C. reinhardtii*, through Inparanoid and OrthoMCL. Then, using *Pantograph*, we projected the iRC1080 model to *N. salina*. This model was curated through a sequence of steps that added reactions from different sources. When the model was not functional, *meneco* provided hints about possible missing reactions. The resulting model was validated against a battery of 32 experimental observations. We used the COBRA Toolbox to manipulate and simulate models. The generated accuracy report guided our manual curators to edit the model, restarting the reconstruction process

growth was achieved if growth rate was greater than 1/3 of maximum growth among all experiments studied [66].

For growth experiments conducted in our laboratory we used *N. salina* obtained from CSIRO (Commonwealth Scientific and Industrial Research Organisation). We used the complete f/2 medium as described by Guillard [55], which contained 75 mg/L of nitrate and 4.41 mg/L of phosphate. Experiments containing urea, nitrate and ammonium in the medium as the only nitrogen source were designed to have the same molar concentrations of nitrogen. Thus, we used a molar concentration of 8.8×10^{-4} for ammonium and nitrate and a molar concentration of 4.4×10^{-4} for urea. Before inoculation, cells were washed with ASW. We inoculated 1 L flasks with

2×10^6 cells/mL, which were maintained at 25 °C, aireation of 0.5 VVM and to 60 μmol m^{-2}s^{-1} of light intensity. For batch cultures growing with different carbon dioxide levels, it was supplied through the gas inlet in concentrations of 2 and 5%, respectively.

In order to diminish the internal reservoir of phosphate in cells for phosphate evaluation, inocula cells came from a culture with diminished phosphate concentration (0.4 mg/L phosphate). All cultures were followed by cell counting (Newbauer chamber and OD750 nm), biomass estimation (dry weight) and in some cultures nitrate consumption, evaluated using a microplate technique as described [67]. All experiments were performed in duplicate.

Table 1 Qualitative validation of iNS934

	Media condition	in vivo	in silico	Gene KO	TFPN	Reference	*Nannochloropsis* species
Luminosity	4 μE (flourecent cool)	-	+		TN	[54]	sp
	40 μE (flourecent cool)	-	+		FP	[54]	sp
	480 μE (flourecent cool)	+	+		TP	[54]	sp
	5 μE (flourecent warm)	+	+		TP	[91]	gaditana
	15 μE (flourecent warm)	+	+		TP	[91]	gaditana
	50 μE (flourecent warm)	+	+		TP	[91]	gaditana
	100 μE (flourecent warm)	+	+		TP	[91]	gaditana
	200 μE (flourecent warm)	+	+		TP	[91]	gaditana
	450 μE (white led)	+	+		TP	[91]	gaditana
	1200 μE (white led)	+	+		TP	[91]	gaditana
	2100 μE (white led)	+	+		TP	[91]	gaditana
	3000 μE (white led)	-	+		FP	[86]	sp
	Red led 673 nm, mixotrophic	+	+		TP	[92]	sp
	Red led 673 nm, autotrophic	+	+		TP	[92]	sp
Carbon	Glucose (mixotrophic)	+	+		TP	[93]	sp
	Glucose (heterotrophic)	+	+		TP	[93]	sp
	Ethanol (heterotrophic)	+	+		TP	[93]	sp
	Ethanol (mixotrophic)	+	+		TP	[93]	sp
	Inorganic carbon sources	+	+		FN	[94]	gaditana
Other	Sodium	+	+		TP	[95]	oculata
	Nitrite	+	+		TP	[96]	sp
	Phosphate	-	-		TP	This work	salina
	Nitrate	+	+		TP	[97]	This work
	Ammonium	+	+		TP	[98]	This work
	Sulfate	+	+		TP	[87]	gaditana
	Urea levels	+	+		TP	[99]	This work
Knockout	Ammonium	+	+	Nitrate reductase	TP	[96]	sp
	Nitrite	+	+	Nitrate reductase	TP	[96]	sp
	Nitrate	-	-	Nitrate reductase	TN	[96]	sp
	Ammonium	+	+	Nitrite reductase	TP	[96]	sp
	Nitrite	-	-	Nitrite reductase	TN	[96]	sp
	Nitrate	-	-	Nitrite reductase	TN	[96]	sp

We compared 32 experiments of growth of *N. salina* under different media conditions, with predictions using our iNS934 model. Four of these conditions were also conducted in our lab. From the 32 experiments, for growth/non-growth comparisons we obtained: TN: 4, FN: 0, TP: 25, FP: 3, with a sensitivity of 1, specificity of 0.57 and an accuracy (geometric mean) of 0.90

Model analysis and validation

To analyze our iNS934 model, we used Flux Balance Analysis (FBA), Flux Variability Analysis (FVA) and dynamic Flux Balance Analysis (dFBA) from the Cobra Toolbox in MATLAB [68]. Flux Coupling Analysis (FCA) was performed using the F2C2 tool [69].

For model validation, we performed FBA with growth rate as the objective function to predict growth in different growth conditions mentioned in the previous section. For each test, we modified the flux boundaries of exchange

reactions to simulate the composition of each growth media. We performed both a qualitative and a quantitative validation. Qualitative validation was used in cases where growth and/or uptake rates could not be obtained from the experimental data gathered from the literature and we only had growth/non-growth data. In these cases, we used an in silico scenario that simulates a rich medium by allowing free uptake of nutrients. The growth rate obtained in this condition was considered the reference maximal growth rate. For each experiment, we established that growth was achieved if the obtained growth rate was greater than 1/3 of the reference value [66, 70]. Then, we generated a confusion matrix and used the geometric mean as a measure of accuracy (See Table 1) to assess the predictive power of iNS934. To have a more accurate evaluation of our model predictions we performed quantitative validations by contrasting predicted growth rates with those obtained in in-house experiments. Prediction of growth rates was conducted using data from two sets of experiments. First, we used data from three batch cultures of *N. salina*: autotrophic growth with nitrate, ammonium and urea. Second, we used data from three batch cultures grown with nitrate at 0.03%, 2% and 5% of CO_2 in the inflow gas. An in silico growth rate was obtained for each FBA simulation using fixed experimental uptake rates of nutrients and setting all other uptake reactions to zero. The error between experimental and predicted growth rates was then calculated.

We estimated uptake of CO_2 by using the formula described previously [71]:

$$CO_2 biofixation = C * P * (MW_{CO_2}/MW_C). \quad (1)$$

Values of biofixation were further transformed to specific uptake rates by dividing them by cell concentrations and the molecular weight of CO_2. The specific uptake rate of CO_2 used to perform simulations was calculated as the average specific uptake rate in exponential phase.

In silico strain optimization

In order to find mutants of *N. salina* which may be useful for lipid overproduction, we developed a method for in silico strain optimization based on reaction knockouts. Our method guarantees the production of a target metabolism, while conserving the functional property, that is, the production of biomass. We iterated 200 times and sorted the resulting knockout sets. The metric to sort the results was defined as $m = \frac{\mu \times r_t}{|k|}$, where μ is the specific growth rate, r_t is the specific production rate of the target t and k is the knockout set.

This method consists of four steps: (1) We randomly traverse the reactions in the iNS934 model removing reactions that are not needed for the production of the target metabolite. (2) Starting again from iNS934, we randomly traverse the model removing all reactions that are not needed for biomass production, while keeping those conserved in step 1. (3) Using FBA we check if the resulting model is able to produce the target metabolite when optimizing growth rate. If so, we continue to step 4. Otherwise we restart from step 1. (4) We try to recover the reactions removed during step 1 one at a time, checking that their inclusion maintains the integrity of the model to produce

biomass and the target metabolite simultaneously. When a reaction breaks this restriction, it is included in the list of reactions to knockout.

Results and discussion

N. salina transcriptome mapping and annotation

We sequenced and assembled transcripts for *N. salina* and then combined them with the reported draft CCMP527 genome [58] to produce a comprehensive gene set for *N. salina*. From this set, 10913 putative genes identified in the transcripts of our *de novo* assembly were not contained in the reference genome (Additional file 5). These, together with the original genes in the genome, makes a total of 17519 putative genes. However, after the annotation, 7205 of these putative genes were not assigned a functional annotation and were therefore not considered in the metabolic reconstruction process. Out of the remaining genes, 3577 were assigned an EC number. In our metabolic model, 490 reactions were associated with putative genes from the transcriptome, which complemented the 1452 reactions predicted using only the CCMP527 genome. The contribution by subsystem of the iNS934 model is depicted in Fig. 2.

iNS934: A genome-scale metabolic model for *Nannochloropsis*

We generated a functional GSMM, able to produce biomass, for the alga *N. salina*, called iNS934. The construction of our model started with an initial draft using reference iRC1080, a genome-scale model of *C. reinhardtii*. Even though models for other algæ have been built, as shown in Table 2, the level of detail varies among them. Only some of them are GSMMs including a genome-scale metabolic network reconstruction with more than a thousand reactions and a mathematical representation suitable for constraint based analysis. iRC1080 stands among them because it is a high quality model whose features include a carefully detailed light-driven algal metabolism, a multi-compartmentalized network and an extensive metabolism of lipids. Moreover, iRC1080 is part of the BIGG database and consequently, offers a controlled vocabulary of reactions and metabolites making it suitable for building GSMMs of standard language. Indeed, these features have been exploited before to construct other models of alga. For example, it was used as the template to build the models of *Chlorella vulgaris* [29], *Chrorella variabilis* [30] and *Phaeodactylum tricornutum* [35], and it was used in the gap-filling process for the model of *Emiliania huxleyi* [43].

An orthology analysis revealed *C. reinhardtii* and *N. salina* shared 2612 orthologs, an amount that allowed us to obtain a reasonable initial draft to begin our reconstruction. This initial draft was improved and tailored to *Nannochloropsis* specific features using the annotation

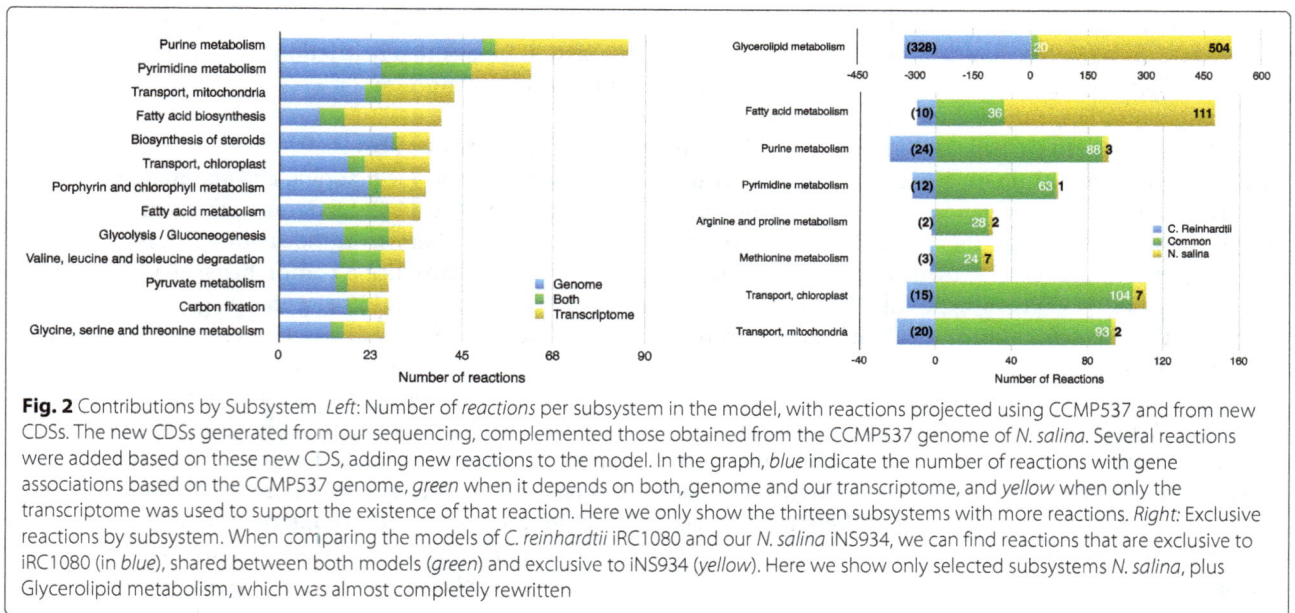

Fig. 2 Contributions by Subsystem *Left*: Number of *reactions* per subsystem in the model, with reactions projected using CCMP537 and from new CDSs. The new CDSs generated from our sequencing, complemented those obtained from the CCMP537 genome of *N. salina*. Several reactions were added based on these new CDS, adding new reactions to the model. In the graph, *blue* indicate the number of reactions with gene associations based on the CCMP537 genome, *green* when it depends on both, genome and our transcriptome, and *yellow* when only the transcriptome was used to support the existence of that reaction. Here we only show the thirteen subsystems with more reactions. *Right*: Exclusive reactions by subsystem. When comparing the models of *C. reinhardtii* iRC1080 and our *N. salina* iNS934, we can find reactions that are exclusive to iRC1080 (in *blue*), shared between both models (*green*) and exclusive to iNS934 (*yellow*). Here we show only selected subsystems *N. salina*, plus Glycerolipid metabolism, which was almost completely rewritten

of the *Nannochloropsis* genome and transcriptome, and manual curation as described in materials and methods. An example of the contributions of manual curation to several subsystems can be seen in Fig. 2.

iNS934 describes 2345 reactions encoded by 934 genes, the 1985 metabolites consumed and produced by those reactions and a biomass function which describes the metabolic requirements for growth in autotrophic and mixotrophic conditions (see Table 3). From the total of reactions, 398 are transport reactions, 95 are exchanges with the media, 1613 are enzymatic reactions with a gene association and 239 without.

The model includes 10 different compartments: extracellular media, cytoplasm, mitochondria, chloroplast, thylakoid lumen, endoplasmic reticulum, peroxisome, Golgi apparatus, lysosome, and nucleus.

For the biomass function of our model, we started with the biomass functions of iRC1080. We adjusted these equations in order to represent the proportions of macronutrients found in *Nannochloropsis* species. In particular, we changed the coefficients related to DNA and RNA production according to previously described methodology [72]. The contribution of glycerolipids to biomass (Additional file 4) was obtained from a study carried out in *Nannochloropsis oceanica* by Li et al. [73].

It is well documented in several microalgæ that an increased lipid accumulation occurs under conditions where there is nitrogen starvation. In *Nannochloropsis* sp. growing in this condition, lipid production increases about one fold [74]. To simulate this behavior, we created a second biomass equation containing new stoichiometric coefficients based on proportions of glycerolipids of

N. oceanica growing in a nitrogen-depleted condition (Additional file 4) [73].

In the following sections, we describe the main metabolic features and particularities of iNS934 with emphasis on lipid and nitrogen metabolism, key processes involved in production of targets with relevance in biodiesel and nutraceutical industries.

Lipids The lipid metabolism in microalgæ is biotechnologically relevant, given that it is key for the production of biodiesel and food additives. Therefore the inclusion of an accurate and species-specific description of lipid pathways into a GSMM is needed if we want to use it as a platform to guide the production of these biotechnological targets. In order to describe the lipid metabolism of *Nannochloropsis*, we first added to iNS934 the biosynthesis pathways of polyunsaturated fatty acids (PUFAs) such as ETA(20:4), ARA(20:4) and EPA(20:5), which were absent in the *C. reinhardtii* model iRC1080 [73, 75]. This was a key step because PUFAs are relevant building blocks for glycerolipids in *Nannochloropsis*. Therefore their inclusion represents a major advance towards in silico simulation of lipid production in this algæ. Additionally, we also added unsaturated fatty acids, such as tetradecenoic acid (14:1) and hexadecadienoic acid (16:2), among others.

Once we added the pathways for all the required fatty acids, we replaced the glycerolipid pathways from the initial draft with 503 new *Nannochloropsis*-specific reactions that define the bioynthesis of TAG, diacylglycerol (DAG), phosphatidylcholine (PC), monogalactosyldiacylglycerol (MGDG), digalactosyldiacylglycerol

Table 2 Metabolic reconstructions of algæ. List of metabolic reconstructions for algæ species

ID	Species	Genome-scale	Detailed glycero-lipid metabolism	Multi-compartment	Availability	in BiGG database	Reference
-	C. prototpecoides	No	No	Yes	SBML	No	[31]
iCS843	C. vulgaris	Yes	Yes	Yes	SBML	No	[29]
iAJ526	C. variabilis	Yes	No	Yes	SBML	No	[30]
-	C. pyrenoidosa	No	No	No	No mathematical model	No	[33]
-	C. sp FC2 IITG	No	No	No	not available	No	[32]
EctoGEM	E. siliculosus	Yes	No	Yes	SBML	No	[42]
iEH410	E. huxleyi	Yes	No	Yes	SBML, MAT	No	[43]
-	C. reinhardtii	Yes	No	No	not available	No	[26]
iRC1080	C. reinhardtii	Yes	Yes	Yes	SBML, MAT	Yes	[22]
ChlamyCyc	C. reinhardtii	Yes	No	No	online	No	[27]
-	C. reinhardtii	No	No	Yes	XLS	No	[25]
-	C. reinhardtii	No	No	Yes	not available	No	[24]
iBD1106	C. reinhardtii	Yes	Yes	Yes	SBML	No	[21]
iCre1355	C. reinhardtii	Yes	Yes	Yes	SBML	No	[20]
AlgaGEM	C. reinhardtii	Yes	No	Yes	SBML	No	[23]
iLB1027	P. tricornutum	Yes	Yes	Yes	SBML	No	[35]
-	P. tricornutum	No	No	Yes	No mathematical model	No	[39]
DiatomCyc	P. tricornutum	Yes	No	No	Online	No	[40]
-	P. tricornutum	No	No	Yes	SBML	No	[41]
-	P. tricornutum	Yes	No	Yes	SBML	No	[38]
-	P. tricornutum	No	No	No	not available	No	[36]
-	P. tricornutum	No	Yes	No	No mathematical model	No	[37]
-	O. lucimarinus	Yes	No	No	SBML	No	[34]
-	O. tauri	Yes	No	No	SBML	No	[34]
-	B. braunii	No	No	No	No mathematical model	No	[28]

(DGDG), sulfoquinovosyl diacylglycerol (SQDG), phosphatidylglycerol (PG), phosphatidylethanolamine (PE), phosphatidylinositol (PI) diacylglyceryl-O-4'-(N, N, N,-trimethyl) homoserine (DGTS) and free fatty acids. In particular, these 503 reactions account for specific proportions of glycerolipids into the biomass of *Nannochloropsis* and specific types of glycerolipids of *Nannochloropsis*

Table 3 Properties of genome-scale metabolic models

	C. reinhardtii (iRC1080)	N. salina (draft)	N. salina (curated)
Genes	1,1146	802	934
Reactions	2,191	1,897	2345
Metabolites	1,706	1,706	1985
Compartments	10	10	10

This table shows the properties of the template model (*C. reinhardtii*), the automatic initial draft produced by *Pantograph* and the results of the manual curation for *N. salina* iNS934

with respect to other algæ. These glycerolipids could represent an important percentage of the *Nannochloropsis* biomass [75]. Given its importance, we created reactions to synthesize each of these glycerolipids and the corresponding stoichiometric coefficients required to account for the proportions in *Nannochloropsis*. The new reactions enable the use of iNS934 as a predictor of lipid metabolism in *Nannochloropsis*.

Furthermore, based on current knowledge of species of chromista [76], we associated the biosynthesis of each glycerolipid to specific compartments, relocating associated metabolites and reactions. We added a compartment for the endoplasmic reticulum, where several glycerolipids are synthesized from fatty acids. Then, the biosynthesis pathways of PC, PE, DGTS and PI were located at the endoplasmic reticulum [77], while pathways for PG, DGDG, MGDG, SQDG were located at the chloroplast. Biosynthesis of TAG was located at both compartments. See Fig. 3 for a schema of our reconstructed lipid subsystem.

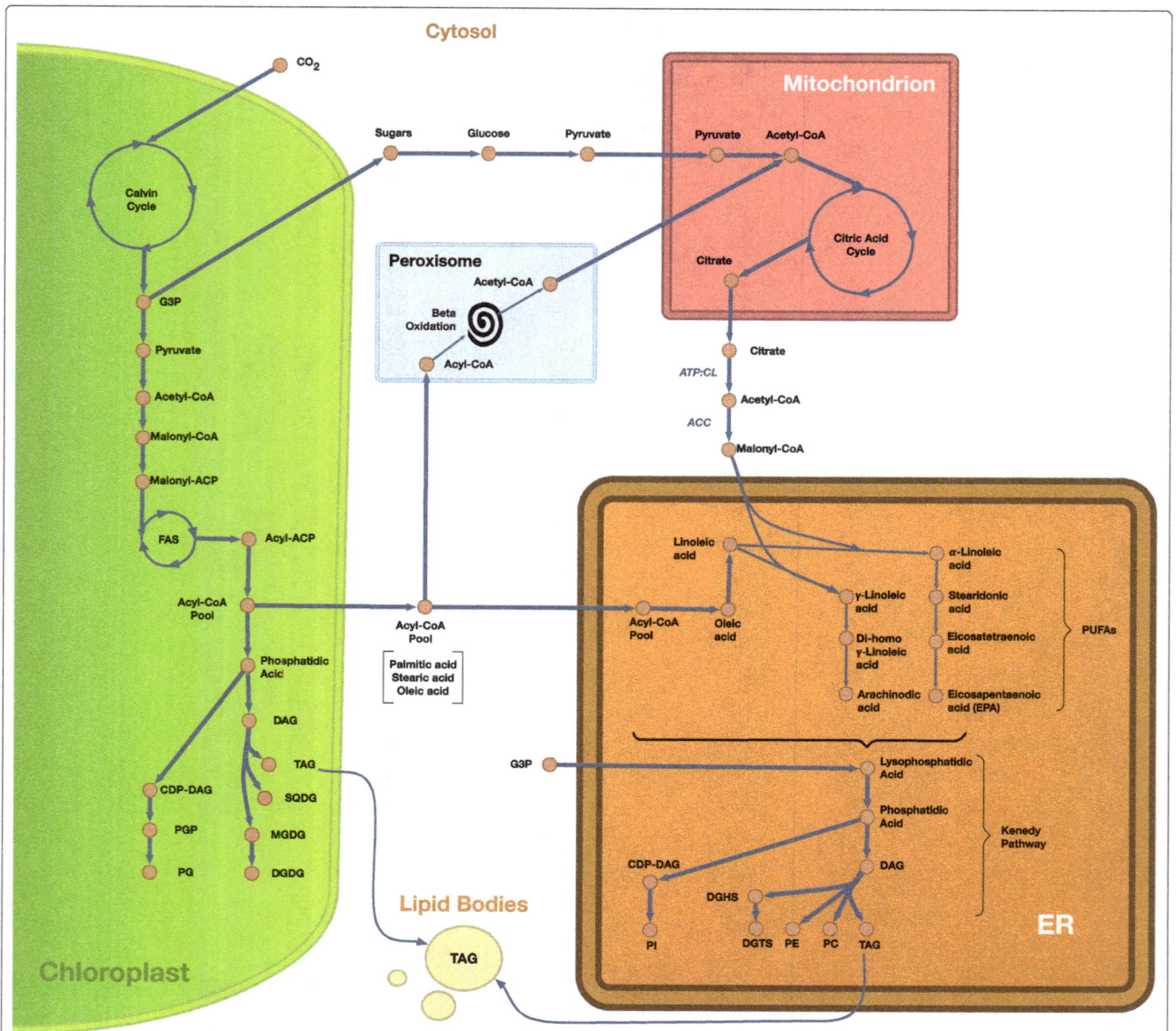

Fig. 3 Diagram of lipid production in *N. salina*, as modeled in iNS934. Starting from CO_2 we follow the chain of reactions until the production of lipids. The biosynthesis of lipids is performed at the Chloroplast and Endoplasmic Reticulum (ER) compartments. The Peroxisome contributes with the splitting of Acyl-CoAs into Acetyl-CoA, while the Mitochondria produce the citrate required Malonyl-CoA, key for the production of PUFAs at the ER. Most of the produced lipids end up in membranes (via the Biomass function), while TAG is also stored in lipid bodies. Inspired by diagrams from [46, 100] and [76]

Urea cycle Nitrogen is quantitatively the most important nutrient affecting growth and lipid accumulation in various algæ [6]. In order to accurately represent the nitrogen metabolism in iNS934, we examined the reactions involved in this process. We found that *N. salina* has a complete urea cycle, including ornithine carbamoyltransferase, argininosuccinate synthase, argininosuccinate lyase, and arginase (Fig. 4). Additionally, we added transporters for nitrate, nitrite, ammonium and urea that

have been identified previously in a *Nannochloropsis* genome [78]. Moreover we have determined experimentally that growth in *N. salina* can be sustained on nitrate, ammonium or urea as sole nitrogen sources (Additional file 5).

Overview of general properties of N. salina GSMM
We performed three analyses in order to find the main network topological features describing the GSMM of *N.*

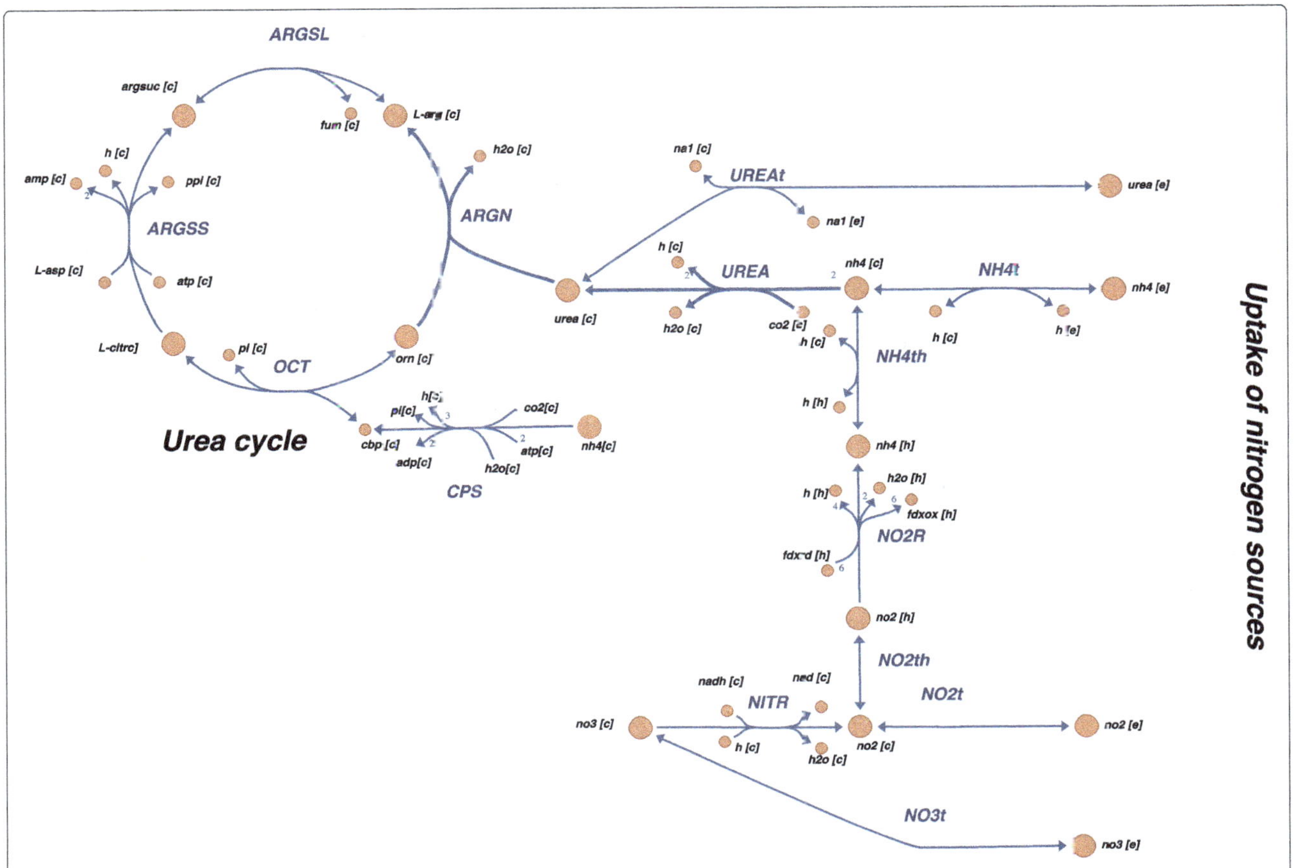

Fig. 4 Nitrogen metabolism: Main reactions and pathways present in iNS934. ARGSL: Argininosuccinate lyase; ARGN:Arginase; ARGSS: Argininosuccinate synthase; CPS:Carbamoyl-phosphate synthase; OCT: Ornithine carbamoyltransferase; UREA: Urease; NO2R: Ammonia:ferredoxin oxidoreductase; NITR: Nitrate reductase; UREAt, N-4t, NO2t, NO3t : Urea, ammonia, nitrite and nitrate transport respectively

salina iNS934. We started performing a FVA by maximizing and minimizing each reaction of the network. We compared this result with the ones obtained for the GSMMs of three other algæ, namely *C. reinhardtii* [22], *C. vulgaris* [29] and *P. tricornutum* [35]. We observed that curve shapes are similar between all networks. Moreover, we found that the amount of reactions having a narrow flux variation is similar: 174, 158, 150 and 379 reactions have a range of 0.01 mmol/gDW hr in *N. salina*, *C. reinhardtii*, *P. tricornutum* and *C. vulgaris* respectively. Additionally, we determined the number of blocked reactions in the network. We found that 38.7% of the reactions in *N. salina* were blocked, while this value decrease to 28.5% for *C. reinhardtii*, 11.9% for *P. tricornutum* and 38.2% for *C. vulgaris*. This result suggests that further refinement of the network is needed in order to accomplish a higher connection and dead-end elimination.

Second, we performed a FCA to determine coupled, partially coupled, directionally coupled and uncoupled reactions. We found that our model has almost the same proportions of types of reactions found in the compared

algæ models. The uncoupled reactions far outnumber other types of reactions, representing nearly 90% of the total reaction pairs. Despite this, a higher percentage of fully and directionally coupled reactions can be observed in *N. salina* compared to the *C. reinhardtii* network, suggesting that the GSMM of *N. salina* is less connected.

Finally, we performed a connectivity and metabolite participation analysis. As is observed in other metabolic networks, we observed that a few metabolites participate in several reactions, meanwhile most metabolites participate in one or two reactions. As an example, we found usual currency metabolites such as H+, ATP, H$_2$O, phosphate and coenzyme A between the 10 most connected metabolites.

Validation of N. salina GSMM

To assess the predictive power of our iNS934 metabolic model, we simulated growth under different conditions using FBA and compared our results to experimental data. We reproduced existing observations of *Nannochloropsis* growth from the literature and from our experiments.

Considering the lack of experimental evidence for *N. salina*, we based part of this validation on experiments from other *Nannochloropsis* species. We complemented this evidence by performing experiments that could help us improve and validate our model.

We modeled in silico autotrophic, mixotrophic and heterotrophic media conditions by setting model constraints according to literature. In particular, for the autotrophic condition, we allowed the model to fix carbon from solar light measured from Earth's surface. Carbon fixation through other types of light were set to zero. A maximum oxygen consumption uptake of 10 mmol/gDW was allowed. Water was allowed to travel only from the chloroplast to the cytosol and not viceversa. No carbon sources were allowed to enter the cell. Additionally, the chloroplastic enzymes divinylprotochlorophyllide vinylreductase, phosphofructokinase, glucose-6-phosphate-1-dehydrogenase and fructose-bisphosphate aldolase were turned off since these enzymes are inactivated with light [79, 80]. To model a mixotrophic condition, we applied the same restrictions except for carbon sources which were allowed to enter the cell. To simulate a heterotrophic condition, acetate, carbon dioxide and oxygen were allowed to enter the cell. Also a dark condition was modeled by inactivating light associated reactions.

Data on *Nannochloropsis* growth under different conditions gathered from the literature was not sufficient to calculate experimental growth and/or uptake rates in all cases, so we used these data for a qualitative validation, where only growth/non-growth prediction accuracy was evaluated. Qualitative assessment of metabolic reconstructions has been previously used to evaluate the biological capabilities of microorganisms on different experimental conditions. In particular, it has been widely used to study the consequences of environmental and genetic parameters that can be experimentally changed [81]. For example, this approach has been used to assess the prediction of nutrient requirements in different strains of *E. coli* [82] as well as other microorganisms [83] and to assess prediction of gene deletions in *S. cerevisiae* [84, 85]. Therefore, we carried out this simple analysis to explore a broader spectrum of scenarios for *Nannochloropsis* growth.

For a quantitative assessment of our model predictions we performed *N. salina* growth experiments using different nitrogen sources and different levels of CO_2 and compared the obtained growth rates with those estimated in silico [85].

Qualitative validation

Table 1 shows the experimental conditions considered for qualitative validation. In all cases, biomass production both experimental and in silico were simplified into binary values (growth/no growth). Corresponding binary results

obtained for all experiments were paired with simulations, with exact agreement in 29 cases (24 true positives and 5 true negatives). Three false positives were observed: *N. salina* growing at 3000 µE m^{-2} s^-1 , 4 µE and 40 µE. In the first case an inhibition of growth was expected [86]. Unfortunately, GSMMs are not yet able to simulate inhibitions. Therefore, this behavior could not be reproduced. In the second and third case, it was expected that growth would be severely affected. However, growth rate was not decreased when compared with cultures grown under control conditions. This result is likely the product of over-optimistic flux simulations and can be reduced through parameter tuning.

The model predicted that phosphate was essential for growth. However, it has been shown that it is not needed to sustain growth experimentally [87]. Based on the model's predictions, we decided to repeat experimentally the result reported in the literature. When we eliminated phosphate from the culture media we obtained growth in the first subculture. However, when we took cells from this first subculture without phosphate, new subcultures did not grow in a media without phosphate, but did grow in a media with phosphate. We presume that *N. salina* cells may accumulate phosphate granules as reservoirs which allowed them to grow in the first subculture and therefore the result reported by Forjan et al. [87] gave a false negative. In light of these new results, we concluded that phosphate was essential to sustain growth of *N. salina*. Therefore the prediction of iNS934 was considered accurate and was classified as a true negative result.

Overall, these growth/no growth simplified comparisons resulted in a prediction accuracy of 0.90 for the 32 evaluated conditions.

Quantitative validation

We assessed iNS934 quantitative predictive power by simulating *N. salina* growth in experimentally tested conditions. These conditions included growth on different nitrogen sources (nitrate, ammonium and urea) and different CO_2 concentrations (0.03%, 2% and 5%) in the gas inlet. Measured uptake rates of nitrogen sources as well as estimated uptake rates of CO_2 used as constraints in the model can be found in Additional file 6. Using these constraints we obtained an average error of 15% in prediction of growth rates (Table 4). This result indicates that iNS934 has a good level of accuracy since, in general, models are considered accurate when they achieve relative errors in growth rate predictions close to 10% [85, 88]. Further refinement of biomass composition as well as experimental measurement of CO_2 uptake rate could improve growth rate predictions.

We also analyzed the inter-compartment fluxes of experimental conditions with different nitrogen sources in order to understand the main metabolic mechanisms

Table 4 Experimental and predicted growth rates of iNS934 Experimental and predicted growth rates for *N. salina* batch cultures growing in six different conditions

CO_2	Nitrogen source	Experimental μ	Predicted μ	Error (%)
0.03%	Nitrate	0.0207	0.0169	22.1%
0.03%	Ammonium	0.0206	0.0156	32.2%
0.03%	Urea	0.0109	0.0098	11.1%
0.03%	Nitrate	0.0203	0.0186	9.3%
2%	Nitrate	0.0183	0.0207	11.8%
5%	Nitrate	0.0185	0.0178	4.2%

Nitrate, ammonium and urea were used independently as nitrogen sources and CO_2 was used as the inorganic carbon source for each batch culture. Air (0.03% of CO_2) and CO_2 enriched air (2% and 5% of CO_2) were used independently in the gas inlet

that are involved in each case. In the first condition, the cell consumes nitrate as the nitrogen source. This is transformed to nitrite by the nitrate redutase in the cytosol. The nitrite is then transported to the chloroplast where it is further transformed to ammonium by the nitrite reductase. This ammonium is used to build some building blocks such as L-serine by the threonine ammonia-lyase (EC: 4.3.1.19). For carbon fixation, carbon dioxide entered the cytosol and was transported to the chloroplast where it participated in the Calvin Cycle. The malonyl-CoA used to build fatty acids is also created from carbon dioxide. Fatty acids such as decanoic acid and PUFAs such as eicosanopentanoic acid were synthesized in the chloroplast. PUFAs leave the chloroplast and travel to the endoplasmic reticulum to synthesize glycerolipids.

In the second condition, *N. salina* consumes ammonium which is transformed in the cytosol to urea (EC: 3.5.1.5) and amino acids such as L-glutamine, L-threonine and glycine. Urea enters the urea cycle and is further transformed to L-arginine. To simulate carbon fluxes, acetyl-CoA, which is synthesized in the cytosol, is transported to the mitochondria in order to generate energy and reducing power through the tricarboxylic acid cycle.

In the third condition *N. salina* consumes urea. The urea is transformed to ammonium by two consecutive reactions (EC: 6.3.4.6 and 3.5.1.54) in the cytosol. Additionally, ammonium is transported to the chloroplast where it is also used to synthesize glutamine and L-serine. The same mechanisms of carbon fixation and biosynthesis of fatty acids observed in experimental condition one was observed for cases where *N. salina* consumed ammonium or urea.

Applications

Simulation of lipid production in nitrogen starvation

It has been shown that lipid content in *Nannochloropsis* changes when cells growing on a nitrogen replete media are transferred to a nitrogen-depleted condition. In particular, the content of TAG increases at least 100 fold [73].

We wanted to test if the iNS934 could predict this behavior at least qualitatively. To do this, using dFBA, we simulated cells growing in a batch culture that faced a sudden change in nitrogen availability. In this simulation we defined three stages. The first stage represented cells growing in a medium with a high availability of nitrate, used as the only nitrogen source. For this purpose we used the biomass equation in a nitrogen-replete condition and we set a maximum value for growth of 0.0045 h^{-1} according to Simionato et al. [75]. At the end of the first stage, we simulated that cells were inoculated into a nitrogen-free media. This represented the beginning of stage 2. In this stage we changed the biomass equation to the one a for nitrogen-depleted condition and we set a maximum value for growth of 0.0036 h^{-1} [75]. At the end of stage 2, cells were once again inoculated in a nitrogen-rich media. In stage three we again used the biomass equation for a nitrogen-replete condition.

As shown in Fig. 5, in the first stage of our simulation, *N. salina* consumed nitrate and generated biomass according to the biomass equation for a nitrogen-replete condition. The glycerolipids were consequently increased as biomass increased. At the second stage, the lipid production increased significantly with respect to stage one. In the third stage, the growth rate and the lipid production were the same as stage one. This preliminary simulation showed that iNS934 is able to accurately describe the behavior in both nitrogen replete and nitrogen depleted conditions. Therefore this a feature that could be further exploited in biotechnological applications related to lipid optimization.

Using iNS934 to guide metabolic engineering of N. salina

As an example for iNS934 use, we focused on TAG production optimization as a case study. We used our model to search for sets of reactions whose blockage resulted in a higher *in silico* rate of TAG production in *N. salina*. This is a powerful and low-cost tool to predict the behavior of *N. salina* in different genetic and media contexts, working as a guide for metabolic engineering efforts. In order to find those in silico mutants, we initially used OptKnock [89] to obtain reactions whose group elimination allowed greater TAG production. However, we performed a FVA which revealed that removal of none of the reaction sets predicted by this tool guaranteed a minimum production of the desired lipids. We also tried OptForce [90], but we could not find reactions whose FVA indicated non-overlapping ranges of values. This is probably caused by the lack of experimental values that could constrain the possible fluxes of reactions in our model. Therefore, we developed an *in house* algorithm to determine sets of reactions that guarantee a minimum desired production, while keeping the

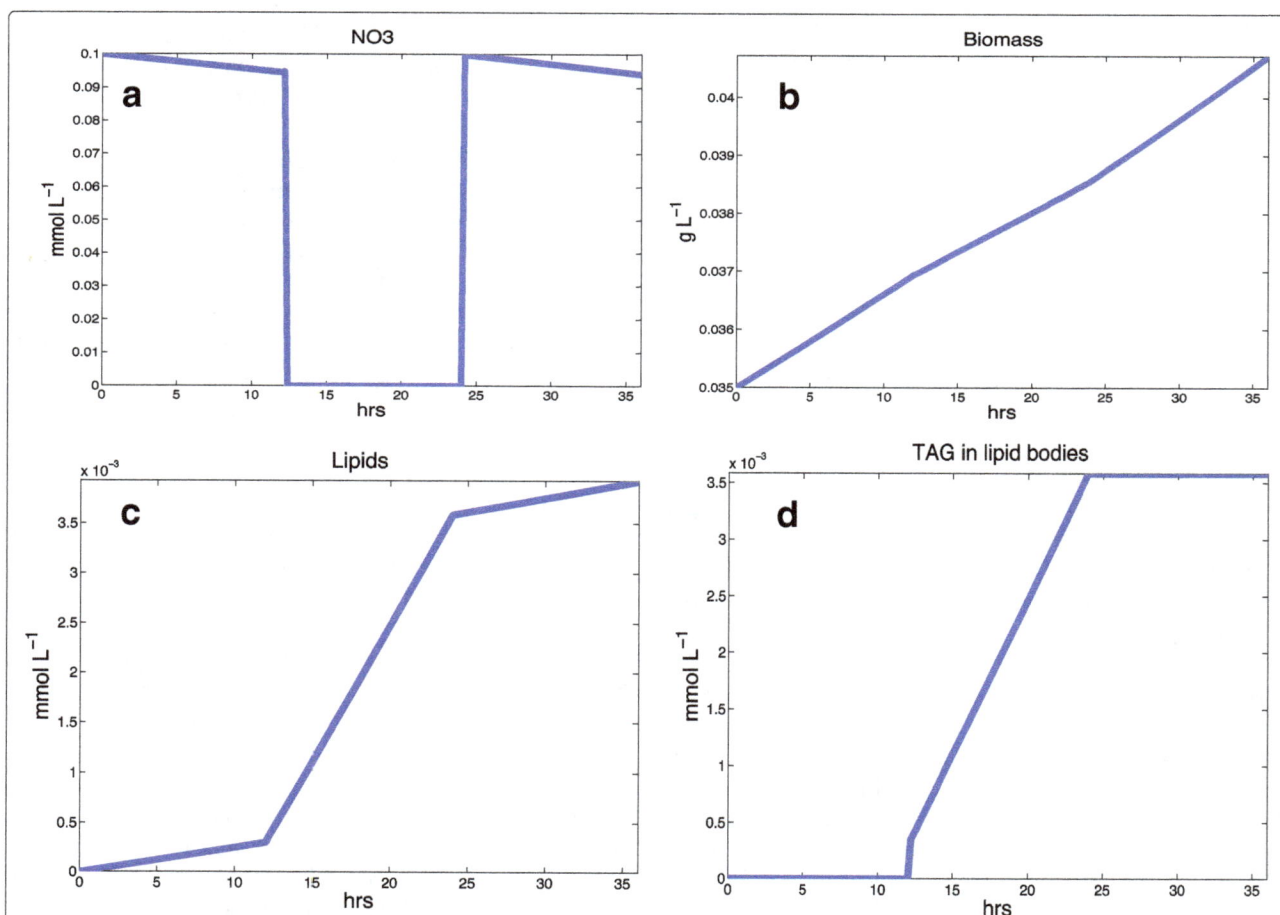

Fig. 5 Dynamic FBA simulation of lipid accumulation in *N. salina* using iNS934. We simulated growth and lipid production using COBRA tools in three stages: NO_3 available, NO_3 depleted and NO_3 available again. **a** NO_3 available in the media. We set an initial amount of NO_3 that was consumed by *N. salina* and eventually depleted. After this event, we added NO_3 to the media, allowing *N. salina* to grow normally again. **b** Biomass accumulation of *N. salina*. After NO_3 depletion, biomass production was reduced in stage 2 and increased again in stage 3. **c** Total lipid production (including TAG). In the first stage, lipid production increases according to the stoichiometric coefficients in the biomass formation reaction. Once NO_3 is depleted, the lipid production increases in a higher rate than stage 1 due to the high amount of TAG produced toward lipid bodies. In stage 3, the lipid production rate is once again as described in stage 1. **d** TAG accumulated on lipid bodies. We created a compartment called "lipid bodies", where the TAG produced, and not consumed by biomass, was stored

capacity to produce biomass (see Methods). We applied this algorithm to optimize our target metabolites.

TAGs, like other glycerolipids, are represented as biomass constituents in GSMMs. This represents a challenge when optimizing these types of compounds because most strain design algorithms are conceived to optimize metabolic products which are secreted from the cell, such as succinate or lactate, not biomass constituents. To simulate an additional production of TAG, we created artificial reactions leading to the production of additional TAG. A demand reaction for that additional TAG was considered as the target objective function. Using our in-house algorithm for strain optimization, we found 82 sets of reactions whose independent elimination guaranteed a minimum TAG production (Additional file 7). We ranked those sets according to

growth rate, minimum TAG guaranteed and number of knockouts. Higher growth rates and TAG productions as well as lower knockout sets lead to better scores.

The first observation we made is that growth rates for mutants were less than the growth obtained for wild-type. This observation makes sense since carbon must be redistributed into the biosynthesis pathways of lipids to produce more lipids, resulting in decreased carbon for use in other pathways. Since the carbon that was used to maximize growth rate in the wild-type is now being used for lipid biosynthesis in the mutant, the growth rate of the mutant must be less than the wild-type. The second observation we made is that, unlike wild-type, the additional TAG production is coupled with growth for all mutants. This explains why a minimum flux is guaranteed for the reaction producing additional TAG.

Most knockout sets involve reactions related to biosynthesis or transport of amino acids between different compartments. For example, one of these sets involves the knockout of a chloroplastic glutamate synthase, an enzyme that consumes glutamine and oxoglutarate to produce glutamate. The glutamate synthase knockout redirects the carbon flux that would be used to synthesize glutamate into the biosynthesis of fatty acids also located in this compartment. The forced flux into the biosynthesis of fatty acid leads to a forced flux into the biosynthesis of TAG. It is worth noting that glutamate is still produced in the cytosol satisfying the requirements for biomass production.

In the case of inter-compartment amino acid transport, the blockage of some transport reactions and the prevention of some transaminations in specific compartments, prevents the interchange of central metabolites, such as pyruvate and oxoglutarate between compartments. This limits the paths in which they can be consumed, which redirects the carbon flux through triacylglicerol biosynthesis.

The results showed in this section demonstrate that iNS934 could be used to propose strategies for optimization of lipid production as well as other biotechnological targets. However, it is worth mentioning that these strategies only involve mass balances and other regulatory mechanisms are not considered. Further experimental information related to reaction knockouts could be used as input in the model to improve the identified strategies.

Conclusions

We have reconstructed iNS934, the first genome-scale metabolic model of the marine algae N. salina. To develop this model we used the alga genome annotation and additionally we generated transcriptomic data that allowed us to identify new putative genes for N. salina. iNS934 contains 1985 metabolites, 2345 reactions, 934 genes and 10 compartments, which in total, achieve a precise description of the metabolism of this alga. We tested iNS934 and we found it was able to make simple growth/nongrowth predictions on 32 different conditions with an accuracy of 90%. These experiments included autotrophic, mixotrophic and heterotrophic conditions as well as key knockouts related to nitrogen metabolism. Moreover, a quantitative estimation of growth rates was achieved with an average error of only 15% for growth experiments with different nitrogen sources and CO_2 supply levels.

iNS934 includes the Nannochloropsis-specific biosynthesis pathways for glycerolipids supported by experimental evidence. It has been shown that Nannochloropsis can grow to where 50% of its biomass is in the form of lipids. Thus, this model could be used to describe and predict the biosynthesis of lipids in high lipid producing species, especially for the biodiesel industry.

We employed iNS934 to find strategies for increasing the production of TAGs. We used a novel approach to handle the optimization of these biomass constituents, which could be used for other new strain optimization algorithms. Additionally, we created an algorithm of strain optimization which allowed us to find 82 sets of knockout reactions whose independent elimination in the network resulted in an improved production of TAG. The results highlight that further experimental information is needed in order to validate the knockout sets experimentally. The incorporation of regulatory mechanisms into GSMMs will probably allow users to predict strategies of strain optimization more accurately.

iNS934 could also be employed for other purposes such as metabolic engineering for improved production of omega-3 and omega-6 or improved production of beta-glucans. Both cases represent important cases of study for producing nutraceuticals with high value for human care.

Endnotes

[1]http://www.phrap.org/phredphrap/phrap.html

[2]http://transdecoder.github.io

[3]http://bigg.ucsd.edu/data_access

Additional files

Additional file 1: CDS. New set of CDS of N. salina used for the reconstruction of iNS934.

Additional file 2: Metabolites in iNS934. List of all metabolites in N. salina model iNS934.

Additional file 3: Model. N. salina iNS934 model. SBML representation of the reconstructed model of N. salina. This file is compatible with SBML Level 3, Version 1, fbc ver. 2. It has been tested with COBRA Toolbox (2.0).

Additional file 4: Extra tables and figures. Tables and figures with details on model reconstruction and analysis.

Additional file 5: ID mapping. Mapping of transcriptome of N. salina to CCMP537 genome assembly [58].

Additional file 6: Constraints used to predict growth rates. Measured uptake rates of nitrogen sources (Nitrate/ammonium/urea) and estimated CO_2 uptake rates for batch cultures of N. salina.

Additional file 7: Strain optimization. Proposed reaction knockouts for improving production of lipids in N. salina.

Abbreviations

ARA: Arachidonic acid; ASW: Artificial sea water; CDS: Coding sequence; CSIRO: Commonwealth Scientific and Industrial Research Organization; DAG: Diacylglycerol; dFBA: Dynamic flux balance analysis; DGDG: Digalactosyldiacylglycerol; DGTS: Diacylglyceryl-O-4'-(N, N, N,-trimethyl) homoserine; DIC: Dissolved inorganic carbon; DLC: Dark, Low CO_2; EC: Enzyme commision; EPA: Eicosapentaenoic acid; ETA: Eicosatetraenoic acid; FBA: Flux balance analysis; FCA: Flux coupling analysis; FVA: Flux variability analysis; GSMM: Genome-scale metabolic model; HLHC: High light, High CO_2; HLLC: High light, Low CO_2; MGDG: Monogalactosyldiacylglycerol; PC: Phosphatidylcholine; PE: Phosphatidylethanolamine; PG: Phosphatidylglycerol; PI: Phosphatidylinositol; PUFAs: Polyunsaturated fatty acids; SBML: Systems biology markup language; SQDG: Sulfoquinovosyldiacylglycerol; TAG: Triacylglycerol

Acknowledgements
This project was supported by grants: Fondap 1509007, Basal program PFB-03, CIRIC-INRIA Chile (line Natural Resources), FONDECYT 11090234, FIA PYT 2016-0339, FONDECYT 3140480 and CCNICYT Doctoral scholarship 21140822. We acknowledge the National Laboratory for High Performance Computing at the Center for Mathematical Modeling (PIA ECM-02 CONICYT).

Authors' contributions
NE, NL and AM conceived the study. MPC, NL, SM and NR reconstructed iNS934. NL and SM analyzed and simulated the model. NE, NG and SM validated the model against experimental evidence. AdG and DT handled genomics and transcriptomics. All authors wrote and approved the final manuscript.

Competing interests
The authors declare that they have no competing interests.

Author details
[1]Mathomics, Center for Mathematical Modeling, Universidad de Chile, Beauchef 851, 7th Floor, Santiago, Chile. [2]Center for Genome Regulation (Fondap 15090007), Universidad de Chile, Blanco Encalada 2085, Santiago, Chile. [3]Centro de Investigación Austral Biotech, Universidad Santo Tomás, Avenida Ejercito 146, Santiago, Chile. [4]Universidad Adolfo Ibáñez, Diagonal Las Torres 2640, Santiago, Chile.

References
1. Ben-Amotz A, Katz A, Avron M. Accumulation of β-carotene in halotolerant alge: purification and characterization of β-carotene-rich globules from Dunaliella bardawil (chlorophyceae). J Phycol. 1982;18(4): 529–37. doi:10.1111/j.1529-8817.1982.tb03219.x.
2. Kleinegris DMM, van Es MA, Janssen M, Brandenburg WA, Wijffels RH. Carotenoid fluorescence in Dunaliella salina. J Appl Phycol. 2010;22(5): 645–9. doi:10.1007/s10811-010-9505-y.
3. Lamers PP, Janssen M, De Vos FCH, Bino RJ, Wijffels RH. Exploring and exploiting carotenoid accumulation in Dunaliella salina for cell-factory applications. Trends Biotechnol. 2008;26(11):631–8. doi:10.1016/j.tibtech.2008.07.002.
4. Chisti Y. Biodiesel from microalgæ. Biotechnol Adv. 2007;25(3):294–306. doi:10.1016/j.biotechadv.2007.02.001.
5. Hu Q, Sommerfeld M, Jarvis E, Ghirardi M, Posewitz M, Seibert M, Darzins A. Microalgal triacylglycerols as feedstocks for biofuel production: perspectives and advances. Plant J Cell Mol Biol. 2008;54(4): 621–39. doi:10.1111/j.1365-313X.2008.03492.x.
6. Griffiths MJ, Harrison STL. Lipid productivity as a key characteristic for choosing algal species for biodiesel production. J Appl Phycol. 2009;21(5):493–507. doi:10.1007/s10811-008-9392-7.
7. Ghirardi M. Microalgæ: a green source of renewable H2. Trends Biotechnol. 2000;18(12):506–11. doi:10.1016/S0167-7799(00)01511-0.
8. Melis A, Zhang L, Forestier M, Ghirardi ML, Seibert M. Sustained Photobiological Hydrogen Gas Production upon Reversible Inactivation of Oxygen Evolution in the Green Alga Chlamydomonas reinhardtii. Plant Physiol. 2000;122(1):127–36.
9. Boyd CE. Fresh-water plants: a potential source of protein. Econ Bot. 1968;22(4):359–68. doi:10.1007/BF02908132.
10. Becker EW. Micro-algæ as a source of protein. Biotechnol Adv. 2007;25(2):207–10. doi:10.1016/j.biotechadv.2006.11.002.
11. Delrue B, Fontaine T, Routier F, Decq A, Wieruszeski JM, Van Den Koornhuyse N, Maddelein ML, Fournet B, Ball S. Waxy Chlamydomonas reinhardtii: monocellular algal mutants defective in amylose biosynthesis and granule-bound starch synthase activity accumulate a structurally modified amylopectin. J Bacteriol. 1992;174(11):3612–20.
12. Slade R, Bauen A. Micro-algæ cultivation for biofuels: Cost, energy balance, environmental impacts and future prospects. Biomass Bioenergy. 2013;53:29–38.
13. Milne CB, Kim PJ, Eddy JA, Price ND. Accomplishments in genome-scale in silico modeling for industrial and medical biotechnology. Biotechnol J. 2009;4(12):1653–70. doi:10.1002/biot.200900234.
14. Smid EJ, van Enckevort FJH, Wegkamp A, Boekhorst J, Molenaar D, Hugenholtz J, Siezen RJ, Teusink B. Metabolic models for rational improvement of lactic acid bacteria as cell factories. J Appl Microbiol. 2005;98(6):1326–31. doi:10.1111/j.1365-2672.2005.02652.x.
15. Carlson R, Srienc F. Fundamental Escherichia coli biochemical pathways for biomass and energy production: creation of overall flux states. Biotech Bioeng. 2004;86(2):149–62. doi:10.1002/bit.20044.
16. Kayser A, Weber J, Hecht V, Rinas U. Metabolic flux analysis of Escherichia coli in glucose-limited continuous culture. I. Growth-rate-dependent metabolic efficiency at steady state. Microbiol (Read Engl). 2005;151(Pt 3):693–706. doi:10.1099/mic.0.27481-0.
17. Förster J, Famili I, Fu P, Palsson BØ, Nielsen J. Genome-scale reconstruction of the Saccharomyces cerevisiae metabolic network. Genome Res. 2003;13(2):244–53. doi:10.1101/gr.234503.
18. Poolman MG, Miguet L, Sweetlove LJ, Fell DA. A genome-scale metabolic model of Arabidopsis and some of its properties. Plant Physiol. 2009;151(3):1570–81. doi:10.1104/pp.109.141267.
19. Baroukh C, Muñoz-Tamayo R, Steyer JP, Bernard O. A state of the art of metabolic networks of unicellular microalgæ and cyanobacteria for biofuel production. Metab Eng. 2015;30:49–60. doi:10.1016/j.ymben.2015.03.019.
20. Imam S, Schäuble S, Valenzuela J, López García De Lomana A, Carter W, Price ND, Baliga NS. A refined genome-scale reconstruction of Chlamydomonas metabolism provides a platform for systems-level analyses. Plant J. 2015;84(6):1239–56. doi:10.1111/tpj.13059.
21. Chaiboonchoe A, Dohai BS, Cai H, Nelson DR, Jijakli K, Salehi-Ashtiani K. Microalgal Metabolic Network Model Refinement through High-Throughput Functional Metabolic Profiling. Frontiers Bioeng Biotechnol. 2014;2(December):68. doi:10.3389/fbioe.2014.00068.
22. Chang RL, Ghamsari L, Manichaikul A, Hom EFY, Balaji S, Fu W, Shen Y, Hao T, Palsson BØ, Salehi-Ashtiani K, Papin JA. Metabolic network reconstruction of Chlamydomonas offers insight into light-driven algal metabolism. Mol Syst Biol. 2011;7:518. doi:10.1038/msb.2011.52.
23. Gomes de Oliveira Dal'molin C, Quek LE, Palfreyman RW, Nielsen LK. AlgaGEM - a genome-scale metabolic reconstruction of algæ based on the Chlamydomonas reinhardtii genome. BMC Genomics. 2011;12 Suppl 4:5. doi:10.1186/1471-2164-12-S4-S5.
24. Kliphuis AMJ, Klok AJ, Martens DE, Lamers PP, Janssen M, Wijffels RH. Metabolic modeling of chlamydomonas reinhardtii: energy requirements for photoautotrophic growth and maintenance. J Appl Phycol. 2012;24(2):253–66. doi:10.1007/s10811-011-9674-3.
25. Boyle NR, Morgan JA. Flux balance analysis of primary metabolism in Chlamydomonas reinhardtii. BMC Syst Biol. 2009;3(1):4. doi:10.1186/1752-0509-3-4.
26. Christian N, May P, Kempa S, Handorf T, Ebenhöh O. An integrative approach towards completing genome-scale metabolic networks. Mol BioSyst. 2009;5(12):1889–903. doi:10.1039/b915913b.
27. May P, Christian JO, Kempa S, Walther D. ChlamyCyc: an integrative systems biology database and web-portal for Chlamydomonas reinhardtii,. BMC Genomics. 2009;10:209. doi:10.1186/1471-2164-10-209.
28. Molnár I, Lopez D, Wisecaver JH, Devarenne TP, Weiss TL, Pellegrini M, Hackett JD. Bio-crude transcriptomics: Gene discovery and metabolic network reconstruction for the biosynthesis of the terpenome of the hydrocarbon oil-producing green alga, botryococcus braunii race b (showa)*. BMC Genomics. 2012;13(1):576. doi:10.1186/1471-2164-13-576.
29. Zuñiga C, Li CT, Huelsman T, Levering J, Zielinski DC, McConnell BO, Long CP, Knoshaug EP, Guarnieri MT, Antoniewicz MR, Betenbaugh MJ, Zengler K. Genome-scale metabolic model for the green alga Chlorella

vulgaris UTEX 395 accurately predicts phenotypes under autotrophic, heterotrophic, and mixotrophic growth conditions. Plant Physiol. 2016;172(1):589–602. doi:10.1104/pp.16.00593.

30. Juneja A, Chaplen FWR, Murthy GS. Genome Scale Metabolic Reconstruction of Chlorella variabilis for exploring its metabolic potential for biofuels. Bioresour Technol. 2016;213:103–10. doi:10.1016/j.biortech.2016.02.118.

31. Wu C, Xiong W, Dai J, Wu Q. Genome-based metabolic mapping and 13C flux analysis reveal systematic properties of an oleaginous microalga Chlorella protothecoides. Plant Physiol. 2015;167(2):586–99. doi:10.1104/pp.114.250688.

32. Muthuraj M, Palabhanvi B, Misra S, Kumar V, Sivalingavasu K, Das D. Flux balance analysis of chlorella sp. fc2 iitg under photoautotrophic and heterotrophic growth conditions. Photosynth Res. 2013;118(1):167–79. doi:10.1007/s11120-013-9943-x.

33. Yang C, Hua Q, Shimizu K. Energetics and carbon metabolism during growth of microalgal cells under photoautotrophic, mixotrophic and cyclic light-autotrophic/dark-heterotroph c conditions. Biochem Eng J. 2000;6(2):87–102. doi:10.1016/S1369-703X(00)00080-2.

34. Krumholz EW, Yang H, Weisenhorn P, Henry CS, Libourel IGL. Genome-wide metabolic network reconstruction of the picoalga ostreococcus. J Exp Bot. 2012;63(6):2353. doi:10.1093/jxb/err407.

35. Levering J, Broddrick J, Dupont CL, Peers G. Beeri K, Mayers J, Gallina AA, Allen AE, Palsson BO, Zengler K. Genome-Scale Model Reveals Metabolic Basis of Biomass Partitioning in a Model Diatom. PLoS ONE. 2016;11(5):0155038. doi:10.1371/journal.pone.0155038.

36. Levitan O, Dinamarca J, Zelzion E, Lun DS, Guerra LT, Kim MK, Kim J, Van Mooy BAS, Bhattacharya D, Falkowsk PG. Remodeling of intermediate metabolism in the diatom Phaeodactylum tricornutum under nitrogen stress. Proc Natl Acad Sci. 2015;112(2):412–7. doi:10.1073/pnas.1419818112.

37. Singh D, Carlson R, Fell D, Poolman M. Modelling metabolism of the diatom Phaeodactylum tricornutum. Biochem Soc Trans. 2015;43(6):1182–6. doi:10.1042/BST20150152.

38. Kim J, Fabris M, Baart G, Kim MK, Goossens A, Vyverman W, Falkowski PG, Lun DS. Flux balance analysis of primary metabolism in the diatom <i>Phaeodactylum tricornutum</i>. Plant J. 2015. doi:10.1111/tpj.13081.

39. Kroth PG, Chiovitti A, Gruber A, Martin-Jezequel V, Mock T, Parker MS, Stanley MS, Kaplan A, Caron L, Weber T, Maheswari U, Armbrust EV, Bowler C. A model for carbohydrate metabolism in the diatom Phaeodactylum tricornutum deduced from comparative whole genome analysis. PLoS ONE. 2008;3(1):1426. doi:10.1371/journal.pone.0001426.

40. Fabris M, Matthijs M, Rombauts S, Vyverman W, Goossens A, Baart GJ. The metabolic blueprint of phaeodactylum tricornutum reveals a eukaryotic entner–doudoroff glycolytic pathway. Plant J. 2012;70(6):1004–14.

41. Hunt KA, Folsom JP, Taffs RL, Carlson RP. Complete enumeration of elementary flux modes through scalable, demand-based subnetwork definition. Bioinformatics. 2014;30(11):021.

42. Prigent S, Collet G, Dittami SM, Delage L, Ethis de Corny F, Dameron O, Eveillard D, Thiele S, Cambefort J, Boyen C, Siegel A, Tonon T. The genome-scale metabolic network of ectocarpus si iculosus (ectogem): a resource to study brown algal physiology and beyond. Plant J. 2014;80(2):367–81. doi:10.1111/tpj.12627.

43. Knies D, Wittmüß P, Appel S, Sawodny O, Ederer M, Feuer R. Modeling and Simulation of Optimal Resource Management during the Diurnal Cycle in Emiliania huxleyi by Genome-Scale Reconstruction and an Extended Flux Balance Analysis Approach. Metabolites. 2015;5(4):659–76. doi:10.3390/metabo5040659.

44. Boussiba S, Vonshak A, Cohen Z, Avissar Y. Richmond A. Microalga Nannochloropsis salina. Biomass. 1987;12:37–47.

45. Emdadi D, Berland B. Variation in lipid class composition during batch growth of Nannochloropsis salina and Pavlova lutheri. Mar Chem. 1989;26(3):215–25. doi:10.1016/0304-4203(39)90004-2.

46. Ma XN, Chen TP, Yang B, Liu J, Chen F. Lipid Production from Nannochloropsis. Mar Drugs. 2016;14(4):. doi:10.3390/md14040061.

47. Lee JH, O'Keefe JH, Lavie CJ, Harris WS. Omega-3 fatty acids: cardiovascular benefits, sources and sustainability Nat Rev Cardiol. 2009;6(12):753–8. doi:10.1038/nrcardio.2009.188.

48. GRYNBERG A. Hypertension prevention: from nutrients to (fortified) foods to dietary patterns. focus on fatty acids. J Hum Hypertens. 2005;19:25–33. doi:10.1038/sj.jhh.1001957.

49. Bauch A, Lindtner O, Mensink GBM, Niemann B. Dietary intake and sources of long-chain n-3 PUFAs in German adults. Eur J Clin Nutr. 2006;60(6) 810–2. doi:10.1038/sj.ejcn.1602399.

50. Klok AJ, Lamers PP, Martens DE, Draaisma FB, Wijffels RH. Edible oils from microalgae: Insights in TAG accumulation. Trends Biotechnol. 2014;32(10):521–8. doi:10.1016/j.tibtech.2014.07.004.

51. Taneja A, Singh H. Challenges for the delivery of long-chain n-3 fatty acids in functional foods. Annu Rev Food Sci Technol. 2012;3(April):105–23. doi:10.1146/annurev-food-022811-101130.

52. Chen B, McClements DJ, Decker EA. Design of foods with bioactive lipids for improved health. Annu Rev Food Sci Technol. 2013;4:35–56. doi:10.1146/annurev-food-032112-135808.

53. Marudhupandi T, Sathishkumar R, Kumar TTA. Heterotrophic cultivation of Nannochloropsis salina for enhancing biomass and lipid production. Biotechnol Rep. 2016;10:8–16. doi:10.1016/j.btre.2016.02.001.

54. Fábregas J, Maseda A, Domínguez A, Otero A. The cell composition of Nannochloropsis sp. changes under different irradiances in semicontinuous culture. World J Microbiol Biotechnol. 2004;20(1):31–5. doi:10.1023/B:WIBI.0000013288.67536.ed.

55. Guillard RRL. Culture of Phytoplankton for Feeding Marine Invertebrates. In: Culture of Marine Invertebrate Animals. Boston: Springer; 1975. p. 29–60. doi:10.1007/978-1-4615-8714-9_3.

56. Chen HL, Li SS, Huang R, Tsai HJ. Conditional production of a functional fish growth hormone in the transgenic line of nannochloropsis oculata (Eustigmatophyceae) 1. J Phycol. 2008;44(3):768–76.

57. White JR, Roberts M, Yorke JA, Pop M. Figaro: a novel statistical method for vector sequence removal. Bioinformatics. 2008;24(4):462–7.

58. Wang D, N ng K, Li J, Hu J, Han D, Wang H, Zeng X, Jing X, Zhou Q, Su X, Chang X, Wang A, Wang W, Jia J, Wei L, Xin Y, Qiao Y, Huang R, Chen J, Han B, Yoon K, Hill RT, Zohar Y, Chen F, Hu Q, Xu J. Nannochloropsis Genomes Reveal Evo ution of Microalgal Oleaginous Traits. PLoS Genet. 2014;10(1):1004094. doi:10.1371/journal.pgen.1004094.s024.

59. Thiele I, Palsson BØ. A protocol for generating a high-quality genome-scale metabolic reconstruction. Nat Protoc. 2010;5(1):93–121. doi:10.1038/nprot.2009.203.

60. Remm M, Storm CE, Sonnhammer ELL. Automatic clustering of orthologs and in-paralogs from pairwise species comparisons. J Mol Biol. 2001;314(5) 1041–52. doi:10.1006/jmbi.2000.5197.

61. Li L, Stoeckert CJ, Roos DS. OrthoMCL: identification of ortholog groups for eukaryotic genomes. Genome Res. 2003;13(9):2178–89. doi:10.1101/gr.1224503.

62. Loira N, Zhukova A, Sherman DJ. Pantograph: A template-based method for genome-scale metabolic model reconstruction. J Bioinforma Comput Biol. 2015;13(02):1550006. doi:10.1142/S0219720015500067.

63. Prigent S, Frioux C, Dittami SM, Thiele S, Larhlimi A, Collet G, Gutknecht F, Got J, Eveillard D, Bourdor J, Plewniak F, Tonon T, Siegel A. Meneco, a topology-based gap-filling tool applicable to degraded genome-wide metabolic networks. PLoS Comput Biol. 2017;13(1):1–32. doi:10.1371/journal.pcbi.1005276.

64. Collet G, Eveillard D, GEBSER M, Prigent S, SCHAUB T, Siegel A, Thiele S. Extending the Metabolic Network of Ectocarpus Siliculosus Using Answer Set Programming. In: Link.springer.com. Berlin, Heidelberg: Springer; 2013. p. 245–56. doi:10.1007/978-3-642-40564-8_25.

65. Bernard T, Bridge A, Morgat A, Moretti S, Xenarios I, Pagni M. Reconciliation of metabolites and biochemical reactions for metabolic networks. Brief Bioinform. 2012;15(1):053.

66. Joyce AR, Reed J, White A, Edwards R, Osterman A, Baba T, Mori H, Lesely SA, Palsson BØ, Agarwalla S. Experimental and computational assessment of conditionally essential genes in Escherichia coli. J Bacteriol. 2006;188(23):8259–71. doi:10.1128/JB.00740-06.

67. Hernandez-Lopez J, Vargas-Albores F. A microplate technique to quantify nutrients (NO2-, NO3-, NH4+ and PO43-) in seawater. Aquacult Res. 2003;34(13):1201–4. doi:10.1046/j.1365-2109.2003.00928.x.

68. Becker SA, Feist AM, Mo ML, Hannum G, Palsson BØ, Herrgard MJ. Quantitative prediction of cellular metabolism with constraint-based models: the COBRA Toolbox,. Nat Protoc. 2007 2(3):727–38. doi:10.1038/nprot.2007.99.

69. Larhlimi A, David L, Selbig J, Bockmayr A. F2c2: a fast tool for the computation of flux coupling in genome-scale metabolic networks. BMC Bioinforma. 2012;13(1):57.

70. Kumar VS, Maranas CD. GrowMatch: an automated method for reconciling in silico/in vivo growth predictions. PLoS Comput Biol. 2009;5(3):1000308. doi:10.1371/journal.pcbi.1000308.

71. Adamczyk M, Lasek J, Skawińska A. Co2 biofixation and growth kinetics of chlorella vulgaris and nannochloropsis gaditana. Appl Biochem Biotechnol. 2016;179(7):1248–61.

72. Suthers PF, Dasika MS, Kumar VS, Denisov G, Glass JI, Maranas CD. A genome-scale metabolic reconstruction of Mycoplasma genitalium, iPS189. PLoS Comput Biol. 2009;5(2):1000285. doi:10.1371/journal.pcbi.1000285.

73. Li J, Han D, Wang D, Ning K, Jia J, Wei L, Jing X, Huang S, Chen J, Li Y, Hu Q, Xu J. Choreography of Transcriptomes and Lipidomes of Nannochloropsis Reveals the Mechanisms of Oil Synthesis in Microalgæ. Plant Cell. 2014;26(4):1645–65. doi:10.1105/tpc.113.121418.

74. Rodolfi L, Chini Zittelli G, Bassi N. Padovani G, Biondi N, Bonini G, Tredici MR. Microalgæ for oil: Strain selection, induction of lipid synthesis and outdoor mass cultivation in a low-cost photobioreactor. Biotech Bioeng. 2009;102(1):100–12. doi:10.1002/bit.22033.

75. Simionato D, Block MA, La Rocca N, Jouhet J, Maréchal E, Finazzi G, Morosinotto T. The response of Nannochloropsis gaditana to nitrogen starvation includes de novo biosynthesis of triacylglycerols, a decrease of chloroplast galactolipids, and reorganization of the photosynthetic apparatus. Eukaryot Cell. 2013;12(5):665–76. doi:10.1128/EC.00363-12.

76. Mühlroth A, Li K, Røkke G, Winge P, Olsen Y, Hohmann-Marriott MF, Vadstein O, Bones AM. Pathways of lipid metabolism in marine algæ, co-expression network, bottlenecks and candidate genes for enhanced production of EPA and DHA in species of Chromista. Mar Drugs. 2013;11(11):4662–97. doi:10.3390/md11114662.

77. Alboresi A, Perin G, Vitulo N, Diretto G, Block M, Jouhet J, Meneghesso A, Valle G, Giuliano G, Maréchal E, Morosinotto T. Light Remodels Lipid Biosynthesis in Nannochloropsis gaditana by Modulating Carbon Partitioning between Organelles. Plant Physiol. 2016;171(4):2468–82. doi:10.1104/pp.16.00599.

78. Radakovits R, Jinkerson RE, Fuerstenberg SI, Tae H, Settlage RE, Boore JL, Posewitz MC. Draft genome sequence and genetic transformation of the oleaginous alga Nannochloropsis gaditana. Nat Commun. 2012;3:686. doi:10.1038/ncomms1688.

79. Plaxton WC. The organization and regulation of plant Glycolysis. Annu Rev Plant Physiol Plant Mol Biol. 1996;47:185–214. doi:10.1146/annurev.arplant.47.1.185.

80. Matsumoto M, Ogawa K. New Insight into the Calvin Cycle Regulation – Glutathionylation of Fructose Bisphosphate Aldolase in Response to Illumination. In: Link.springer.com. Dordrecht: Springer; 2008. p. 871–4. doi:10.1007/978-1-4020-6709-9_193.

81. O'Brien EJ, Monk JM, Palsson BO. Using genome-scale models to predict biological capabilities. Cell. 2015;161(5):971–87. doi:10.1016/j.cell.2015.05.019.

82. Monk JM, Charusanti P, Aziz RK. Lerman JA, Premyodhin N, Orth JD, Feist AM, Palsson BO. Genome-scale metabolic reconstructions of multiple escherichia coli strains highlight strain-specific adaptations to nutritional environments. Proc Natl Acad Sci. 2013;110(50):20338–43. doi:10.1073/pnas.1307797110. http://www.pnas.org/content/110/50/20338.full.pdf.

83. Teusink B, van Enckevort FH, Francke C, Wiersma A, Wegkamp A, Smid EJ, Siezen RJ. In silico reconstruction of the metabolic pathways of lactobacillus plantarum: comparing predictions of nutrient requirements with those from growth experiments. Appl Environ Microbiol. 2005;71(11):7253–62.

84. Kuepfer L, Sauer U, Blank LM. Metabolic functions of duplicate genes in saccharomyces cerevisiae. Genome Res. 2005;15(10):1421–30.

85. Nookaew I, Jewett MC, Meechai A, Thammarongtham C, Laoteng K, Cheevadhanarak S, Nielsen J, Bhumiratana S. The genome-scale metabolic model iin800 of saccharomyces cerevisiae and its validation: a scaffold to query lipid metabolism. BMC Syst Biol. 2008;2(1):71.

86. Zou N, Zhang C, Cohen Z, Richmond A. Production of cell mass and eicosapentaenoic acid (EPA) in ultrahigh cell density cultures of Nannochloropsis sp. (Eustigmatophyceae). Eur J Phycol. 2000;35(2):127–33. doi:10.1080/09670260010001735711.

87. Forján Lozano E, Garbayo Nores I, Casal Bejarano C, Vílchez Lobato C. Enhancement of carotenoid production in Nannochloropsis by phosphate and sulphur limitation In: Mendez-Vilas, editor.

Communicating current research and educational topics and trends in applied microbiology. Formatex, Badajoz. Formatex Research Center; 2007. p. 356–64.

88. Herrgård MJ, Fong SS, Palsson BØ. Identification of genome-scale metabolic network models using experimentally measured flux profiles. PLoS Comput Biol. 2006;2(7):72.

89. Burgard AP, Pharkya P, Maranas CD. Optknock: a bilevel programming framework for identifying gene knockout strategies for microbial strain optimization. Biotech Bioeng. 2003;84(6):647–57. doi:10.1002/bit.10803.

90. Ranganathan S, Suthers PF, Maranas CD. OptForce: an optimization procedure for identifying all genetic manipulations leading to targeted overproductions. PLoS Comput Biol. 2010;6(4):1000744. doi:10.1371/journal.pcbi.1000744.

91. Simionato D, Sforza E, Corteggiani Carpinelli E, Bertucco A, Giacometti GM, Morosinotto T. Acclimation of Nannochloropsis gaditana to different illumination regimes: effects on lipids accumulation. Bioresour Technol. 2011;102(10):6026–32. doi:10.1016/j.biortech.2011.02.100.

92. Das P, Lei W, Aziz SS, Obbard JP. Enhanced algæ growth in both phototrophic and mixotrophic culture under blue light. Bioresour Technol. 2011;102(4):3883–7. doi:10.1016/j.biortech.2010.11.102.

93. Fang X, Wei C, Zhao-Ling C, Fan O. Effects of organic carbon sources on cell growth and eicosapentaenoic acid content of Nannochloropsis sp. J Appl Phycol. 2004;16(6):499–503. doi:10.1007/s10811-004-5520-1.

94. Huertas E, Montero O, Lubián LM. Effects of dissolved inorganic carbon availability on growth, nutrient uptake and chlorophyll fluorescence of two species of marine microalgæ. Aquac Eng. 2000;22(3):181–97. doi:10.1016/S0144-8609(99)00038-2.

95. Su CH, Chien LJ, Gomes J, Lin YS, Yu YK, Liou JS, Syu RJ. Factors affecting lipid accumulation by Nannochloropsis oculata in a two-stage cultivation process. J Appl Phycol. 2010;23(5):903–8. doi:10.1007/s10811-010-9609-4.

96. Kilian O, Benemann CSE, Niyogi KK, Vick B. High-efficiency homologous recombination in the oil-producing alga Nannochloropsis sp. Proc Natl Acad Sci. 2011;108(52):21265–9. doi:10.1073/pnas.1105861108.

97. Alsull M, Omar WMW. Responses of Tetraselmis sp. and Nannochloropsis sp. Isolated from Penang National Park Coastal Waters, Malaysia, to the Combined Influences of Salinity, Light and Nitrogen Limitation. In: International Conference on Chemical, Ecology and Environmental Sciences (ICEES'2012), Bangkok; 2012. p. 142–5.

98. Hii YS, Soo CL, Chuah TS, Mohd-Azmi A, Abol-Munafi AB. Interactive effect of ammonia and nitrate on the nitrogen uptake by Nannochloropsis sp. J Sustain Sci Manag. 2011;6(1):60–8.

99. Rocha JMS, Garcia JEC, Henriques MHF. Growth aspects of the marine microalga Nannochloropsis gaditana. Biomol Eng. 2003;20(4-6):237–42. doi:10.1016/S1389-0344(03)00061-3.

100. Bellou S, Baeshen MN, Elazzazy AM, Aggeli D, Sayegh F, Aggelis G. Microalgal lipids biochemistry and biotechnological perspectives. Biotechnol Adv. 2014;32(8):1476–93. doi:10.1016/j.biotechadv.2014.10.003.

PERMISSIONS

All chapters in this book were first published in SB, by BioMed Central; hereby published with permission under the Creative Commons Attribution License or equivalent. Every chapter published in this book has been scrutinized by our experts. Their significance has been extensively debated. The topics covered herein carry significant findings which will fuel the growth of the discipline. They may even be implemented as practical applications or may be referred to as a beginning point for another development.

The contributors of this book come from diverse backgrounds, making this book a truly international effort. This book will bring forth new frontiers with its revolutionizing research information and detailed analysis of the nascent developments around the world.

We would like to thank all the contributing authors for lending their expertise to make the book truly unique. They have played a crucial role in the development of this book. Without their invaluable contributions this book wouldn't have been possible. They have made vital efforts to compile up to date information on the varied aspects of this subject to make this book a valuable addition to the collection of many professionals and students.

This book was conceptualized with the vision of imparting up-to-date information and advanced data in this field. To ensure the same, a matchless editorial board was set up. Every individual on the board went through rigorous rounds of assessment to prove their worth. After which they invested a large part of their time researching and compiling the most relevant data for our readers.

The editorial board has been involved in producing this book since its inception. They have spent rigorous hours researching and exploring the diverse topics which have resulted in the successful publishing of this book. They have passed on their knowledge of decades through this book. To expedite this challenging task, the publisher supported the team at every step. A small team of assistant editors was also appointed to further simplify the editing procedure and attain best results for the readers.

Apart from the editorial board, the designing team has also invested a significant amount of their time in understanding the subject and creating the most relevant covers. They scrutinized every image to scout for the most suitable representation of the subject and create an appropriate cover for the book.

The publishing team has been an ardent support to the editorial, designing and production team. Their endless efforts to recruit the best for this project, has resulted in the accomplishment of this book. They are a veteran in the field of academics and their pool of knowledge is as vast as their experience in printing. Their expertise and guidance has proved useful at every step. Their uncompromising quality standards have made this book an exceptional effort. Their encouragement from time to time has been an inspiration for everyone.

The publisher and the editorial board hope that this book will prove to be a valuable piece of knowledge for researchers, students, practitioners and scholars across the globe.

LIST OF CONTRIBUTORS

Zhanzhan Cheng and Kai Huang
School of Computer Science, Fudan University, Handan Road, 200433 Shanghai, China

Shuigeng Zhou
School of Computer Science, Fudan University, Handan Road, 200433 Shanghai, China
The Bioinformatics Lab at Changzhou NO. 7 People's Hospital, Changzhou, 213011 Jiangsu, China

Hui Liu
The Bioinformatics Lab at Changzhou NO. 7 People's Hospital, Changzhou, 213011 Jiangsu, China
Lab of Information Management, Changzhou University, 213164 Changzhou, China

Jihong Guan
Department of Computer Science and Technology, Tongji University, 201804 Shanghai, China

Yang Wang
School of Computer Science, Jiangxi Normal University, 330022 Nanchang, China

Anna Doloman, Charles D. Miller
Department of Biological Engineering, Utah State University, Old Main Hill 4105, 84322-4105 Logan, UT, USA

Honey Varghese, and Nicholas S. Flann
Department of Computer Science, Utah State University, Old Main Hill 420, 84322-4205 Logan, UT, USA

Tao Peng, Adam L MacLean and Qing Nie
Department of Mathematics, Center for Complex Biological Systems, and Center for Mathematical and Computational Biology, University of California, Irvine, CA 92697, USA

Linan Liu, Chi Wut Wong and Weian Zhao
Department of Pharmaceutical Sciences, Department of Biomedical Engineering, Department of Biological Chemistry, Sue and Bill Gross Stem Cell Research Center, Chao Family Comprehensive Cancer Center & Edwards Life sciences Center for Advanced Cardiovascular Technology, University of California, 845 Health Sciences Road, Irvine, CA 92697, USA

Luis Miguel Serrano-Bermúdez and Dolly Montoya
Bioprocesses and Bioprospecting Group, Universidad Nacional de Colombia. Ciudad Universitaria, Carrera 30 No. 45-03, Bogotá, D.C, Colombia

Andrés Fernando González Barrios
Grupo de Diseño de Productos y Procesos (GDPP), Departamento de Ingeniería Química, Universidad de los Andes, Carrera 1 N.° 18A – 12, Bogotá, Colombia

Costas D. Maranas
Department of Chemical Engineering, The Pennsylvania State University, University Park, PA 16802, USA

Runxuan Zhang
Information and Computational Sciences, The James Hutton Institute, Invergowrie, Dundee, Scotland DD2 5DA, UK

Wenbin Guo
Information and Computational Sciences, The James Hutton Institute, Invergowrie, Dundee, Scotland DD2 5DA, UK
Plant Sciences Division, School of Life Sciences, University of Dundee, Invergowrie, Dundee, Scotland DD25DA, UK

Cristiane P. G. Calixto and Nikoleta Tzioutziou
Plant Sciences Division, School of Life Sciences, University of Dundee, Invergowrie, Dundee, Scotland DD25DA, UK

Robbie Waugh and John W. S. Brown
Plant Sciences Division, School of Life Sciences, University of Dundee, Invergowrie, Dundee, Scotland DD25DA, UK
Cell and Molecular Sciences, The James Hutton Institute, Invergowrie, Dundee, Scotland DD2 5DA, UK

Ping Lin
Division of Mathematics, University of Dundee, Nethergate, Dundee, Scotland DD1 4HN, UK

Irena Kuzmanovska Jan Mikelson and Mustafa Khammash
Department of Biosystems Science and Engineering, ETH Zurich, Mattenstrasse 26, 4058 Basel, Switzerland

Andreas Milias-Argeitis
Department of Biosystems Science and Engineering, ETH Zurich, Mattenstrasse 26, 4058 Basel, Switzerland
Groningen Biomolecular Sciences and Biotechnology, University of Groningen, Nijenborgh 4, 9747 AG Groningen, Netherlands

Christoph Zechner
Plant Sciences Division, School of Life Sciences, University of Dundee, Invergowrie, Dundee, Scotland DD25DA, UK
Max Planck Institute of Molecular Cell Biology and Genetics and Center for Systems Biology, Pfotenhauerstrasse 108, 01307 Dresden, Germany

Zu-Guo Yu
School of Mathematics and Computational Science, Xiangtan University, Xiangtan 411105, China

Changhe Fu
School of Mathematics and Computational Science, Xiangtan University, Xiangtan 411105, China
School of Mathematics and System Science, Shenyang Normal University, Shenyang 110034, China

Su Deng and Xinxin Wang
School of Mathematics and System Science, Shenyang Normal University, Shenyang 110034, China

Guangxu Jin
Center of Systems Biology and Bioinformatics, Wake Forest School of Medicine, Winston-Salem, NC 27157, USA

Niek Welkenhuysen, Mattias Backman and Loubna Bendrioua
Department of Chemistry and Molecular Biology, University of Gothenburg, SE-412 96 Gothenburg, Sweden

Stefan Hohmann
Department of Chemistry and Molecular Biology, University of Gothenburg, SE-412 96 Gothenburg, Sweden
Department of Biology and Biological Engineering, Chalmers University of Technology, SE-412 96 Gothenburg, Sweden

Johannes Borgqvist and Marija Cvijovic
Department of Mathematical Sciences, Chalmers University of Technology and the University of Gothenburg, SE-412 96 Gothenburg, Sweden

Mattias Goksör and Caroline B Adiels
Department of Physics, University of Gothenburg, SE-412 96 Gothenburg, Sweden

Deshun Sun
Control and Simulation Center, Harbin Institute of Technology, West Dazhi Street 92, 150001 Harbin, People's Republic of China

Fei Liu
Control and Simulation Center, Harbin Institute of Technology, West Dazhi Street 92, 150001 Harbin, People's Republic of China
School of Software Engineering, South China University of Technology, Building B7, 510006 Guangzhou, People's Republic of China

Ryutaro Murakami and Hiroshi Matsuno
Faculty of Science, Yamaguchi University, Yoshida 1677-1, 753-8512 Yamaguchi, Japan

Yulin Wang
School of Computer Science and Engineering, University of Electronic Science and Technology of China, Chengdu, Sichuan, China

Hongyu Miao
Department of Biostatistics, School of Public Health, University of Texas Health Science Center at Houston, Houston, TX 77030, USA

Yijie Wang and Xiaoning Qian
Department of Electrical & Computer Engineering, Texas A and M University, MS 3128, TAMU, College Station, TX, USA

Yuzhen Guo and Fengying Tao
Department of Mathematics, Nanjing University of Aeronautics and Astronautics, 210000 Nanjing, People's Republic of China

Yong Wang
National Center for Mathematics and Interdisciplinary Sciences, Academy of Mathematics and Systems Science, Chinese Academy of Sciences, 100190 Beijing, People's Republic of China
University of Chinese Academy of Sciences, 100049 Beijing, People's Republic of China

ZikaiWu
University of Shanghai for Science and Technology, 200433 Shanghai, People's Republic of China
Shanghai Key Laboratory of Intelligent Information Processing, Fudan University, 200433 Shanghai, People's Republic of China

Sabine Hug
Helmholtz Zentrum München - German Research
Center for Environmental Health, Institute of
Computational Biology, Ingolstädter Landstraße 1,
85764 Neuherberg, Germany

**Benjamin Ballnus, Jan Hasenauer and Fabian J.
Theis**
Helmholtz Zentrum München - German Research
Center for Environmental Health, Institute of
Computational Biology, Ingolstädter Landstraße 1,
85764 Neuherberg, Germany
Technische Universität München, Center for
Mathematics, Chair of Mathematical Modeling
of Biological Systems, Boltzmannstraße 15, 85748
Garching, Germany

Kathrin Hatz and Linus Görlitz
Bayer AG, Engineering & Technologies, Applied
Mathematics,   Kaiser-Wilhelm-Allee,   51368
Leverkusen, Germany

Alejandro F. Villaverde and Julio R. Banga
BioProcess Engineering Group, IIM-CSIC, Eduardo
Cabello 6, 36208 Vigo, Spain

Attila Gábor
BioProcess Engineering Group, IIM-CSIC, Eduardo
Cabello 6, 36208 Vigo, Spain
JRC-COMBINE,   RWTH   Aachen   University,
Photonics Cluster, Level 4, Campus-Boulevard 79,
52074 Aachen, Germany

**Nicolás Loira, Sebastian Mendoza, Dante
Travisany, Alex Di Genova, and Alejandro Maass**
Mathomics, Center for Mathematical Modeling,
Universidad de Chile, Beauchef 851, 7th Floor,
Santiago, Chile
Center for Genome Regulation (Fondap 15090007),
Universidad de Chile, Blanco Encalada 2085,
Santiago, Chile

María Paz Cortés
Mathomics, Center for Mathematical Modeling,
Universidad de Chile, Beauchef 851, 7th Floor,
Santiago, Chile
Center for Genome Regulation (Fondap 15090007),
Universidad de Chile, Blanco Encalada 2085,
Santiago, Chile
Universidad Adolfo Ibáñez, Diagonal LasTorres
2640, Santiago, Chile

Natalia Rojas
Center for Genome Regulation (Fondap 15090007),
Universidad de Chile, Blanco Encalada 2085,
Santiago, Chile

Natalia Gajardo and Nicole Ehrenfeld
Centro   de   Investigación   Austral   Biotech,
Universidad Santo Tomás, Avenida Ejercito 146,
Santiago, Chile

Index